To The Worl[illegible]

Second Greatest

Polymer Chemist

With Love

Rudy

Polymer Structure, Properties and Applications

Polymer Structure, Properties and Applications

RUDOLPH D. DEANIN, Ph.D.
Professor of Plastics
Lowell Technological Institute

Sponsored by
Plastics World, a Cahners publication

CAHNERS BOOKS, *Division of Cahners Publishing Company, Inc.*
89 Franklin Street, Boston, Massachusetts 02110

International Standard Book Number: 0-8436-1202-9
Library of Congress Catalog Card Number: 70-156478

Printed in the United States of America.
The Maple Press Company, York, Pennsylvania, U.S.A.

Contents

Preface

The relationships between polymer structure, properties, and applications are of obvious importance to chemists and engineers working in the polymer industries—plastics, rubber, fibers, coatings, and adhesives. Most technical people in these industries feel a need for greater understanding of these relationships. Most courses and literature on the subject are concerned mainly with the theory of mechanical and thermal properties; while most of the practical knowledge is dispersed throughout the industry in technical bulletins or unpublished form.

In view of the great need and interest in this area, I have tried to collect both theoretical and practical information on these relationships, and organize it into a system which will make it more available and useful to specialists in the field. Since the original scientific literature is too extensive and too scattered, I have worked mainly from recent textbooks and industrial experience.

The reader should have a reasonable background in the field, either through industrial experience or through introductory courses in polymer science or technology. The primary purpose of this book is to collect and organize our present understanding of the important relationships between structure, properties, and applications. Given this starting point, the reader should then be able to build on it more intensively through further reading and experience.

The complexity of the subject suggests that it might best be examined successively in three separate books, starting from the separate viewpoints of the chemist (structure), engineer (properties), and product designer (applications). Most specialists will want to read just one of these three books, whichever is closest to his specialty and point of view. The more interested reader may want to consider two or even all three points of view. The present volume examines the subject from the structural point of view. I hope to offer separate volumes from the property and application points of view within the next few years.

This book would never have been possible without the help of many friends. I am forever indebted first of all to Carl S. Marvel and John M. DeBell for their teaching, guidance, and inspiration. In the preparation of this book, I have received much valuable help from Professor Henry E. Thomas and my students in the Plastics Department at Lowell Technological Institute. In the actual labor of taping, typing, editing, and copying, I am primarily indebted to my wife, Joan, for her labor, encouragement, and patience; and we are both indebted to Mrs. Sarah Rabinovic and Mrs. Peggy Orroth for their further assistance. To all of them, I extend my gratitude and thanks.

Part One

INTRODUCTION

Chapter One

The Relationship between Structure, Properties, and Applications

Although polymer science, as applied to plastics, began its major growth almost a half-century ago, the new materials which it produced first began to find major practical uses during World War II in answer to the technological demands for new materials needed in military applications. During the war the nation mobilized its resources to back the vast research, development, and plant construction efforts required to perfect and produce these new polymers in large quantities.

As World War II drew to a close, the industries which had developed this large capacity for producing new polymers recognized that they would no longer have military markets for them. They began an aggressive search to find peacetime consumer applications for the polymers that they were able to produce, and soon introduced them into quite a variety of new products. The rapid appearance of these new materials on the consumer market proved quite dazzling, both to the consumer and to the scientist.

It was at this point that farseeing polymer scientists first began to talk of tailoring molecular structure to give any needed properties for any desired end product. At that early stage in polymer science, this concept was more optimistic hope than practical reality. Polymer chemists and chemical engineers knew how to make several types of polymer structures, and a few variations on some of them. These structures produced several limited ranges of properties, and an extremely limited choice in the balance of such properties; and these properties were measured only by rather inappropriate tests developed for older conventional materials. Practical applications at this time consisted mainly of cheap substitutes for established materials—metal, ceramic, wood, leather, cloth, and paper—and usually inferior substitutes at that.

During the past 25 years, polymer chemists and chemical engineers have greatly improved their ability to make broad ranges of polymer structures. Polymer physicists and mechanical engineers have learned to measure the significant and critical properties of these polymers. Product engineers and industrial designers have learned to use these materials in thousands of applications, most of them better than the older conventional materials they replaced, and many of them quite impossible before these new materials became available.

During the past 25 years, there have also been quite a number of researches on the relationship between structure and properties, and a few on the relationship between properties and applications. These researches have naturally been highly specialized in nature, and have been published in a wide variety of scientific journals, leaving their findings generally unavailable to the average polymer scientist.

There were some early attempts to develop a simple overall correlation between structure and properties (Table 1-1 and Figure 1-1). In recent years, several general textbooks on polymer science, such as those by Billmeyer (ref 3), Brydson (ref 4), Golding (ref 5), and Schmidt (ref 6), have included one or more chapters summarizing current knowledge of the relationship between polymer structure and properties. Several recent monographs, notably those by Alfrey (ref 7), Baer (ref 8), Miller (ref 9), and Nielsen (ref 10) have given very good summaries of the existing research and the knowledge it has produced.

All of these researches and publications are best on the relationship between structure and fundamental properties, particularly mechanical and thermal properties. Knowledge of processability, and of electrical and chemical properties, is less complete and less generally available. Acoustic and optical properties have received much less attention. And studies of the relationship between properties and practical applications, although they have obviously produced thousands of commercial successes, have rarely been treated thoroughly in the scientific literature.

The average chemist, chemical engineer, physicist, mechanical engineer, product engineer, and industrial designer doing practical work with polymers still must work primarily by trial and error. This is the biggest missing link in current polymer science and technology. When organic polymer chemists and chemical engineers synthesize a new polymer, they may predict that it will have properties which will lead to certain appli-

Table 1-1. Relation between Properties and Structural and Chemical Characteristics of High Polymers (ref 1)

Property	Average Degree of Polymerization	Branching	Cross-linkage	Non-polar Structures	Polar Structures	Flexibility of Chains	Readiness of Chains to Crystallize	Orientation of Crystallites
Modulus of elasticity	+	?	+	?	?	−	+	+
Ultimate tensile strength	+	?	?	−	+	?	+	+
Ultimate impact strength	+	?	−	+	?	+	−	−
Elongation at break	+	?	−	+	?	+	−	−
Range of reversible extensibility	+	−	−	+	?	+	?	?
Tendency to contract	+	?	−	+	−	+	+	?
Surface hardness	+	?	+	−	+	−	+	+
Resistance to temperature	+	?	+	−	+	−	+	?
Electrical resistance	?	?	+	+	−	?	?	?
Dielectric constant	?	?	?	−	+	?	?	?
Resistance to swelling	+	+	+	?	?	−	+	+
Adhesive power	?	+	−	?	+	+	−	−
Resistance to cold flow or creep	+	?	+	−	+	−	+	+
Resistance to moisture	+	+	+	?	?	−	+	+
Resistance to acids	?	?	+	+	?	?	+	?
Resistance to alkalies	?	?	+	+	?	?	+	?

A plus sign means that the property under consideration increases; a minus sign means that it decreases. A question mark indicates that the effect on the property is not known.
Mainly after
Mark, H., *Ind. Eng. Chem.* 34, 1343 (1942).

cations, but then often find that the polymer they have produced actually has properties leading to quite different applications.

When polymer physicists and mechanical engineers are considering specific types of properties, they may predict that these will result from

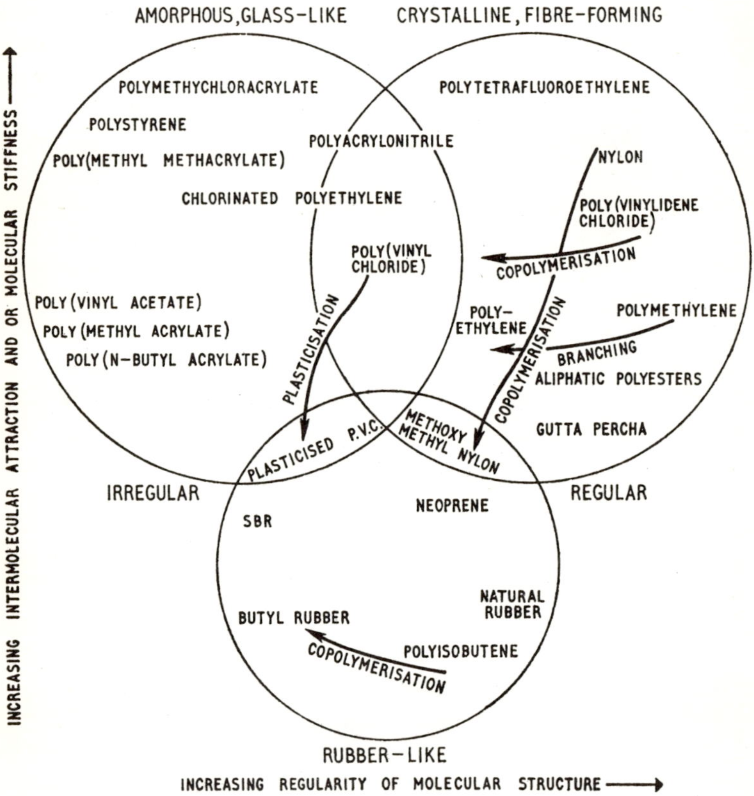

Figure 1-1. Effect of structure regularity, molecular stiffness and intermolecular attraction on polymer properties (after Swallow, ref 2).

certain types of polymer structure and lead to specific practical applications, but when they actually complete their study they often find that these properties were obtained from quite different types of polymer structure and led to quite different types of practical applications.

When the product engineer and industrial designer want to make a new product, they may predict that it should have certain properties and that these should be obtained from particular polymer structures; but, when they have finally completed the successful commercial development of the new product, they may find that it ultimately required entirely different properties which were obtained from wholly different structures.

Even with this limited knowledge and understanding, new polymeric materials offer such fertile practical potential that they have enough successes to more than pay for all of the errors and still provide a profitable growth rate of about 15 percent a year, which means that the industry doubles every 5 years. This is a very healthy rate of commercial growth, and shows every sign of continuing into the foreseeable future.

It should be possible to collect and organize the information now available on the relationships between polymer structure, properties, and applications. Such a collection and organization would make the information more readily available to the polymer industries. It should also make the data more meaningful and thus provide scientists and engineers working in these industries with a much fuller understanding and therefore a much greater ability to predict and control their future developments. This should reduce the cost of research, development, and commercial failures involved in introducing useful new materials and products; and it should thus make polymeric products available more rapidly and at lower cost, and thus produce an even faster rate of growth for the industry. Such an organization of existing knowledge, leading to greater availability, understanding, and control, should do for polymer science and technology, as applied to plastics, what Mendeleeff's periodic table did a century ago for inorganic chemistry.

This book is an attempt to undertake such a collection and organization of our knowledge about the relationship between polymer structure, properties, and applications. It is based both upon the fundamental research published and summarized in the recent scientific literature, and also upon commercial experience—some of it reported in manufacturers' bulletins, much of it available only by direct contact with commercial experience.

In approaching such a complex three-dimensional subject, it should be considered from each of the three points of view in succession, in order to get a full, well-rounded view of the entire subject. For the present approach, we start with the fundamental aspects of polymer structure, and consider how each of these factors affects properties and applications. Proceeding from the smallest to the largest units in polymer structure, we look first at chemical composition—the atoms and functional groups within the polymer molecule; and then in succession at the size and flexibility of individual polymer molecules; intermolecular order and bonding between groups of adjacent polymer molecules; and supermolecular structures involving larger areas in the total product.

Table 1-2. Structural Features of Commercial Polymers

Polymer	Structure
High-density polyethylene	$—(CH_2CH_2)—$
Low-density polyethylene	$—(CH_2CH_2)—(CH_2CH(C_2H_5))—(CH_2CH(C_4H_9))—(CH_2CH(C_nH_{2n+1}))—$
Ionomer	$—(CH_2CH_2)—(CH_2C(CH_3)(C(=O)—O^- Na^+))—$
Ethylene/vinyl acetate copolymer	$—(CH_2CH_2)—(CH_2CH(OC(=O)CH_3))—$
Ethylene/ethyl acrylate copolymer	$—(CH_2CH_2)—(CH_2CH(C(=O)—OC_2H_5))—$
Chlorosulfonated polyethylene	$—(CH_2CH_2)—(CH_2CH(Cl))—(CH_2CH(S(=O)_2Cl))—$
Ethylene/propylene copolymer	$—(CH_2CH_2)—(CH_2CH(CH_3))—$
Polypropylene	$—(CH_2CH(CH_3))—$

Poly-1-butene	$—(CH_2\underset{\displaystyle C_2H_5}{\underset{\mid}{C}}H)—$
Polyisobutylene	$—(CH_2\overset{\displaystyle CH_3}{\overset{\mid}{\underset{\displaystyle CH_3}{\underset{\mid}{C}}}})—$
Poly(4-methylpentene-1)	$—(CH_2\underset{\displaystyle \underset{\displaystyle CH_3CHCH_3}{\underset{\mid}{CH_2}}}{\underset{\mid}{C}}H)—$
Polybutadiene	$—(CH_2CH{=}CHCH_2)—$
Polyisoprene (natural rubber)	$—(CH_2\underset{\displaystyle CH_3}{\underset{\mid}{C}}{=}CHCH_2)—$
Butadiene/styrene rubber (SBR or GR-S)	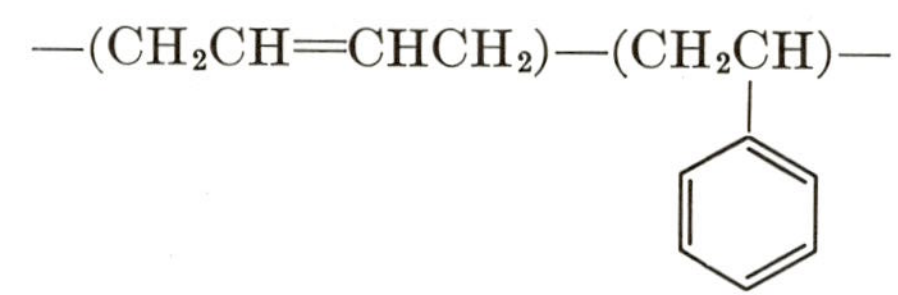
Butadiene/acrylonitrile rubber (nitrile rubber or NBR)	$—(CH_2CH{=}CHCH_2)—(CH_2\underset{\displaystyle C{\equiv}N}{\underset{\mid}{C}}H)—$
Polychloroprene (neoprene)	$—(CH_2\underset{\displaystyle Cl}{\underset{\mid}{C}}{=}CHCH_2)—$

Table 1-2. Structural Features of Commercial Polymers (*Continued*)

Polystyrene	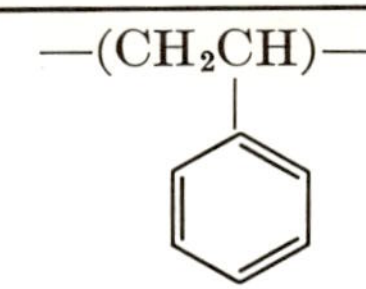
Styrene/acrylonitrile copolymer	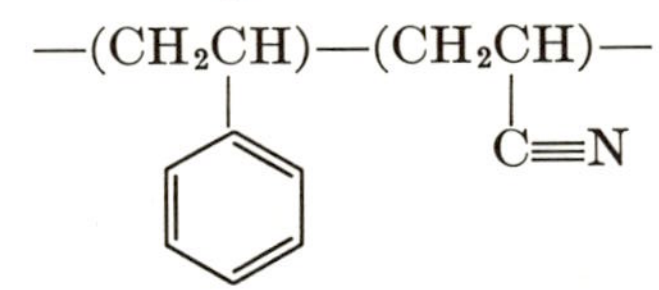
Impact styrene	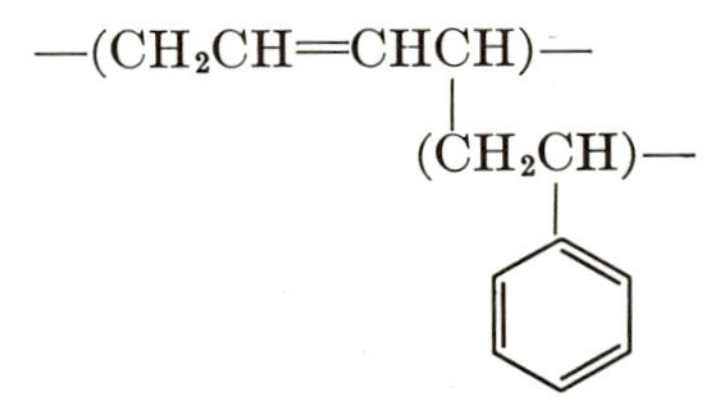
Acrylonitrile/butadiene/styrene terpolymer (ABS)	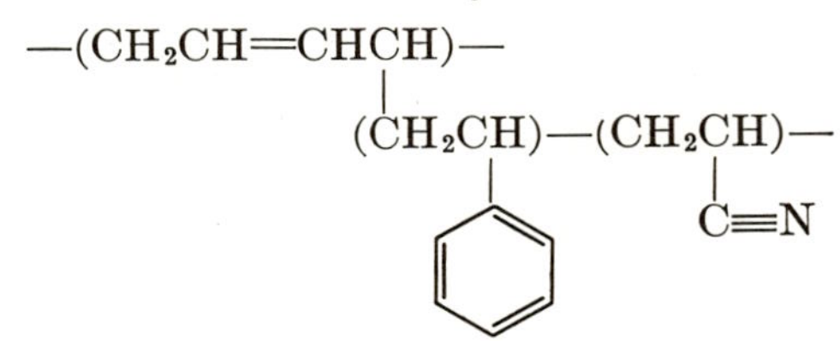
Poly-p-xylylene	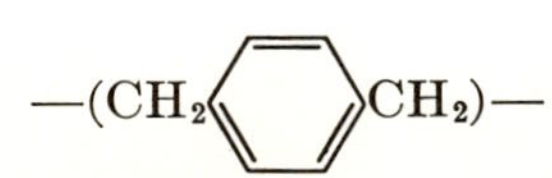

Polymer	Structure
Polyvinyl chloride	$—(\mathrm{CH_2}\underset{\mathrm{Cl}}{\mathrm{CH}})—$
Vinyl chloride/vinyl acetate copolymer	$—(\mathrm{CH_2}\underset{\mathrm{Cl}}{\mathrm{CH}})—(\mathrm{CH_2}\underset{\mathrm{OCCH_3}\ (\mathrm{C{=}O})}{\mathrm{CH}})—$
Polyvinylidene chloride	$—(\mathrm{CH_2}\overset{\mathrm{Cl}}{\underset{\mathrm{Cl}}{\mathrm{C}}})—$
Polytetrafluoroethylene	$—(\mathrm{CF_2CF_2})—$
Fluorinated ethylene/propylene copolymer	$—(\mathrm{CF_2CF_2})—(\mathrm{CF_2}\underset{\mathrm{CF_3}}{\mathrm{CF}})—$
Polychlorotrifluoroethylene	$—(\mathrm{CF_2}\underset{\mathrm{Cl}}{\mathrm{CF}})—$
Polyvinylidene fluoride	$—(\mathrm{CH_2CF_2})—$
Polyvinyl fluoride	$—(\mathrm{CH_2}\underset{\mathrm{F}}{\mathrm{CH}})—$
Vinylidene fluoride/hexafluoropropylene copolymer	$—(\mathrm{CH_2CF_2})—(\mathrm{CF_2}\underset{\mathrm{CF_3}}{\mathrm{CF}})—$

Table 1-2. Structural Features of Commercial Polymers (*Continued*)

Polyvinyl alcohol	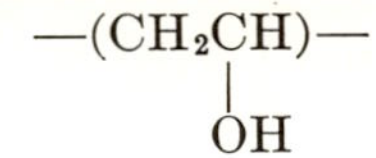
Polyvinyl acetate	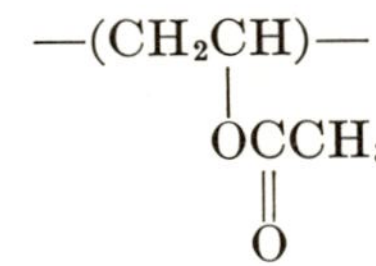
Polyvinyl formal	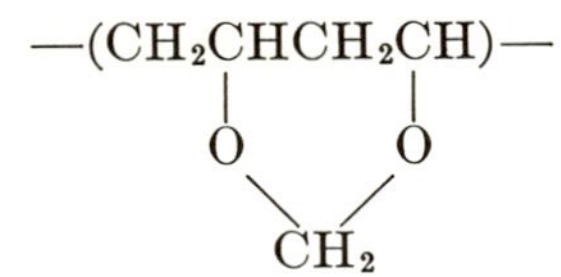
Polyvinyl butyral	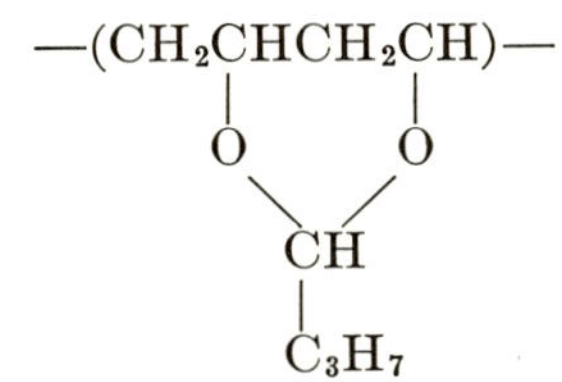
Polyvinyl ethyl ether	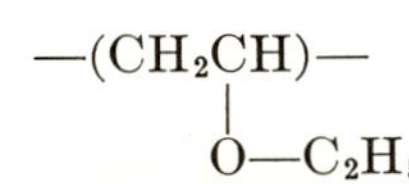
Polyethyl acrylate	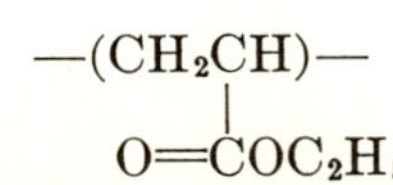

Polymethyl methacrylate	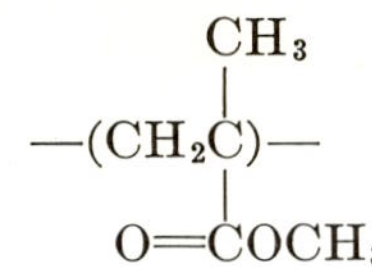
Polyacrylonitrile	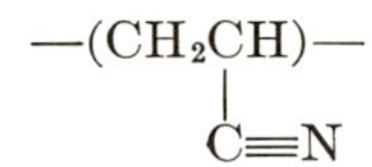
Polyvinyl pyrrolidone	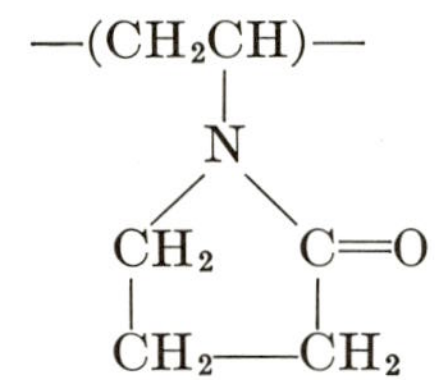
Polyethylene oxide	$—(CH_2CH_2O)—$
Polypropylene oxide	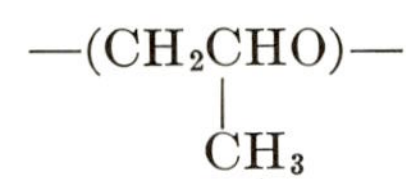
Polyepichlorohydrin	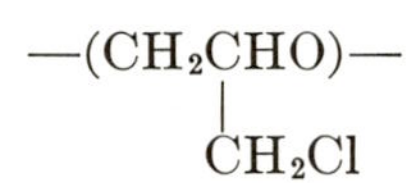
Poly-3,3-bis(chloromethyl) oxetane (Penton)	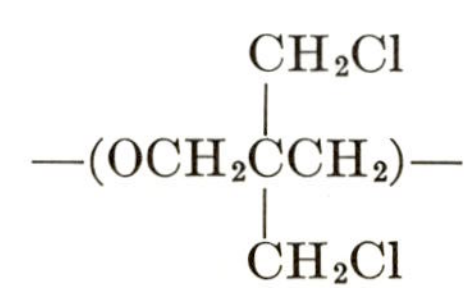

Table 1-2. Structural Features of Commercial Polymers (*Continued*)

Polymer	Structure
Phenoxy resin	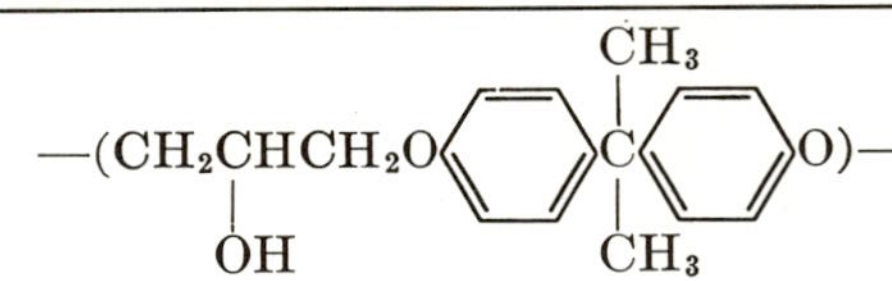
Epoxy resin	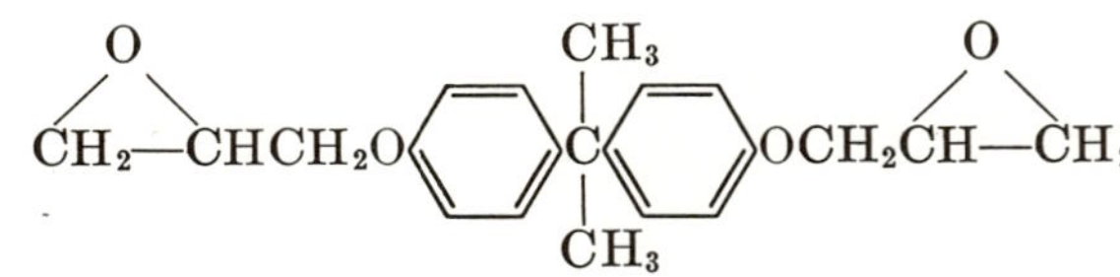
Poly(2,6-dimethylphenylene oxide) (PPO)	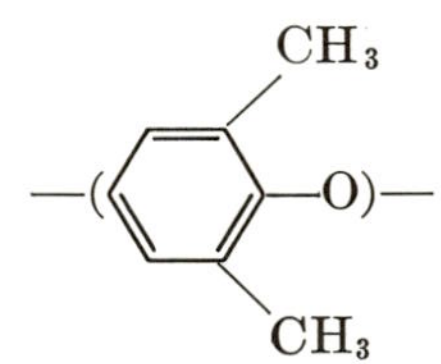
Polysulfone	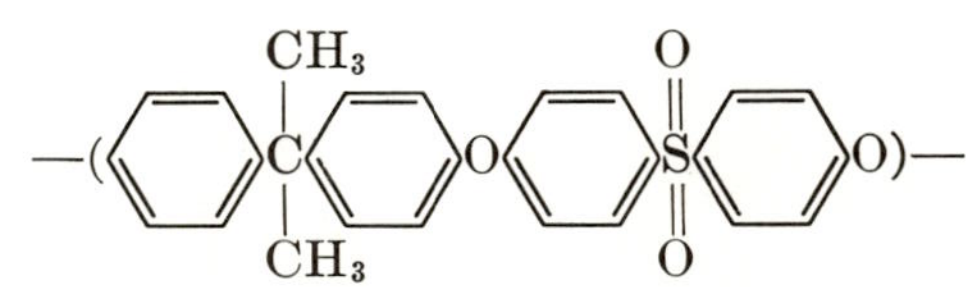
Polysulfide elastomer	$—(CH_2CH_2OCH_2OCH_2CH_2SSCH_2CH_2SS)—$
Polyoxymethylene (acetal)	$—(CH_2O)—$

Cellulose	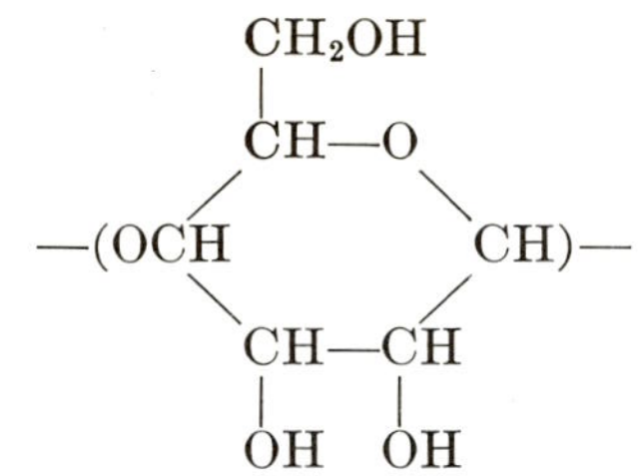
Cellulose triacetate	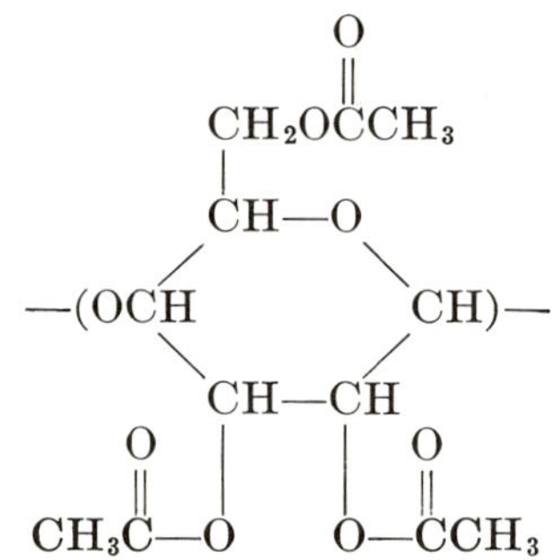
Cellulose nitrate	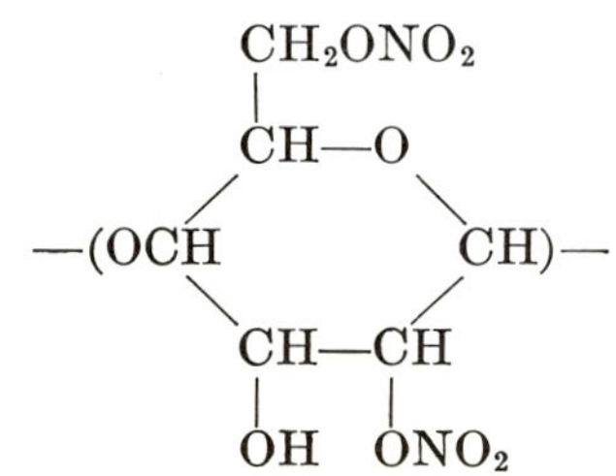

Table 1-2. Structural Features of Commercial Polymers (*Continued*)

Polymer	Structure
Ethyl cellulose	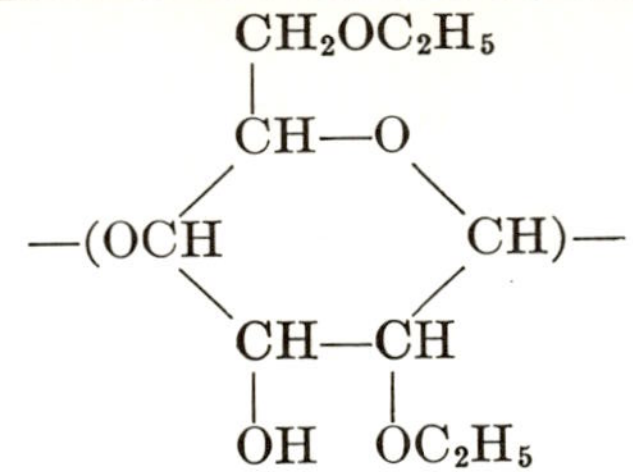
Sodium carboxymethyl cellulose (CMC)	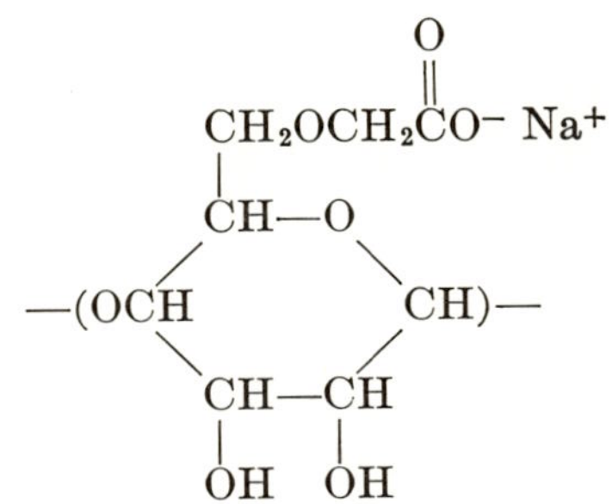
Polycarbonate	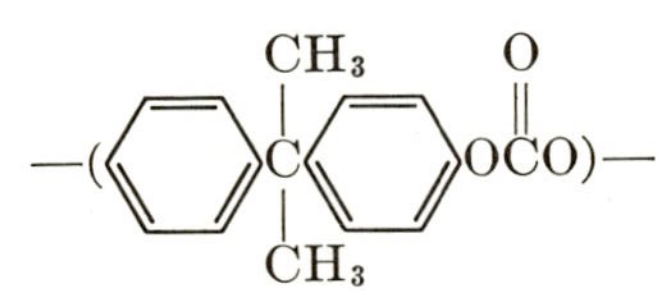
Poly(ethylene terephthalate)	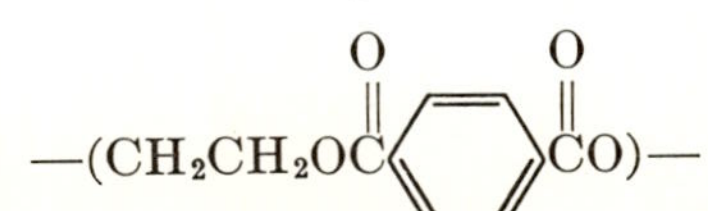

Unsaturated polyester (cured)

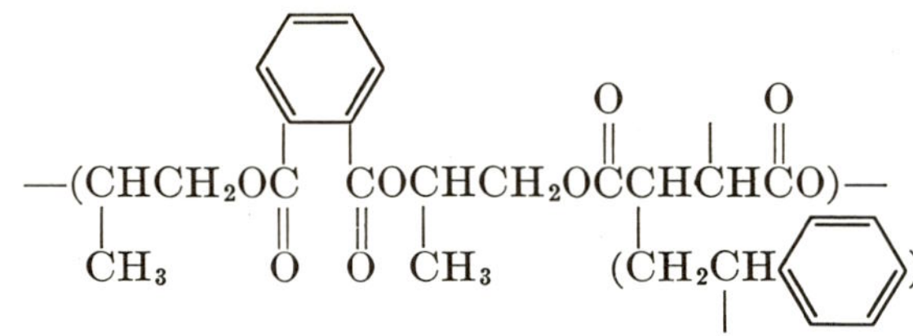

Diallyl phthalate

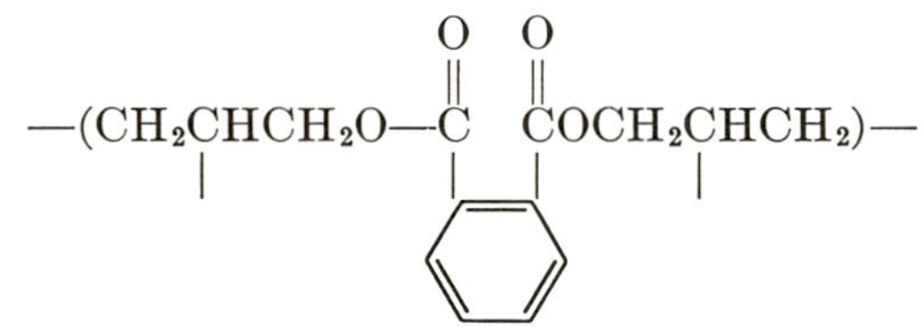

Drying oil

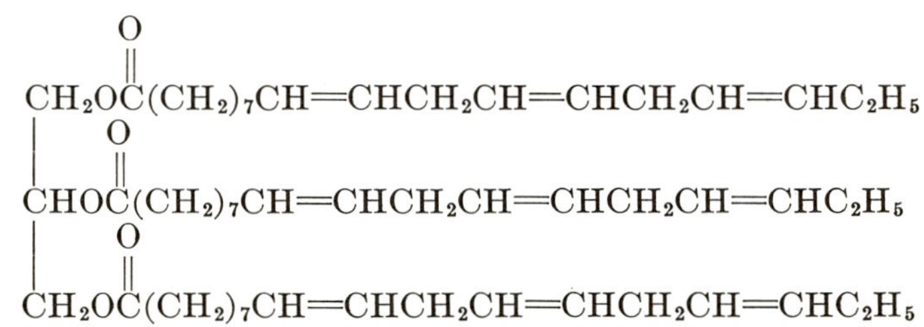

Alkyd

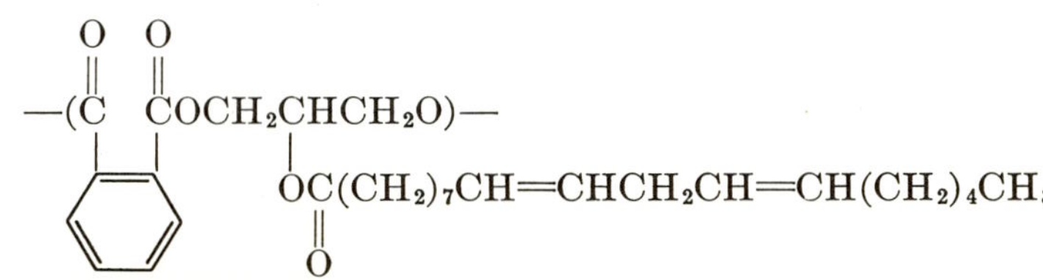

Table 1-2. Structural Features of Commercial Polymers (*Continued*)

Polyether polyurethane	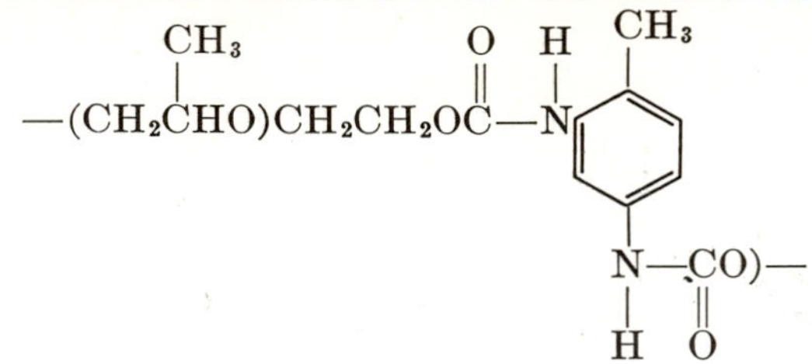
Polyester polyurethane	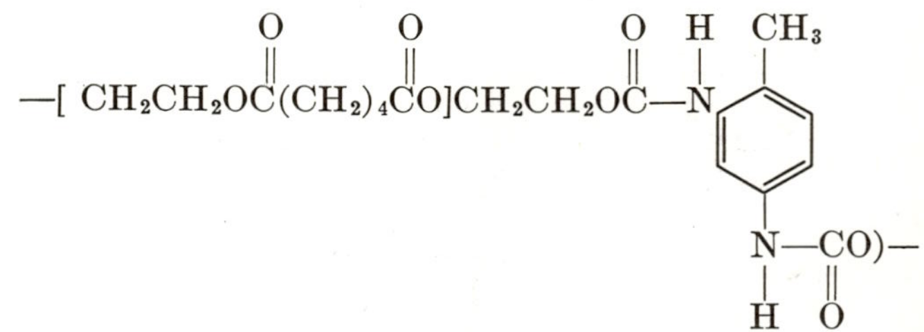
Nylon 6	$\text{—[(CH}_2)_5\text{C(=O)—N(H)]—}$
Nylon 66	$\text{—[(CH}_2)_6\text{N(H)—C(=O)(CH}_2)_4\text{C(=O)—N(H)]—}$
Nylon 610	$\text{—[(CH}_2)_6\text{N(H)—C(=O)(CH}_2)_8\text{C(=O)—N(H)]—}$
Nylon 11	$\text{—[(CH}_2)_{10}\text{C(=O)—N(H)]—}$

Nylon 12	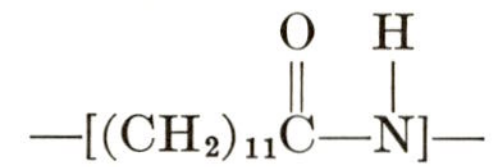
Polyimide	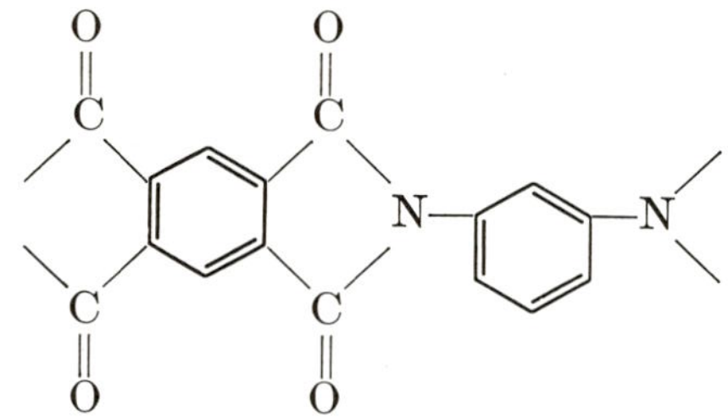
Polybenzimidazole	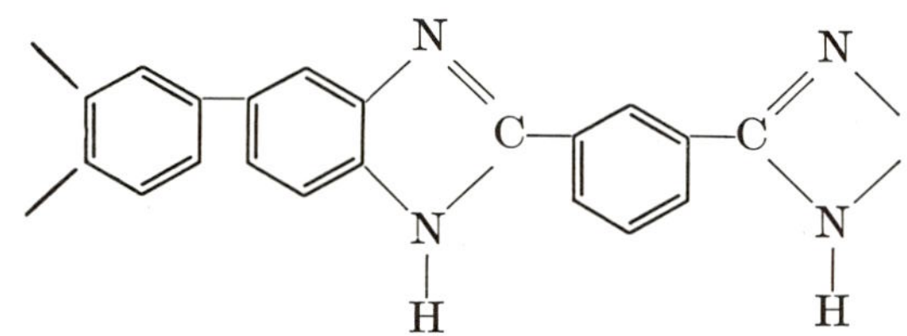
Urea-formaldehyde	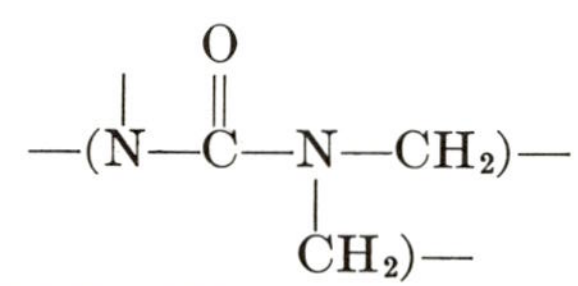

Table 1-2. Structural Features of Commercial Polymers (*Continued*)

Polymer	Structure
Melamine-formaldehyde	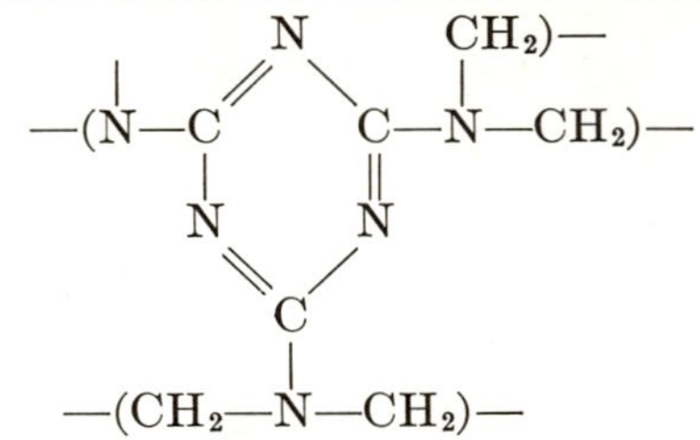
Phenol-formaldehyde	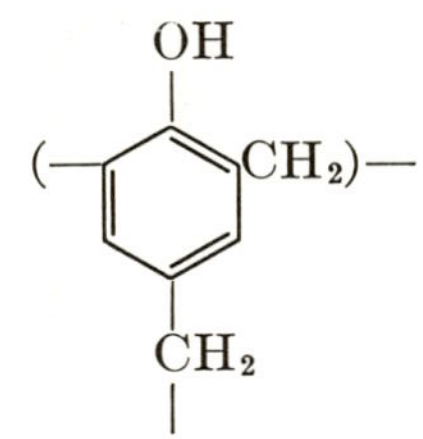
Silicone	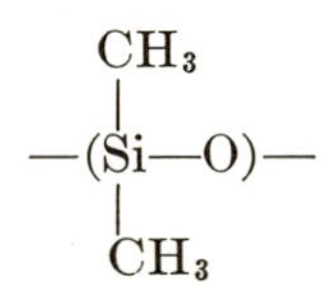

A second approach should consider the different types of polymer properties and the way in which each depends upon structure and leads to applications. Polymer properties can be classified into a series of more or less conventional groups, starting with processability, and continuing on to mechanical, acoustic, thermal, electrical, optical, and chemical properties, as well as cost. This second approach could be completed by considering the additional, more complex problem of achieving a combination and balance of properties for specific applications.

A third approach should consider the different major types of applications of polymers, and for each consider the properties required and the structures that will produce them. Lacking any established technical method of classifying polymer applications, a classification can be developed based upon the gross physical dimensions and qualities of final products: rigid solids, flexible solids, foams, films, coatings, adhesives, fibers, and miscellaneous applications in discontinuous form.

Such a thorough three-dimensional approach should produce a much more complete and more useful understanding of the complex interrelationships among polymer structure, properties, and applications. For the present, this volume will simply undertake the first approach, organizing present knowledge on the basis of the fundamental aspects of polymer structure. The second and third approaches must await full development at a later date.

Before beginning, the reader should be thoroughly familiar with the structural features of commercial polymers (Table 1-2), as these will provide the basis for discussion throughout the entire study. In addition, two structural features should be mentioned briefly at this point because, while they are of major importance, they cut across the organization of this study, and thus occur in every section where they are important, without ever being selected for unified study themselves:

Plasticizers are monomeric or low-molecular-weight polymeric chemicals which are frequently added to polymers to improve their processing or soften their properties. In the terms of the present study, they change the Chemical Composition (Chapter 2) of the polymeric material; introduce lower Molecular Size (Chapter 3) species; are subject to the same considerations of Molecular Flexibility (Chapter 4) as the polymer itself; reduce Intermolecular Order (Chapter 5) and Intermolecular Bonding (Chapter 6); and, when not completely compatible, even affect the Supermolecular Structure (Chapter 7) of the total system. Thus their effects are discussed in each of these places, but not collected into a single section. The same is true of moisture absorbed by polar polymers, which acts as a plasticizer in most of these ways.

Copolymerization combines the structural features of two or more homopolymers into a single polymer molecule, and thus affects Chemical Composition (Chapter 2), Molecular Flexibility (Chapter 4), Intermolecular Order (Chapter 5), and Intermolecular Bonding (Chapter 6). If, in addition, the copolymer is not a random structure, but contains organized groups of monomer units, as in block and graft copolymers (Figure 1-2) the effects on all of those structural features may be further modified; and aggregation of similar regions in such block and graft copolymers may even affect Supermolecular Structure (Chapter 7). Thus the effects of copolymerization are discussed in each of these places, but not collected into a single section.

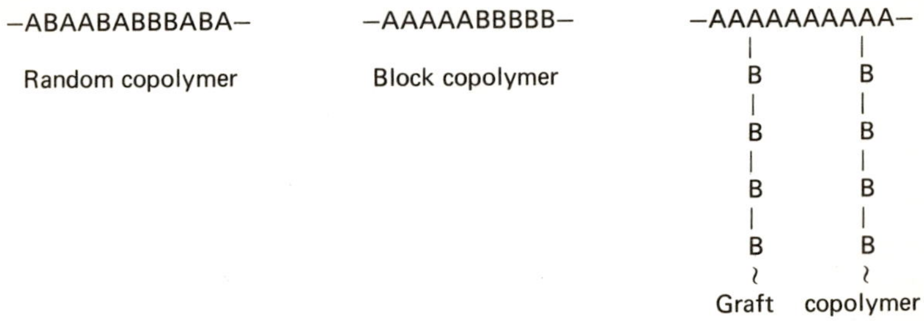

Figure 1-2. Organized groups of monomer units

With these two exceptions, most of the other significant structural features of polymers appear to lend themselves well to the organization of this study, and can each be considered in its proper place.

REFERENCES

1. Schmidt, A. X., & Marlies, C. A., *Principles of High-Polymer Theory and Practice,* McGraw-Hill, 1948, p 95.
2. Brydson, J. A., *Plastics Materials,* Van Nostrand, 1966, p 46.
3. Billmeyer, F. W., Jr., *Textbook of Polymer Science,* Interscience, 1962.
4. Brydson, *op. cit.*
5. Golding, Brage, *Polymers and Resins,* Van Nostrand, 1959.
6. Schmidt, *op. cit.*
7. Alfrey, Turner & Gurnee, E. F., *Organic Polymers,* Prentice-Hall, 1967.
8. Baer, Eric, *Engineering Design for Plastics,* Reinhold, 1964.
9. Miller, M. L., *The Structure of Polymers,* Reinhold, 1966.
10. Nielson, L. E., *Mechanical Properties of Polymers,* Reinhold, 1962.

Part Two

SUBMOLECULAR STRUCTURE

Chapter Two

Chemical Composition

Polymer molecules are distinguished from other chemical compositions primarily by their tremendous size, and most of their properties result from this unusual size. But even far below the molecular size, many of the properties and applications of polymers are greatly influenced by the individual atoms and functional groups within the polymer molecule. In a thorough study of the relationship between polymer structure, properties, and applications, it is therefore important to begin by considering these submolecular aspects of chemical composition.

These relationships may be approached in a very general way by listing some of the chemical structural features which are most important to each major class of polymer properties (Table 2-1). Such a tabulation forms a useful starting point for this study, by indicating the general types of relationships which will be involved. In itself, however, it does not provide the specific practical utility which we are seeking.

For a more specific, detailed study of the effects of chemical composition upon polymer properties and applications, it is best to consider first the effects of the atoms and functional groups in the polymer molecule itself, and later the effects of the chemical composition of impurities remaining from the polymerization reaction, and of other monomeric materials purposely added during compounding.

Table 2-1. General Relationships between Chemical Composition and Polymer Properties

General features of chemical composition	Effects on polymer properties
	Processability
Reactivity	Pot life, polymerization, & cure
Bond strength	Thermal stability
Similarity of atoms, groups, polarity, & H-bonding	Solubility
Volatile impurities	Bubbles & toxicity
	Mechanical properties
Volatile impurities	Low rigidity, strength, & dimensional stability
Surface energy	Friction
	Acoustic properties
Atomic weights	Sound absorption
	Thermal properties
Atomic constants	Specific heat
Bond strength	Thermal stability
Specific elements	Flame resistance
	Electrical properties
Polarity of bonds	Dielectric constant
Polar & ionic impurities	Conduction
Conjugated unsaturation	Conduction
	Optical properties
Polarity of bonds	Refractive index, spectroscopic absorption
Conjugated unsaturation	Color, ultraviolet light sensitivity
	Chemical properties
Similarity of atoms, groups, polarity, & H-bonding	Solubility & permeability
Bond strength	Reactivity & stability
Surface energy	Adhesion
Specific elements & functional groups	Toxicity
	Price
Elements & groups	Cost of synthesis
Atomic weights	Specific gravity

2-1 CHEMICAL COMPOSITION OF THE POLYMER MOLECULE

The effects of atoms and functional groups in the polymer molecule are best understood by reference to the position of their elements in the periodic table (Table 2-2). Their electropositivity or electronegativity (Table 2-3) affects the strength and type of bonds between pairs of atoms, which in turn affect polymer reactivity, thermal stability, ultraviolet light stability, weather resistance, and aging in general. The electropositivity or electronegativity of the atoms forming a bond also determines the polarity of the bond, which in turn determines polymer electrical properties, refractive index (ref 2), solubility, permeability, surface energy, surface tension, adhesion, and friction. Atomic weight is a major factor determining polymer specific gravity, and other atomic constants may be directly additive in determining properties such as specific heat.

Table 2-2. Periodic Arrangement of Elements Commonly Found in Polymers

	Group						
Period	I	II	III	IV	V	VI	VII
2				C	N	O	F
3	Na	Mg		Si	P	S	Cl
4		Ca					Br

Table 2-3. Electronegativity of Elements Commonly Found in Polymers (ref 1)

Sodium	0.9	Phosphorus	2.1	Nitrogen	3.0
Calcium	1.0	Carbon	2.5	Chlorine	3.0
Magnesium	1.2	Sulfur	2.5	Oxygen	3.5
Silicon	1.8	Bromine	2.8	Fluorine	4.0
Hydrogen	2.1				

While a rigorous approach might proceed directly through the periodic table, it is better for our present purposes to consider the elements in order of their importance in polymer composition.

Carbon

Carbon is the most important element in organic chemistry and polymer chemistry, and forms all or most of the main chain in most polymer molecules. Carbon itself is neither electropositive nor electronegative; and the C—C bond by itself is completely neutral and very stable, as we see in diamond and graphite. The energy and strength of the C—C bond are quite sufficient for thermal stability and for ultraviolet light stability (ref 3). The instability of the quaternary carbon atom in polymer structure (ref 4) is probably due to steric hindrance which favors depolymerization.

Attachment of other elements to the carbon atom introduces polarity into the C—C bond and thus lowers its stability. Most of the effects of chemical composition upon polymer properties are due to the attachment of these other elements to carbon.

In the C═C double bond the two pairs of electrons between two carbon atoms are more than can be accommodated electrically with good stability. This makes the C═C double bond very reactive with oxygen, ozone, halogens, and hydrogen halides (ref 5), reactions which are important in the utilization of rubber. When the C═C double bond is in the alpha position at the end of the molecule, it is very often polymerizable by free radical or ionic mechanisms, particularly with application of appropriate initiators. This is very important in monomer casting, particularly of methyl methacrylate, diallyl phthalate and diethyleneglycol bis(allyl carbonate) for glazing and lenses. Even internal C═C double bonds, in the interior of the polymer molecule, may be sufficiently activated by adjacent groups to become readily polymerizable, for example in the cure of unsaturated polyesters made from maleic or fumaric acids, particularly for reinforced plastics. When there are a large number of C═C or other double bonds on alternating pairs of atoms in the polymer molecule, such multiple conjugated unsaturation produces chromophoric structures which are responsible for color (ref 6), for example in the deterioration of polyvinyl chloride. When such multiple conjugated unsaturation is extended very broadly throughout the polymer structure, it may give rise to a fairly free-floating cloud of electrons which may even permit electrical conductivity (Figure 2-1) (ref 8).

In aromatic structures, on the other hand, the unsaturation of the C═C double bond is completely stabilized by resonance, and these aromatic structures generally provide high thermal stability in polymer molecules, especially in the newer high-temperature resins.

Hydrogen

After carbon, hydrogen is the next most common element in organic chemistry and in polymer structure. While it behaves fairly neutrally, forming covalent bonds with carbon, these bonds are somewhat more reactive than the perfectly balanced C—C bond, and often form the point of attack which produces chemical reactivity or instability in the

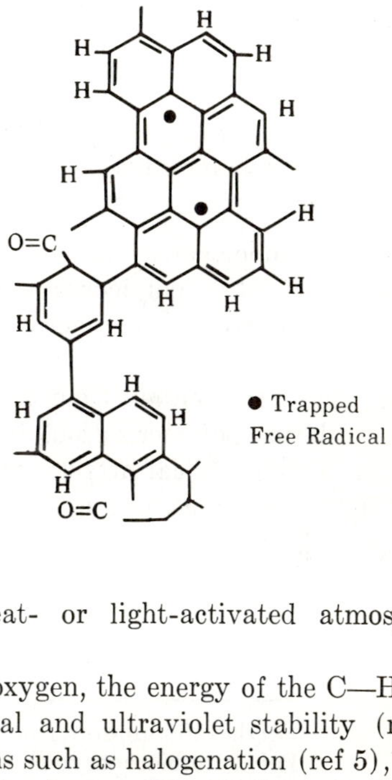

Figure 2-1. Structure of semiconductive pyrolyzed poly(divinyl benzene) (ref 7).

polymer molecule, particularly heat- or light-activated atmospheric oxidation.

In the absence of atmospheric oxygen, the energy of the C—H bond is sufficient to provide good thermal and ultraviolet stability (ref 3). It does undergo substitution reactions such as halogenation (ref 5), which are important in forming chlorinated polymers. The aliphatic C—H bond also has low surface energy, producing low surface tension (Table 2-4), low adhesion (Table 2-5), and often low friction (Table 2-6) (ref 10).

The aromatic C—H bond is quite stable, but still permits substitution reactions (ref 5). It has moderate surface tension, producing somewhat less lubricity and higher adhesion (ref 12). When the aromatic ring is activated by groups such as phenolic hydroxyl or amine, these greatly activate C—H bonds in the ortho and para positions, and this

Table 2-4. Surface Tension of Polymers (ref 9)

Polymer	Surface tension (dynes/cm at 20°C)
Polyfluoroester of acrylic and methacrylic acid	10–12
Polyhexafluoropropylene	16.2
Polytetrafluoroethylene	18.5
Polytrifluoroethylene	22
Polyvinylidene fluoride	25
Polyvinyl fluoride	28
Polyethylene	31
Polytrifluorochloroethylene	31
Polystyrene	33
Polyvinyl alcohol	37
Polyvinyl chloride	40
Polyvinylidene chloride	40
Polyethylene terephthalate	43
Polyhexamethylene adipamide	46

reactivity is important in the polymerization and cure of phenolic resins, and in the oxidative polymerization of phenols to polyphenylene oxides, and even of benzene to polyphenylene.

Tertiary hydrogen atoms, on carbon attached to three different groups, can result either from monomer structure or from branching reactions during polymerization. Such tertiary hydrogen atoms are quite sensitive to free radical attack, producing oxidation (ref 5) and ultraviolet light scission, as well as substitution and cross-linking reactions.

Table 2-5. Adhesive Strength of Polymers (ref 10)

Polymer	Work of adhesion (ergs/cm^2)
Polyhexafluoropropylene	52.5
Polytetrafluoroethylene	58.0
Polytrifluoroethylene	72.5
Polystyrene	78.0
Polyethylene	82.0
Polyethylene terephthalate	88.0
Polyvinylidene fluoride	88.5
Polyvinyl fluoride	92.0
Polyvinyl chloride	92.0
Polyvinylidene chloride	92.0
Polyhexamethylene adipamide	97.0

When the hydrogen atom is on a carbon atom adjacent to an activating group, such an alpha hydrogen may be even more reactive, particularly toward ultraviolet light and cross-linking (ref 3). This is especially significant in allyl structures, where the allylic hydrogen alpha to the C=C double bond becomes very reactive to free radical and oxidative attack;

$$\sim CH{=}CH{-}\overset{\overset{\displaystyle H}{|}}{C}H\sim \;\rightarrow\; \sim CH{=}CH{-}CH\sim$$

this is used in vulcanization of rubber and in oxidative cure of drying oils. Such reactions can also be observed on C—H bonds adjacent to carboxyl groups, such as polyacrylic esters.

Table 2-6. Coefficients of Friction for Some Common Plastics (ref 11)

Polymer	Steel on polymer Static	Steel on polymer Kinetic	Polymer on polymer Static	Polymer on polymer Kinetic
PTFE ("Teflon") (polytetrafluoroethylene)	0.10	0.05	0.04	0.04
PTFE-HFP copolymer (FEP "Teflon") (tetrafluoroethylene/ hexafluoropropylene)	0.25	0.18	—	—
Polyethylene (low density)	0.27	0.26	0.33	0.33
Polyethylene (high density)	0.18	0.08–0.12	0.12	0.11
Acetal resin ("Delrin")	0.14	0.13	—	—
Polyvinylidene fluoride	0.33	0.25	—	—
Polycarbonate	0.60	0.53	—	—
PET("Mylar") (polyethylene terephthalate)	0.29	0.28	0.27*	0.20*
Nylon (polyhexamethylene adipamide)	0.37	0.34	0.42*	0.35*
PFCE("Kel-F") (polytrifluorochloroethylene)	0.45*	0.33*	0.43*	0.32*
PVC(polyvinyl chloride)	0.45*	0.40*	0.50*	0.40*
PDVC (polyvinylidene chloride)	0.68*	0.45*	0.90*	0.52*

* "Stick-slip" (intermittent motion).

When hydrogen is attached to elements other than carbon, it becomes much more reactive. In particular the hydrogen attached to oxygen in hydroxyl groups may be highly reactive (ref 5); this is important in formation of cellulose esters, polyesters, and polyvinyl acetals, and in the reactivity of methylol groups on urea, melamine, and phenolic resins during polymerization and cure. In finished polymers such as polyvinyl alcohol and cellulose, it also produces high polarity and hydrogen bonding, increasing dielectric properties and water absorption.

Oxygen

Oxygen is the third most common element in organic chemistry and in polymer structure. It is much more electronegative than carbon and hydrogen, and the bonds it forms with carbon and hydrogen are much more polar and reactive. After it has formed these bonds, the oxygen atom still has two more pairs of electrons fairly available, and generally lends these for the formation of fairly strong hydrogen bonds to adjacent molecules (Figure 2-2). The presence of oxygen atoms in polymer structure also frequently produces high surface energy and surface tension (ref 13).

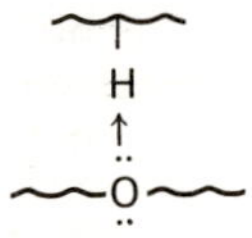

Figure 2-2.

The stability of the C—O—C bond depends very much upon the groups to which it is attached: in aromatic ethers it is extremely stable; in aliphatic ethers it activates the adjacent C—H bond, causing easy peroxidation of the alpha hydrogen. In acetals the C—O—C—O—C bond is particularly sensitive to acid hydrolysis, which is a significant reason for instability in polyvinyl acetals, polyformaldehyde, and cellulosic polymers. In epoxy resins, particularly glycidyl aromatic ethers, the presence of oxygen in a highly-strained three-membered ring produces very high reactivity, which permits the easy cure of epoxy resins, even at room temperature.

The C=O carbonyl bond is a resonating structure which affects the polarity and reactivity of adjacent groups. It also absorbs ultraviolet light strongly in the range of 2800–3200 Å, and this can cause instability of the polymer toward ultraviolet light aging (ref 14). On the other hand, proper use of the carbonyl bond is important in the design of ultraviolet light absorbers, which are added during compounding to stabilize the polymer against ultraviolet light aging.

The —CO_2— group in polyesters, and in the side chain of polymers made from ester type monomers, is quite sensitive to hydrolysis (ref 5). This reactivity is particularly useful in the conversion of polyvinyl acetate to polyvinyl alcohol, but is a problem in the use of reinforced polyesters for chemical process equipment. The —CO_3— group in polycarbonates is also sensitive to hydrolysis, particularly in aliphatic structures, but even in aromatic structures under alkaline conditions.

The —CO_2H group in carboxylic acids can be useful for polyesterification and polyamidification in the cure of low-molecular-weight prepolymers. The presence of carboxylic acid groups, in the side chains of polymer molecules, can be very useful for producing adhesion of coatings and adhesives to substrates, dispersability and dispersion-stability of latex, solubility in alkali, and thermoplastic ionic cross-linking by metals in ionomers. On the other hand, the presence of residual carboxylic acid groups in polyesters can have very harmful effects on electrical properties (ref 15).

The O—O bond of peroxides is very unstable. This is useful in peroxide initiation of free radical polymerization and cure. It can be very detrimental when peroxides appear in the structure of the final polymer, either through copolymerization with oxygen impurities during synthesis, or through post-polymerization atmospheric peroxidation of tertiary hydrogens, alpha hydrogen, or C═C double bonds in the polymer structure. The presence of such weak O—O peroxide bonds in the polymer structure makes it particularly sensitive to thermal degradation.

Nitrogen

Nitrogen is the fourth most common element in organic chemistry and polymers. It is intermediate between carbon and oxygen in the periodic table, and therefore in its electronegativity. Bonds between carbon and nitrogen are covalent and fairly strong, but also fairly polar, providing strong adhesion to polar substrates. When nitrogen has formed its normal three covalent bonds, it still has a pair of unshared electrons, which it lends readily to form strong hydrogen bonds to adjacent molecules (Figure 2-3). Although the C—N bond is generally quite stable in quaternary ammonium compounds such as ion-exchange resins and cationic detergents, in antistatic structures it is sensitive to thermal decomposition, forming volatile tertiary amines which cause a very undesirable fishy odor in some plastics processes.

The —C≡N: nitrile group, which is present in ABS and SAN resins,

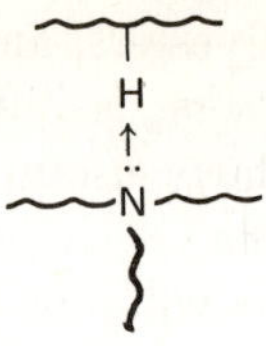

Figure 2-3.

acrylic fiber, and nitrile rubber, is strongly polar and the unshared pair of electrons on the nitrogen is highly exposed and available for very strong hydrogen bonding. A series of adjacent side-chain nitrile groups, such as are found in polyacrylonitrile acrylic fiber, tend to react with each other, particularly under the influence of heat and ionic impurities, to form a ladder-type polymer with multiple conjugated unsaturation which causes discoloration (Figure 2-4); when carried to extremes by oxidative pyrolysis, this highly conjugated unsaturation can even produce electrical conductivity (ref 16). When acrylonitrile polymers are burned, it is possible that HCN gas may be released, forming very toxic conditions; and this has occasionally caused concern to underwriters and industrial designers in textile, building, and aircraft applications.

The N—H bond is fairly reactive, which is important in the polymerization of polyamides and in the cure of epoxy, urethane, urea, and

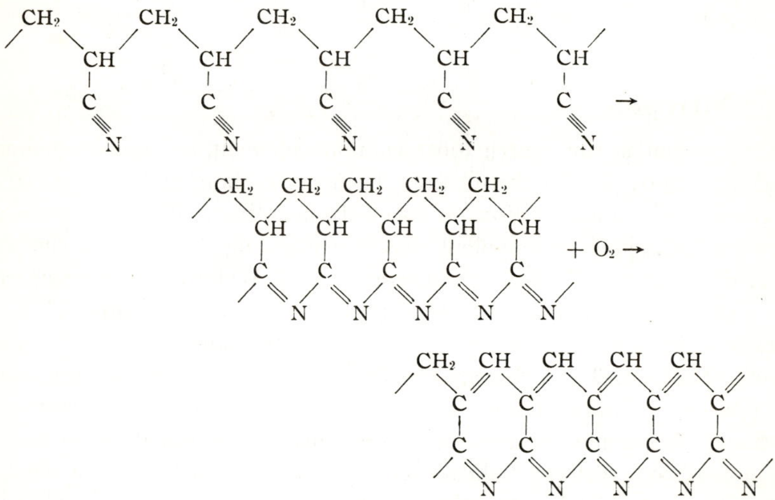

Figure 2-4. Pyrolyzed polyacrylonitrile (ref 16).

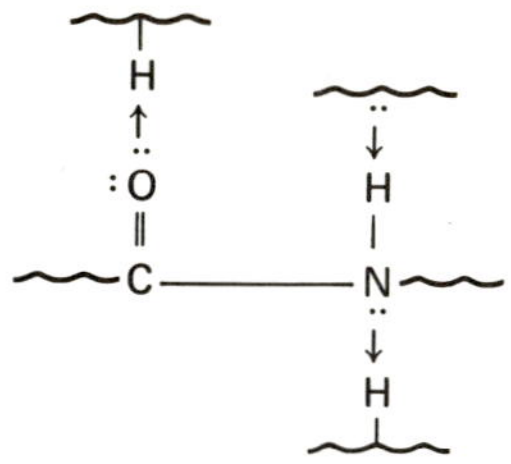

Figure 2-5.

melamine resins, as well as in the modification of the properties of nylons.

The $\overset{\overset{O}{\|}}{C}$—$\overset{\overset{H}{|}}{N}$ bond in polyamides introduces polarity and powerful hydrogen bonding (Figure 2-5). This is responsible for water absorption even in moderately moist atmospheres (Figure 2-6), which causes dimensional instability (Figure 2-7) and, under more drastic conditions, hydrolysis and degradation as well (ref 5). It is also responsible for very high surface energy and surface tension which are useful in adhesives (ref 12).

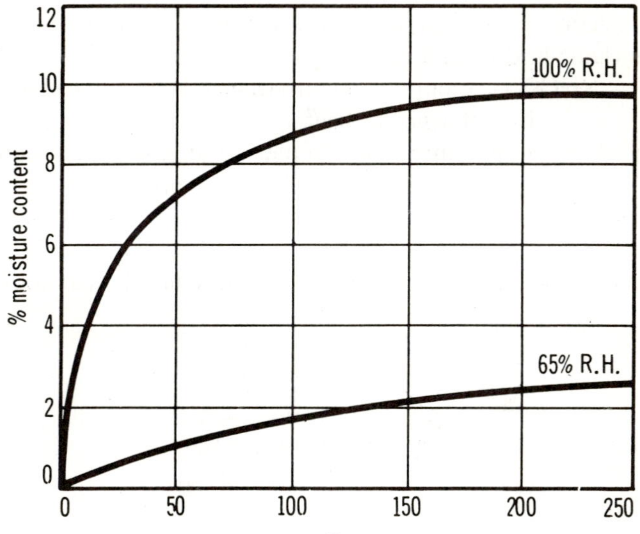

Figure 2-6. Moisture absorption of nylon 6 (ref 17).

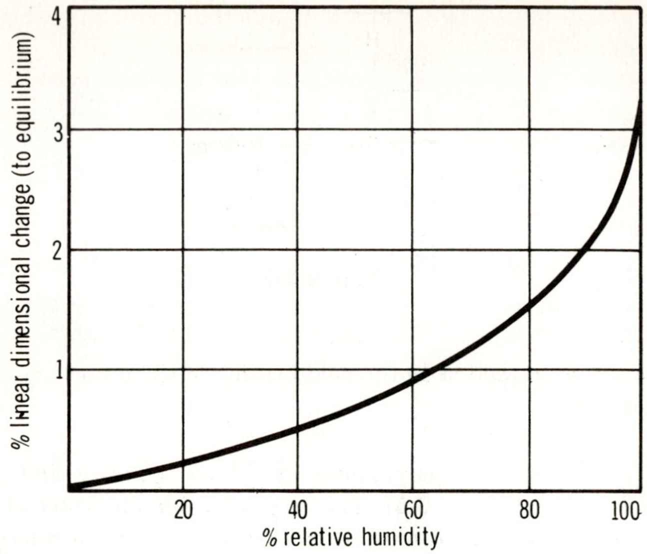

Figure 2-7. Effect of humidity on dimensional stability of nylon 6 (ref 17).

The —N=C=O group in isocyanate prepolymers is highly reactive and polymerizable (Figure 2-8), and this is used both in the casting of polyurethane elastomers and adhesives and in the foaming and cure of polyurethane flexible and rigid foams.

The —$NHCO_2$— group in polyurethanes is fairly stable, but may decompose under the influence of heat.

In heterocyclic polymers the alternation of single and double bonds, between carbon and nitrogen and sometimes other elements, provides for high resonance stabilization which produces extremely high thermal stability, and is therefore very useful in high-temperature thermosetting resins such as melamine-formaldehyde and triallyl cyanurate cure of

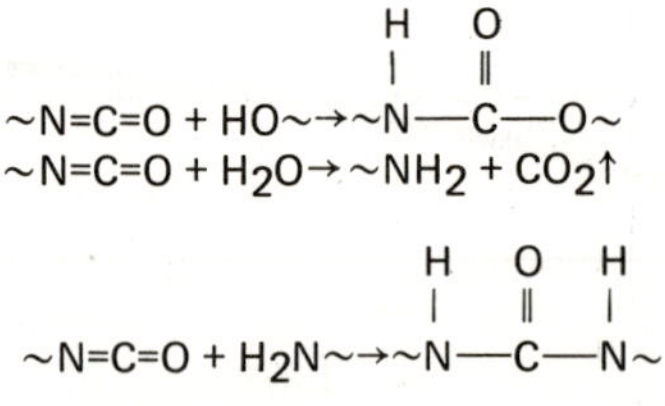

Figure 2-8.

polyesters, and is most dramatic in many of the newest ultra-high-temperature resins which are currently being developed.

The —ONO_2 group in nitrocellulose acts like a powerful built-in oxidizing agent, producing useful explosive action in gun cotton, but undesirably high flammability in plastics and films.

Halogens

The halogen elements in Group VII of the periodic table are the most negative elements known, and also have the smallest atomic radii. It has been reported that halogenated polyethylenes are the only polymers which are transparent to infrared (ref 18).

Fluorine

Fluorine is the most negative and smallest of the halogens, almost as small as hydrogen. The C—F bond has very low surface energy, giving low surface tension, high lubricity, and nonwetting properties to a high degree (ref 12). These properties are particularly useful in stain-proofing fabrics, nonstick cooking utensils, mold lubricants, and self-lubricating bearing surfaces.

The C—F bond is chemically very stable, especially in perfluorinated structures. It has high thermal and ultraviolet light stability (ref 19). This stability is useful in gaskets, and in high-temperature plastics and elastomers, especially for instruments and aerospace applications. It has been reported that when fluorine and hydrogen are present on adjacent carbon atoms, such structures may tend to split out HF. In elastomeric fluorocarbon copolymers, the presence of hydrogen or chlorine atoms attached to some of the carbon atoms provides sufficient reactivity for vulcanization (ref 5). In certain cases thermal or explosive decomposition of fluoropolymers has been known to produce highly toxic gaseous decomposition products, and this is occasionally a cause for concern in certain specialized applications.

Chlorine

Chlorine is slightly less negative and larger than fluorine, but still forms quite strong covalent bonds with carbon. The presence of chlorine in the polymer structure is very effective in providing flame-resistance, one of the major advantages of polyvinyl chloride in building, electrical, and transportation applications. The chlorine creates strong polarity and

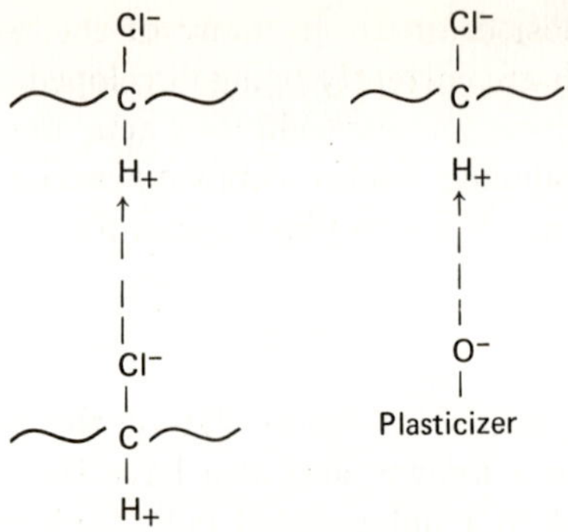

Figure 2-9.

perhaps some hydrogen-bonding (ref 32) (Figure 2-9), thus providing both strength and plasticizer compatibility in polyvinyl chloride. Because of its higher atomic weight, it also produces polymers of higher specific gravity. The high polarity and surface energy of the C—Cl bond also produce high surface tension and adhesion (ref 12).

The existence of chlorine and hydrogen atoms on adjacent carbon atoms in the polymer molecule is a source of thermal, ultraviolet, and chemical instability, particularly when the chlorine is on a tertiary or allylic carbon atom (Figure 2-10). Hydrogen and chlorine atoms have a tendency to split out of the polymer molecule and form HCl, leaving behind an unsaturated C═C bond in the polymer molecule. The HCl formed then autocatalyzes further degradation, and the C═C double bonds formed cause allylic activation of the same mechanism, leading to discoloration (Section on carbon).

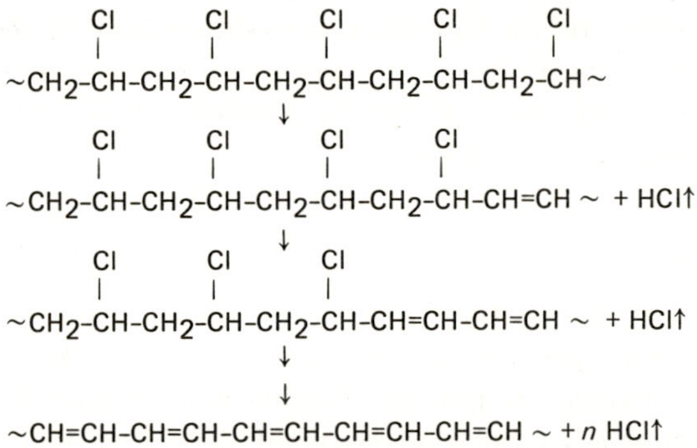

Figure 2-10.

Bromine and Iodine

Bromine is considerably larger and less electronegative than chlorine. It forms fairly stable bonds with carbon, but splits out fairly easily, particularly under the influence of heat or ultraviolet light, usually causing serious discoloration. Its higher atomic weight produces higher density in the resulting polymers. Its most useful effect is a high degree of flame resistance when even small amounts of bromine are incorporated into the polymer composition.

Iodine is the largest and least electronegative of the halogens, and the bonds it forms with carbon are quite unstable, so that it is hardly ever found in polymer molecules. If it could be firmly bound, it would probably provide high density, high flame-resistance, and certain biocidal effects which might be quite useful.

Sulfur

Sulfur is fairly similar to oxygen in its chemical properties, but less electronegative and larger in size. The sulfide link in aliphatic polymers is quite sensitive to atmospheric or chemical oxidation. The sulfide

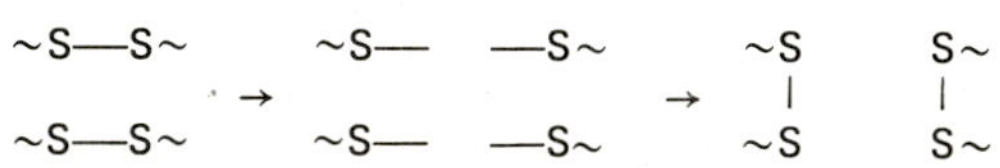

Figure 2-11.

link in aromatic polymers, however, has proved useful in high-temperature resins, suggesting that the resonance and steric hindrance of the benzene rings are more than sufficient to protect the sulfide group against oxidation.

The —S_x— group of polysulfide rubbers is fairly weakly bonded and fluid, providing the chemical reactivity which is useful in equilibration and vulcanization, but also permitting stress-relaxation during aging under tension (ref 20) (Figure 2-11).

The —SO_2— group in polymeric sulfones is itself quite stable, and the oxygens projecting from it each have two pairs of unshared electrons to donate for strong hydrogen bonding to adjacent molecules. In aromatic polymers the sulfone group participates in conjugated resonance and provides high thermal stability, but is sensitive to ultraviolet light degradation. In aliphatic vinyl sulfone polymers, on the other hand, thermal stability is so low that such polymers often depolymerize back to their monomers even at room temperature or below, and have never found practical utility.

Silicon

Silicon is somewhat more electropositive and larger than carbon. This makes Si—Si bonds very unstable, so that they are of no practical value in the formation of polymers. The Si—H bond is also quite unstable and very reactive, and this is useful in preparing silicon derivatives and in delicate control of the cure of silicone polymers. The Si—C bond in organosilanes is very stable, making silicone polymers useful in high-temperature elastomers. The methyl groups attached to silicon contain C—H bonds which are quite stable, but are still sufficiently active to permit some vulcanization by peroxides under vigorous conditions. The projecting methyl groups also provide surfaces of low energy and surface tension, and high lubricity, which are useful in surfactants and in release coatings on paper, mold surfaces, and cooking utensils.

The Si—OH silanol bond is very reactive, and polymerizes rapidly to form silicones. The Si—O—R organosiloxane bond is also quite reactive and polymerizes or reacts with substrates fairly readily. The Si—O—Si bond in both glass and silicone polymers is very stable to heat. It is labile to acid and alkali, and this reaction is useful in the preparation of silicone polymers by equilibration. All of these reactions are used in the polymerization, casting, vulcanization and cure of silicone elastomers, resins, and coatings.

Phosphorus

Phosporus is slightly more electropositive than nitrogen. Its presence in polymers invariably produces the greatest flame-retardant effects known, probably due to the formation of phosphorus-oxygen glasses at the surface of a burning material, and it is most often used for these effects. The P—O—C bond often provides useful structures for incorporation into polyethers, polyesters and polyurethanes. The direct P—C bond may sometimes be accompanied by very high toxicity, and is not commonly used in production of polymers.

Metals

A variety of metallic elements are occasionally useful in polymer structures. Sodium ion is commonly used in ionomers to provide fairly

Table 2-7. Effect of Chemical Composition on Prices of Polymers (ref 22–24)

Elements	Polymers	Prices
C,H	Low-density polyethylene	$0.1125
	Polystyrene	0.155
	Impact styrene	0.175
	High-density polyethylene	0.18
	Polybutadiene	0.20
	Polypropylene	0.21
	Butadiene/styrene rubber	0.21
C,H,Cl	Polyvinyl chloride	0.15
	Polyvinylidene chloride	0.34
C,H,O	Phenol-formaldehyde	0.215
	Polyester	0.22
	Polyvinyl acetate	0.29
	Cellulose acetate	0.40
	Polyvinyl alcohol	0.43
	Polymethyl methacrylate	0.455
	Epoxy	0.49
	Acetal	0.65
C,H,(O),N	Styrene/acrylonitrile	0.26
	Acrylonitrile/butadiene/styrene	0.32
	Urea-formaldehyde	0.32
	Melamine-formaldehyde	0.36
	Butadiene/acrylonitrile rubber	0.45
	Polyurethane	0.59
	Nylon	0.75
C,H,(O),S	Polysulfide rubber	0.85
	Polysulfone	1.00
C,H,O,Si	Silicone resin	1.90
	Silicone rubber	3.66
C,F	Polytetrafluoroethylene	3.25
	Perfluorinated ethylene/propylene	4.95

strong thermoplastic cross-linking (ref 21) which combines processability and toughness in the final polymer. Polymers containing high concentrations of sodium carboxylate side groups are soluble in water, and may be used for thickening foods (carboxymethyl cellulose), latex paints (polyacrylic acid), and other aqueous materials.

Several Group II metals are frequently used in vulcanization of elastomers, particularly magnesium, calcium, and zinc. Calcium also provides cross-linking when calcium acrylate is polymerized for soil stabilization.

Several of the transition metals have been examined in basic research on the preparation of chelated structures for high-temperature thermal stability, but these have not yet led to any successful commercial applications.

It is appropriate at this point to note briefly the effect of chemical composition upon the last and often most important property of polymers, their price (Table 2-7). In general, polymers based simply upon carbon and hydrogen are readily produced from very low-cost raw materials, and are the most economical and widely used of all plastic materials. Introduction of other elements into the polymer molecules generally requires a larger number of more complicated synthetic reactions to produce the starting monomer, and therefore increases the cost of the final polymer. In order of increasing costs, in general, we find chlorine, oxygen, nitrogen, sulfur, silicon and fluorine producing polymers of successively increasing cost. This increase in cost must obviously be accompanied by more valuable properties in order to justify the commercial production of such polymers.

2-2 MONOMERIC INGREDIENTS IN THE FINAL POLYMER COMPOSITION

In addition to the chemical composition of the polymer molecule itself, the final commercial composition often contains monomeric ingredients whose chemical composition also has important effects upon properties and applications. These monomeric ingredients may be either: (1) Impurities remaining from the original polymerization reaction, or (2) Monomeric additives purposely introduced during compounding to modify composition and final properties.

Impurities

The polymerization reaction may leave a number of impurities in the final composition, and these may have a variety of effects upon its properties and uses. In general, volatilization or leaching of such impurities may be responsible for shrinkage or crazing (ref 25). In addition, each type of impurity has specific effects which deserve individual consideration.

Monomer

Residual unpolymerized monomer often remains tenaciously in the final polymer. This generally softens its mechanical and thermal properties (nylon 6), causes poor color stability during aging (ethyl maleate), and produces undesirable odors (styrene and acrylic esters). Some residual monomers may even be very toxic, particularly isocyanates, vinyl carbazole, and chloroacrylate (ref 26).

Solvent

Residual solvent generally softens mechanical and thermal properties and causes odor in the finished plastic. Some aromatic hydrocarbon solvents may also cause toxicity. These problems are particularly troublesome after certain solution polymerizations and coating operations.

Water

Water may remain from the polymerization and washing steps in the preparation of the polymer, or it may be absorbed from the atmosphere or immersion during storage and use. In any case, the presence of moisture in many polymers causes volatilization problems during processing; it also softens mechanical properties, lowers dimensional stability and electrical properties, and is the starting point for hydrolytic degradation of the polymer molecule. This is particularly important in polyamides (previously covered in section on nitrogen).

Chloride Ion

The presence of residual chloride ion as an impurity in the polymer can be responsible for poor electrical properties (ref 15). In combination with acidic conditions, the resulting HCl can also produce serious corrosion. In general, the presence of mobile ions can contribute to electrical conductivity (ref 27), and thus lower insulating ability.

Catalysts and Initiators

Residual catalysts, initiators, or their decomposition products in the final polymer, most often peroxides or transition metal compounds, may be harmful to thermal stability, electrical properties, and weathering. Some initiator residues may also be toxic, particularly those from organic azo compounds.

Dispersant

Surface active agents, used in dispersing the polymerization reaction, can hurt thermal stability, electrical properties, and weatherability. Certain dispersants, particularly methyl silicones, can also be very detrimental to postdecoration operations such as painting. On the other hand, surfactants may be of crucial value in many foaming and coating operations.

Additives in Compounding

The addition of monomeric additives in compounding, to produce the final commercial composition and properties, is a very common and useful technique. At the same time, it should be recognized that such monomeric additives will often have lower chemical resistance than the polymer itself. In addition each type of additive has certain important specific effects which should be considered.

Plasticizer

The addition of monomeric liquid plasticizers, to improve processability or flexibility of polymers, is very common practice, especially in polyvinyl chloride and cellulose esters. These must be sufficiently similar to the polymer, in polarity and hydrogen-bonding, to be miscible and compatible with it, and low enough in volatility and diffusability to remain fairly permanently in it.

Monomeric plasticizers are relatively nonvolatile, but still somewhat fugitive when heated or aged for long periods of time (Table 2-8). Most

Table 2-8. Loss of Plasticizer from Plasticized Polyvinyl Chloride (ref 28). Reprinted by permission of Modern Plastics Encyclopedia, McGraw-Hill, Inc.

		Percent loss from 10-mil film			
Plasticizer	Percent of plasticizer in total composition	Activated carbon volatility 24 hr/90°C	Oil extraction 10 dy/25°C	Hexane extraction 2 hr/25°C	1% soap extraction 24 hr/90°C
Di-2-ethylhexyl adipate	32	15.5	17.5	23.5	13.8
Diisooctyl adipate	33	14.0	17.5	24	16
Dioctyl phthalate	35	7.6	7.7	21.8	7.8
n-Octyl, *n*-decyl adipate	34	5.1	15.5	25.8	5.4
Diisodecyl adipate	38	4.8	12.5	31.5	2.1
Di-2-ethylhexyl sebacate	35	4.1	15.9	28.0	1.0
Epoxidized soybean oil	36.5	0.6	3.9	10.8	0.2
High MW polyester	40.5	0.6	+0.2	0.9	0.3

of them are less solvent- and chemically-resistant than the polymer to which they are added, and many are extractable by soapy water, gasoline, motor oil, foods, or dry-cleaning solvents. Most of them are combustible, and lower the flame resistance of the total composition. Also many of them contain structures which are less resistant to aging than the polymers to which they are added. And finally, some of them are fairly toxic, and should be used with caution where they could be extracted by foods, drugs, or cosmetics.

Stabilizers

Substances which can help to stabilize the polymer against heat or ultraviolet light degradation are extremely useful in many polymer

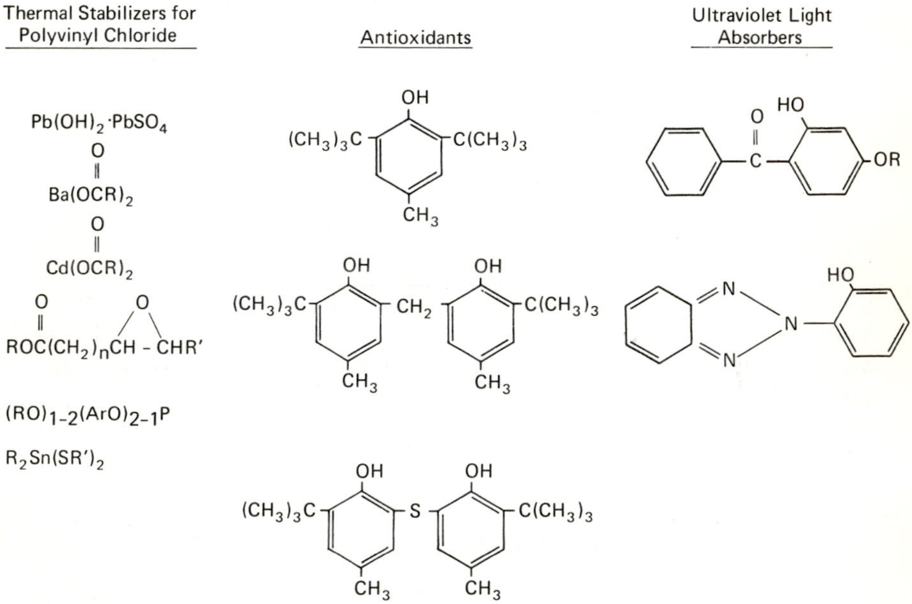

Figure 2-12. Stabilizers for polymers.

systems and are very often added in small concentrations for these purposes (Figure 2-12). Their effect on electrical properties may be good or bad, depending upon the specific stabilizer; for example, lead stabilizers in vinyl electrical insulation, although they are relatively polar and ionic in nature, actually improve the volume resistivity of the resulting insulation, probably by adsorbing ionic impurities which otherwise would have conducted electricity (Table 2-9).

Phenolic type antioxidants may stabilize the polymer against heat, and yet be sensitive to ultraviolet light, and thus sensitize the polymer toward ultraviolet light degradation (ref 14). Such additives may also interfere with clarity, if they should happen to be incompatible with the polymer. Some of them affect color, particularly certain ultraviolet light stabilizers whose conjugated carbonyl structures may cause yellow discoloration. And carbon black, which is the strongest known ultraviolet light stabilizer, of course produces opaque black products in every case. In addition certain stabilizers may introduce problems of toxicity; lead, cadmium, and some tin compounds used to stabilize polyvinyl chloride are particularly important in this consideration. Cost of thermal stabilizers is not usually a problem; but ultraviolet light stabilizers are extremely expensive, and are therefore used insufficiently in commercial practice.

Table 2-9. Effect of Lead Stabilizer on Electrical Resistance of Plasticized Polyvinyl Chloride (ref 29)

Tribasic lead sulfate, PHR*	Volume resistivity, ohm—cm
7	1.3×10^{14}
0.7	0.32×10^{14}
0.07	0.15×10^{14}
0.00	0.099×10^{14}

* Parts per hundred of resin.

Reinforcements and Fillers

Solid inorganic powdered fillers are often added to increase rigidity and decrease cost; and fibrous reinforcements are also added to increase strength and dimensional stability. Addition of fibrous reinforcing agents causes serious processability problems and increases processing costs considerably, particularly when long fibers, woven fabrics, and filament winding techniques are used. Alkaline fillers tend to hurt electrical properties (ref 15). Clay fillers may actually improve volume resistivity radically, for example in vinyl electrical insulation; this is probably due to their ability to adsorb ionic impurities which otherwise would have been mobile enough to conduct current. Carbon black, graphite, copper, and other metal fillers can actually be used in high concentration to produce electrical conductivity (ref 30). In addition most reinforcements and fillers have refractive indices different enough from the polymer to make it quite opaque, and many of them also introduce colors or at least tinting effects.

Cross-linking Agents

The use of curing agents for thermosetting resins, cross-linking agents for some thermoplastic resins, and vulcanizing agents for elastomers, is very important for their final mechanical, thermal, and chemical properties, and depends upon a delicate balance between pot life versus rate and extent of cure; they must be carefully chosen to optimize this balance. Most of them cause opacity and some of them cause color. In addition some of them are toxic and must be avoided in applications where they would be detrimental.

Foaming Agents

The use of chemical foaming agents, which decompose to liberate nitrogen gas, is very important in the production of some types of foams

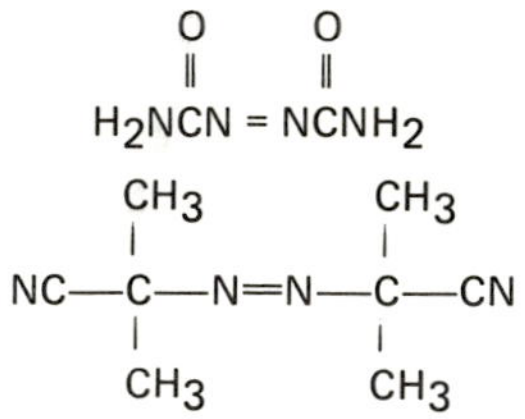

Figure 2-13.

(Figure 2-13). These are quite expensive, and some of them also produce toxic by-products which must be avoided in certain applications. Very toxic foaming agents cannot even be tolerated in normal processing techniques, and require particularly good ventilation.

Flame Retarders

Organic phosphates and halogen compounds, often synergized by antimony oxide, are frequently added to polymers to increase their flame resistance. At the same time, these additives may lower thermal and ultraviolet light stability, particularly if they use organic halogen compounds. Many flame-retarders cause opacity or toxicity problems as well, and most increase cost.

Lubricants

Lubricants are often added to improve processibility and to reduce friction in the final product. They may be liquids or low-melting waxes which exude to the surface, in which case they may produce poor surface appearance or cause painting problems. Or they may be solids like MoS_2 or brass, dispersed throughout the polymer, in which case they cause opacity and sometimes loss of strength.

Antistatics

Ionic materials such as quaternary ammonium compounds, or hydrophilic materials such as polyethoxylated organics, are often coated on finished products, or added to the polymer composition during processing, to reduce its tendency to develop and hold a static electrical charge during processing or use. Those which are simply postcoated on the surface are impermanent. Those which are added during compounding and processing must have precisely the right amount of semicompatibility so that they will continue to migrate to the surface and replace the superficial material which has worn away during handling and use. Neither the mechanism nor the technology of their action is well understood. In addition some of the best antistatic quaternary ammonium compounds have less than sufficient thermal stability, and tend to decompose to volatile tertiary amines and produce very undesirable fishy odors during high-temperature processing.

Biocides

The addition of biocides to protect polymers against bacteria, fungus, rot, and rodents must not simultaneously produce serious toxicity to humans, either during processing or during long-term use. This requires a very delicate balance of toxic properties, which is often hard to achieve. Ideally they should be semicompatible and migrate gradually to the surface during use, so that they may continue to replenish the surface layer which is being worn away. Some of them harm the color-stability of the polymer, particularly during thermal processing. Many of them are expensive as well. Use of elements or groups such as lead, phenols, aromatic hydrocarbons, and aromatic amines introduces problems of toxicity to humans, which must be considered and guarded very carefully (ref 26).

Colors

The addition of colors to polymers is a fine art of blending and color matching. For maximum thermal stability, inorganic colors are generally used; and since these are generally incompatible with the organic polymer, they produce only opaque colors. Development of clear, thermally-stable organic colors is much more difficult. Certain colors such as cadmium compounds introduce problems of toxicity. Many colors break down chemically, particularly during thermal processing or ultraviolet light aging, causing discoloration in use (ref 31).

REFERENCES

1. Pauling, Linus, *The Nature of the Chemical Bond,* Cornell, 1942, p 64.
2. Brydson, J. A., *Plastics Materials,* Iliffe, 1966, pp 94–96.
3. *Ibid.,* pp 77–79.
4. Baer, Eric, *Engineering Design for Plastics*. Reinhold, 1964, p 416.
5. Brydson, *op. cit.,* p 75.
6. Schmidt, A. X., & Marlies, C. A., *Principles of High Polymer Theory & Practice,* McGraw-Hill, 1948, p 458.
7. Miller, M. L., *The Structure of Polymers,* Reinhold, 1966, p 676.
8. *Ibid.,* pp 675–79.
9. Baer, *op. cit.,* p 702.
10. *Ibid.,* p 714.
11. *Ibid.,* p 736.
12. *Ibid.,* Chapter 10.
13. *Ibid.,* pp 702–03.
14. Brydson, *op. cit.,* p 78.
15. Volk, M. C., Lefforge, J. W., & Stetson, Russell, *Electrical Encapsulation,* Reinhold, 1962, p 21.
16. Fettes, E. M., *Chemical Reactions of Polymers,* Interscience, 1964, p 111.
17. Gulf Oil Corporation, Technical Bulletin, *Effects of Moisture on Physical Properties of Gulf Nylon.*
18. Baer, *op. cit.,* p 592.
19. Brydson, *op. cit.,* pp 75–79.
20. Tobolsky, A. V., *Properties & Structure of Polymers,* Wiley, 1960, p 206–11.
21. *Modern Plastics Encyclopedia 1968,* McGraw-Hill, 1967, p 119.

22. *Modern Plastics, 45* (5), 86, 234 (January, 1968); *46* (1), 38 (January, 1969).
23. *Plastics Technology, 13* (10), 11 (October, 1967).
24. Thiokol Chemical Corporation, Price Quotation.
25. Baer, *op. cit.*, p 603.
26. Brydson, *op. cit.*, p 84.
27. Billmeyer, F. W., Jr., *Textbook of Polymer Science,* Interscience, 1962, p 114.
28. *Modern Plastics Encyclopedia, op. cit.*, p 449.
29. Deanin, R. D., Skowronski, B. J., & Wixon, F. R., *Plasticization & Plasticizer Processes,* Advances in Chemistry Series 48. American Chemical Society, 1965, p 142.
30. Miller, *op. cit.*, p 671.
31. Baer, *op. cit.*, pp 603–04.
32. Miller, *op. cit.*, p 170.

Part Three

MOLECULAR STRUCTURE

Many of the properties of polymers are inherent in the structure of the individual polymer molecules. Most polymer properties depend largely upon the behavior of the individual molecule, plus the modifying effects of larger structural features. Thus a detailed study of molecular structure can both explain many polymer properties, and also lay the groundwork for analyzing most of the more complex properties.

Molecular structure is conveniently divided into two major aspects: (1) Molecular Size and Shape includes molecular weight, molecular weight distribution, and branching resulting from intra- or intermolecular chain transfer reactions during polymerization. (2) Molecular Flexibility describes the ability of individual polymer molecules to coil and uncoil in response to external and internal forces, and is determined primarily by freedom or restriction of rotation around the interatomic bonds in the main-chain of the polymer molecule.

Chapter Three

Molecular Size and Shape

The presence of temperature reveals that atoms and molecules are in constant motion. Small molecules move freely as gases, down to quite low temperatures. Larger molecules move fairly freely as liquids. Still larger molecules are held together by forces of attraction between them and are solid at ordinary temperatures, but these forces are not sufficient to provide high mechanical strength; under a high mechanical stress such molecules slide over each other and come apart quite easily (Table 3-1). Only at much higher molecular weight does substantial cohesive strength develop.

The polymers used in plastics, rubbers, coatings, adhesives, and fibers are all characterized primarily by their extremely high molecular weight. The entanglement and attraction between such large molecules provide the substantial mechanical strength which makes them useful as structural materials. The large size of these polymer molecules has particularly striking effects upon their processability and upon their mechanical, thermal, and chemical properties.

Thus most properties of polymers depend upon their molecular weight. Some properties such as melt processability, hot strength, and solvent resistance depend upon molecular disentanglement, and change continuously with molecular weight up to infinity (Figure 3-1). Most properties change as molecular weight increases from low to medium size, either because the number of polymer end groups is decreasing, or because the total intermolecular attractive forces are increasing with the size of the molecule; such properties will tend to level off gradually

Table 3-1. Effect of Molecular Weight upon Character of High-pressure Polyethylene (ref 1)

Number of —(CH_2—CH_2)— Units	Molecular Weight	Softening Temperature (°C)	Character of Polymer at 25°C
1	30	−169*	Gas
6	170	−12*	Liquid
35	1,000	37	Grease
140	4,000	93	Wax
250	7,000	98	Hard wax
430	12,000	104	Hard resin
750	21,000	110	Hard resin
1,350	38,000	112	Hard resin

* Melting point

as molecular size becomes very large, often following an equation of the type $P = a + b/M$ (ref 2). In this equation a and b are constants which depend upon the property being measured and upon the dimension in which it is measured. The sign of b may be positive or negative. The molecular weight M is a more complex function and requires further discussion.

The reactions by which polymers are prepared do not produce a single species of uniform molecular weight. Addition polymerization reactions may chain-transfer or terminate whenever the growing polymer molecule happens to meet a molecule of initiator, solvent, chain-transfer agent, or another growing polymer molecule; and such meetings may occur when the growing polymer molecule is either small or large, thus producing a broad spectrum of different size polymer molecules. Condensation polymers are formed by stepwise increase in molecular weight, and such stepwise reactions occur occasionally at any molecular size, thus producing a broad spectrum of final molecular weights. Addition and substitution reactions on polymer molecules also attack them fairly randomly, and often produce cross-linking or cleavage in a random fashion which thus also produces a broad range of molecular weights. Thus no matter how the polymer is produced, it will have a spectrum of molecular weights rather than a single homogeneous molecular weight (Figure 3-2).

The exact shape of this molecular weight distribution will vary considerably from one polymer to another, depending upon the monomers polymerized and the conditions of the polymerization reaction. These

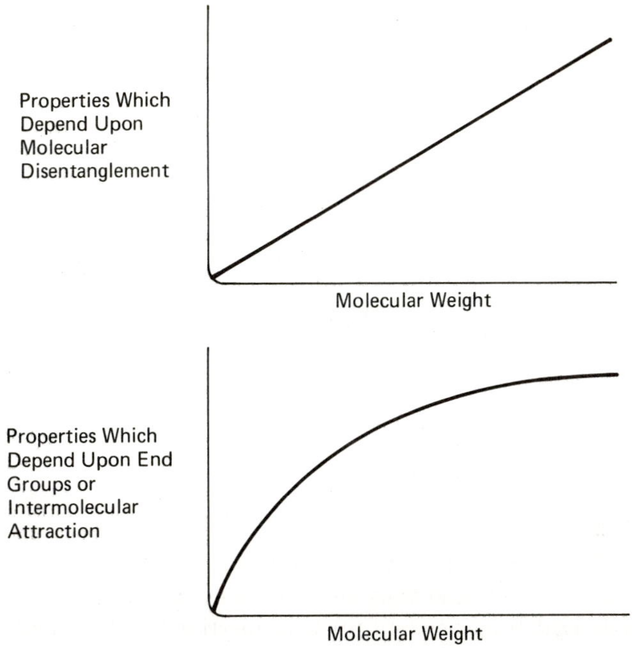

Figure 3-1. Effects of molecular weight upon polymer properties.

variables can shift the curve to lower or higher molecular weights, make it narrower or broader, or skew it in various ways. Such differences are easily represented graphically, particularly by use of analytical tools such as gel permeation chromatography. They are much harder to describe by simple mathematical values which can be used for tabulation and correlation. Present mathematical treatment is based primarily

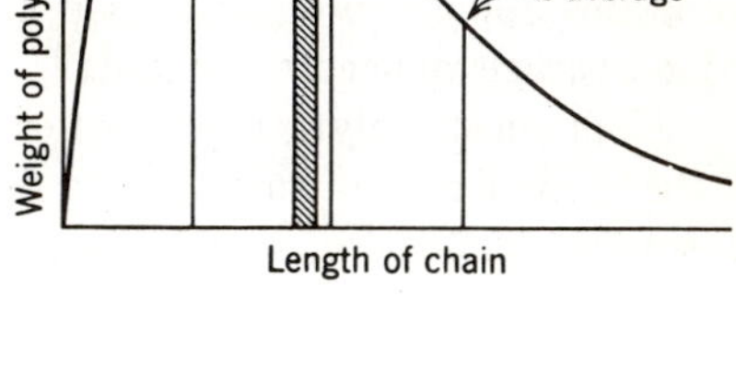

Figure 3-2. Molecular weight distribution and molecular weight averages in a typical polymer (ref 3)

upon calculating an average molecular weight by one or more techniques. The number average molecular weight $\bar{M}_n$

$$\bar{M}_n = \frac{w}{\sum_{i=1}^{\infty} N_i} = \frac{\sum_{i=1}^{\infty} M_i N_i}{\sum_{i=1}^{\infty} N_i} \qquad \text{(ref 4)}$$

simply divides the total weight of material present by the number of molecules in the material, and is particularly useful for properties which depend primarily upon the smaller molecules. The weight average molecular weight (designated $\bar{M}_w$) in the equation

$$\bar{M}_w = \frac{\sum_{i=1}^{\infty} w_i M_i}{w} = \frac{\sum_{i=1}^{\infty} N_i M_i^2}{\sum_{i=1}^{\infty} N_i M_i}$$

considers the weight of material of each size rather than the number of molecules of each size, and thus places greater emphasis upon the higher-molecular-weight fractions and upon properties which depend primarily upon the larger molecules. The weight average molecular weight is thus always greater than the number average molecular weight. In a homogeneous polymer of narrow molecular weight distribution, this difference may not be very great and the ratio $\bar{M}_w/\bar{M}_n$ may be only 1.5–2.0; whereas in a polymer of very broad molecular weight distribution, the difference may be very large and the ratio may be as high as 20–50 (ref 3). Thus this ratio may be used as a simple expression of the broadness or polydispersity of the molecular weight distribution. Occasionally even higher-order average molecular weights such as the *z*-average molecular weight $\bar{M}_z$

$$\bar{M}_z = \frac{\sum_{i=1}^{\infty} w_i M_i^2}{\sum_{i=1}^{\infty} w_i M_i} = \frac{\sum_{i=1}^{\infty} N_i M_i^3}{\sum_{i=1}^{\infty} N_i M_i^2}$$

may be measured by techniques such as ultracentrifugation, and some properties appear to depend upon and correlate better with these higher-order average molecular weights.

Thus most polymer properties depend upon the molecular weight of the polymer; but some properties are more affected by the number of smaller molecules, while other properties are more affected by the size of the larger molecules. In seeking the relationship between a property and molecular weight, it is important to determine which type of average molecular weight is most appropriate and will correlate best. When the property is plotted against an inappropriate average molecular weight, it will produce a family of curves; whereas when it is plotted against the appropriate average molecular weight, it will produce a single unified curve (Figure 3-3).

In addition to the size of the polymer molecule itself, two other factors should be considered here:

(1) *Plasticizers* are monomeric, or low-molecular-weight polymeric, materials which are added to polymers to improve their processability

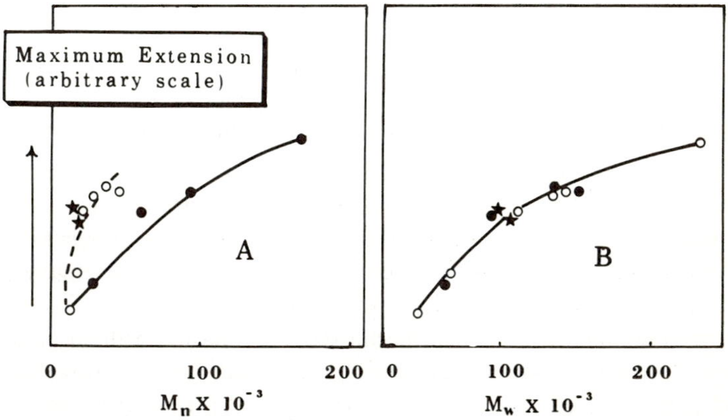

Figure 3-3. Correlation of average molecular weights with ultimate elongations of acrylic films. ● fractions, ○ whole polymers,* blends (ref 7).

or to soften their final properties. When they improve processability, they appear to function simply by increasing the low-molecular-weight fraction of the total material. When they soften final properties, their effect may be considered either: (*a*) as the effect of a low-molecular-weight fraction in the total material, in which case they are properly considered in the present discussion; or (*b*) their effect may be due to their ability to reduce attractive forces between polymer molecules, in which case they are best considered at a later point in the discussion (Chapter 6).

(2) *Cross-linking* joins polymer molecules together by permanent covalent bonds. While a discussion of such intermolecular bonding will be reserved primarily for later, cross-linking does in fact convert moderate-size polymer molecules into infinite-size polymer molecules; and

the present discussion would be incomplete if it did not take occasional note of such ultimate effects of molecular size.

The relationship between molecular weight and polymer properties is best examined by considering each type or group of properties separately. We will begin with processability; continue to mechanical, thermal, electrical, optical, and chemical properties; and finally consider the effects of molecular weight upon price and upon overall balance of important properties.

3-1 PROCESSABILITY

At risk of gross oversimplification, it is convenient to start with the common observation that processability is best at low molecular weight, while properties of the finished product are best at high molecular weight. Even though there are many exceptions to this simple rule, it is sufficiently important to play a major role in the design of polymers for practical use. (1) The largest volume of polymer usage, particularly in plastics and synthetic fibers, is based upon thermoplastic processing of a polymer at constant molecular weight, and this molecular weight is chosen to represent the best compromise between processability and final properties. (2) On the other hand, the greatest variety of polymers and processes, particularly in thermosetting plastics, rubber, and coatings, start with a low-molecular-weight material for easy processing and convert it into a high-molecular-weight material for best ultimate properties. The significance of molecular weight in these two different types of processing is best considered separately.

Thermoplastic Processing at Constant Molecular Weight

In general the largest volume of thermoplastic polymers used in plastics and synthetic fibers is based upon rapid low-cost melt processing to produce the finished products. This in turn depends upon low molecular weight to produce low melt viscosity. For example low-molecular-weight polyethylene is preferred for high draw-down rates in extrusion coating of paper and for injection molding (ref 8,9). Low-molecular-weight polyvinyl chloride is preferred for high-speed calendering of plasticized film and sheet, and for molding and extrusion of rigid pipe, bottles, and phonograph records. Low-molecular-weight methyl methacrylate resins permit the fastest injection molding into the most intricate shapes. In these and other thermoplastic polymers, low molecular weight is used to provide low melt viscosity which supports rapid economical processing.

General Relationships

The relationship between molecular weight and thermoplastic processability has been quite well studied and understood. In the thermoplastic processing of linear amorphous polymers, the viscosity of the molten polymer at low shear rates is directly related to the weight average molecular weight (ref 10). In general, $\eta = KM^q$ (ref 2). More specifically for polyamides, $\log \eta = A + CM^{\frac{1}{2}}$ (ref 2); and for polyesters, $\log \eta = A\bar{M}_w^{\frac{1}{2}} + K$ (ref 11). At relatively low molecular weights, in general, $\log \eta = \log M + \log K$ (ref 12). As weight average molecular

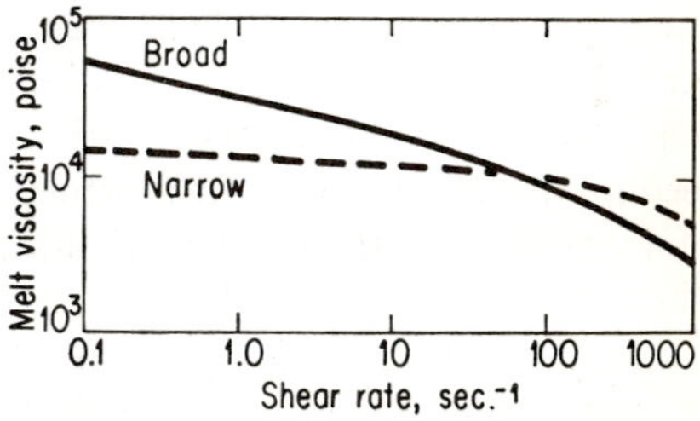

Figure 3-4. Effect of polyethylene molecular weight distribution upon pseudoplasticity in melt processing (ref 14).

Reprinted by permission of Modern Plastics Encyclopedia, McGraw-Hill, Inc.

weight increases, it passes a critical value M_c beyond which entanglement of molecules becomes important. Beyond this point then $\log \eta = 3.4 \log \bar{M}_w + I$ (ref 11). The close agreement between theoretical calculation and experimental observation indicates the fundamental understanding which has been developed of this phenomenon (ref 13).

Complex Relationships

Under certain circumstances the relationships may become more complex and the theory less understood. At high rates of shear, or at broad molecular weight distributions, viscosity may be related to a different average molecular weight rather than $\bar{M}_w$ (ref 10). Long-chain branching of the polyethylene molecule, producing a broad molecular weight distribution, provides a more pseudoplastic melt (Figure 3-4) which supports higher draw-down rates in extrusion coating, reduces neck-in (shrinkage) and beading (thick edges) of extrusion coating, and broadens heat-seal range for film and coatings (ref 9). At the other extreme, polyethylene of narrow molecular weight distribution and low thixotropy is preferred for low warpage in thin-wall injection moldings, extrusion of film, and rotational molding (ref 8).

Processes Requiring High Molecular Weight

While processability is generally considered best at low molecular weight, there are many processes in which the relationship between molecular weight and processability is more complex. In injection molding, a polymer whose melt is too fluid may exude excessively through flash and vent openings. Whereas a low-molecular-weight material may be injection molded into a rigid mold which will support the low-strength molten product, extrusion of a continuous profile into the open air requires a material with sufficient melt strength to support itself as it exits from the die, and thus extrusion generally requires higher molecular weights than injection molding (ref 15). Similarly the spinnability and drawability of aliphatic polyester fibers increase directly with molecular weight (ref 16). In postforming operations, particularly the thermoforming, stretching, and orientation of thermoplastic sheet, film, and fiber, the material must have sufficiently high molecular weight and strength to withstand the high stresses and strains involved in such process techniques. In the production of thermoplastic foam, the melt viscosity of the polymer must be low enough to permit blowing and expansion but the melt strength of the blown material must be high enough to prevent rupture of the thin cell walls which are being produced. And occasionally the unusually excellent properties of ultra-high-molecular-weight polymers such as polyethylene warrant the development of unusual process techniques such as sintering and high-pressure compaction in order to cope with the difficult processability of such ultra-high-molecular weights (ref 17,146).

Plasticizers

Plasticizers are monomeric or low-molecular-weight polymeric substances which are often added to high-molecular-weight polymers to improve their processability. Fundamentally, they function by broadening the molecular weight distribution and increasing the low-molecular-weight fraction of the total composition. In plastics they generally lower the melt viscosity (ref 18,19) and processing temperature; most common are the use of phthalate esters in polyvinyl chloride and cellulose esters. In rubber they reduce the processing temperature and the power required for mixing, increase the plasticity of the composition, and also improve related processing properties such as tackiness and wetting and dispersion of pigments (ref 20,21); for these reasons they are often called softeners or tackifiers; they are most commonly petroleum hydrocarbons. In adhesives they also improve processability and tack (ref 22).

In some materials plasticizers are also referred to as lubricants, but this term is more properly applied to low concentrations of semicompatible monomeric materials of low surface tension, which separate from the polymer during processing, migrate to the interface between the polymer and the process equipment, and lower the adhesion between the polymer and the process equipment, thus reducing the tendency to stick and retard processing in this way. Stearic acid, stearates, oils, and waxes are typical of such materials.

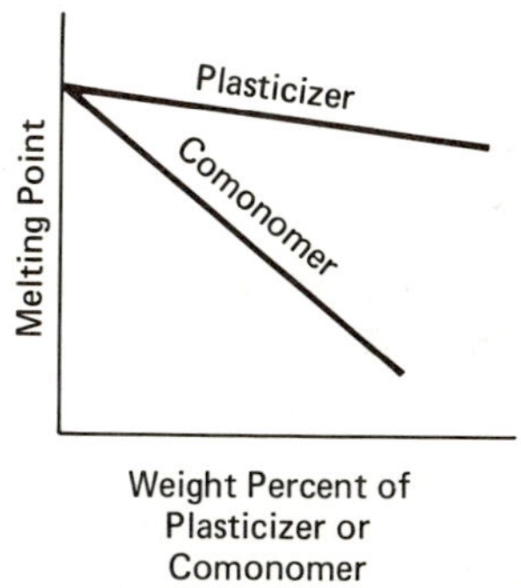

Figure 3-5. Effect of plasticizer and comonomer on melting point (ref 24).

Crystalline Polymers

Crystalline polymers introduce more complex phenomena and will be considered primarily later, but still deserve brief mention. Above their melting points, processability is similar to amorphous polymers, and increasing molecular weight produces increasing viscosity, from mobile liquids to viscous liquids to rubbery melts (ref 23). At lower temperatures, however, the sharp melting point of the crystalline phase produces a break in the normal Viscosity vs Temperature curve. Low-molecular-weight fractions introduce high concentrations of end groups which do not fit neatly into the crystalline lattice, and can often lower the perfection of crystallinity and the melting point of the crystalline phase; increasing molecular weight then produces a rise in melting point as well as an increase in melt viscosity, and ultimately may cause harder processing (ref 145). Addition of low-molecular-weight plasticizer can also lower the melting point and thus improve processability (comonomer units occupy less volume than plasticizer molecules and are therefore even more effective in this respect: Figure 3-5) (ref 25).

On the other hand, lower-molecular-weight polymers have lower melt viscosity and can therefore crystallize more readily and sometimes to a higher degree (ref 26). Hence addition of a trace of plasticizer may similarly speed crystallization to equilibrium and to higher practical levels of crystallinity (ref 27,28), and when it does it is sometimes called an antiplasticizer.

Thus molecular weight may correlate either directly or inversely with melting point and melt processability in individual polymer systems.

Conversion to High Molecular Weight during Processing

Unlike the major thermoplastic polymers considered above, most polymers and most applications do not permit such an easy compromise between the low molecular weight required for processing and the high molecular weight required for optimum properties. In most cases a wide variety of techniques are used to process the polymer at low molecular weight and then convert it to high molecular weight during the terminal stages of processing, thus combining the easy processability of the initial low molecular weight with the optimum properties of the final molecular weight. For convenience, these techniques are best classified according to whether: (1) the material is used at 100% total solids and the entire composition is converted to high molecular weight; or (2) the material contains a fugitive low-molecular-weight fraction which is removed during processing to leave only the high-molecular-weight polymer in the final product.

Conversion of 100% Total Solids from Low to High Molecular Weight

Whenever possible, a processor would prefer to convert the entire composition into the final product, and not have to remove, recover, or discard a fugitive low-molecular-weight fraction during processing. Many practical systems approach or even achieve such 100% conversion into final product.

THERMOSETTING RESINS Most of the thermosetting polymers used in molded and laminated plastics, such as unsaturated polyesters, urea, melamine, and phenolic resins, are first polymerized to a low molecular weight A-stage or B-stage polymer which is still soluble, fusible, and highly reactive during final processing. They are generally catalyzed and/or heated to finish the polymerization reaction, cross-linking them up to high and almost infinite molecular weights. This step is often referred to as cure, and will be considered more fully later. Epoxy, polyurethane and silicone polymers are normally polymerized first to a low-molecular-weight viscous liquid or soluble low-melting prepolymer, which is then mixed with a curing agent and/or catalyst, poured or spread into the desired form, and polymerized rapidly up to high or even infinite molecular weights in the final plastic, elastomeric, coating, or adhesive product. In all these techniques the original A-stage or B-stage resin or prepolymer is made at very low molecular weight to retain maximum processability, but must be converted to much higher molecular weight during final processing in order to produce the desired final properties.

MONOMER CASTING Monomer casting is an even more extreme version of the same technique. Monomers such as methyl methacrylate, diethylene glycol bis(allyl carbonate), and molten caprolactam may be mixed with polymerization initiators or catalysts and poured in their highly fluid form into the mold in which they are then heated and polymerized to the final hard product. In the polymerization of methyl methacrylate such monomer casting techniques produce large, optically perfect sheets, which could not be obtained by conventional thermoplastic processing, and also permit embedment of delicate specimens which would be damaged by the high melt viscosity of conventional thermoplastic processing techniques. Diethylene glycol bis(allyl carbonate) can be cast to optically perfect hard thermoset lenses. Caprolactam can be cast into larger and higher-molecular-weight nylon 6 products, containing less residual plasticizing monomer, than would be possible by conventional thermoplastic molding techniques; such castings are particularly useful for making large gears, buckets, rolls, and similar engineering products. In all of these cases, casting a highly fluid monomer into the final form and then polymerizing to high molecular weight has significant advantages.

DRYING OILS The natural and synthetic drying oils used in most paints and coatings are liquid monomeric or low polymeric materials which can be flowed uniformly onto the substrate surface, wet and penetrate all the irregularities of that surface, and then polymerize by atmospheric oxygen with or without the benefit of added catalysts such as transition metal soaps. The polymerization reaction converts them from liquid monomers to fully-cured and cross-linked polymeric coatings, and may actually improve adhesion to the substrate further by reacting directly with it during drying and cure.

Removal of Fugitive Low Molecular Weight Fractions during Processing

In the processing of man-made fibers, coatings, and adhesives, high-molecular-weight polymers are often dissolved in low-molecular-weight fugitive solvents to lower the average molecular weight and viscosity of the material during processing, and the fugitive solvents are removed again during the later stages of processing, to regenerate the high-molecular-weight polymer with all its ultimate properties. Such techniques only work when the thickness of the final product is low enough so that solvent molecules need not travel very far to reach the surface.

EVAPORATION Most common solvents are fugitive simply by virtue of their volatility. Most coatings, many adhesives, and some fibers are produced from solution in volatile solvents, using the low-molecular-weight solvent to lower the viscosity of the high-molecular-weight polymer sufficiently for spreading on a surface or spinning through an orifice, and then evaporating the solvent to leave the high-molecular-weight polymer in finished form. Some ramifications of this technique in coatings are particularly interesting.

In dilute solutions each polymer molecule is a swollen sphere travelling independently of the others. In the concentrated solutions required for practical processing, however, these swollen spheres may tend to entangle somewhat with each other, producing considerably higher viscosities (ref 29). Even in solution, excessively high-molecular-weight polymers may produce higher viscosities than can be used in coating applications. Dilution to lower viscosity produces thinner coats, requiring multiple coating operations to produce the desired coating thickness, and thus increasing cost (ref 30).

Choice of solvent is often very critical in coating solutions. Evaporation of solvent must be fast enough to be economical yet slow enough for the polymer molecules to intercoil and produce substantial continuity for optimum final properties. Addition of diluents which are poor solvents serves to lower the viscosity and the cost but also complicates this delicate evaporation process (ref 31).

If the polymer is dissolved in a mixture of good and poor solvents, evaporation of the good solvent before the poor one produces a gel, and, during further evaporation and drying, this produces orientation and increases strength and toughness; but if the good solvent evaporates too soon or the poor solvent is too weak, it may actually cause precipitation of the polymer and produce a grainy weak product (ref 32).

COAGULATION Some polymers are too infusible to be processed by conventional thermoplastic processing techniques and too insoluble to be dissolved in low-boiling-temperature solvents, but they may still be soluble in high-boiling-temperature nonvolatile or reactive solvents. These solutions in low-molecular-weight solvents provide all the fluidity needed for processing, but the solvents must then be removed by coagulation in order to regain the pure high-molecular-weight polymer.

Thus polyacrylonitrile is dissolved in a high-boiling-temperature solvent such as dimethylformamide and extruded through the fine holes of a spinnerette to form fibers, but the solution must be extruded into nonsolvent coagulants to gel the polyacrylonitrile fibers and separate them from the solvent. In this process, low-molecular-weight polymer provides fluidity for faster and smoother extrusion of fiber, but high-molecular-weight is required to provide wet tensile strength during drawing of the fiber; so the optimum molecular weight must be a compromise between these two factors, and is also best at very specific types of molecular weight distribution (Figure 3-6) (ref 34).

Cellulose is infusible and insoluble in conventional solvents, but can be swollen and softened in concentrated aqueous zinc chloride solution; this gel can be pressed into sheets and then leached with excess water to extract the zinc chloride, thus reprecipitating the cellulose from the water in sheet form and producing tough rigid plastics known as vulcanized fiber. Cellulose can also be dissolved in aqueous alkali by reaction with carbon disulfide to form the xanthate, and when this solution is extruded in film or fiber form into aqueous acid, the acid destroys the soluble xanthate and reprecipitates the high-molecular-weight cellulose in the form of cellophane film or rayon fiber.

In all of these techniques, high-molecular-weight polymer is dissolved in low-molecular-weight solvent to form a fluid solution and the polymer is then coagulated from the solution to separate it from the low-molecular-weight solvent and regenerate the high-molecular-weight polymer in its finished form.

Conversion of low-molecular-weight to high-molecular-weight polymer during processing, by either 100% solids or fugitive solvent techniques, is particularly important in processes specifically requiring the

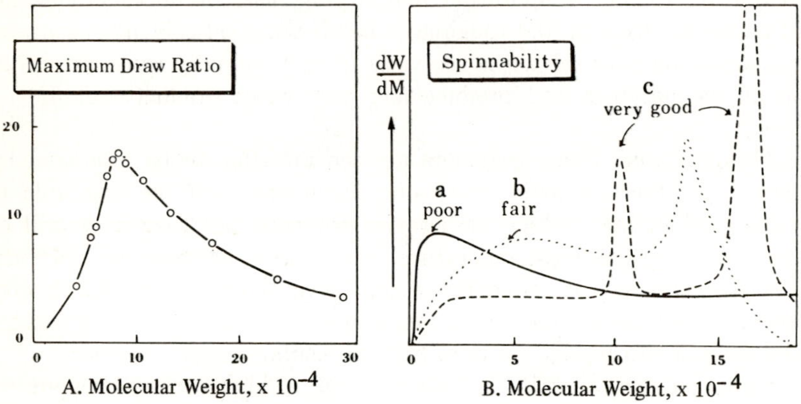

Figure 3-6. Effect of molecular weight and distribution upon solution spinnability of polyacrylonitrile (ref 33).

A. Relationship between maximum draw ratio and weight-average molecular weight for fractions of polyacrylonitrile. B. Molecular-weight distributions of three samples of polyacrylonitrile, a, b and c, that have nearly the same weight-average molecular weight but differ in molecular-weight distribution and in spinnability.

high fluidity of low-molecular-weight material and is therefore most useful in applications such as foam, impregnation, coating, and adhesives.

3-2 MECHANICAL PROPERTIES

The effects of molecular weight upon mechanical properties, and upon physical properties in general, are among the most frequently studied of the many structure → properties relationships, both theoretical and applied. In the most fundamental and general approach to this relationship, the thermal energy of atoms and molecules keeps them in constant vibratory motion. If the entire molecule is very small, and the attractive forces between molecules are not too great, these molecules move freely throughout space in the form of a gas. If the molecules are somewhat larger and/or the attractive forces between them are somewhat greater, they still move fairly freely in the form of a liquid. As the size of the molecules becomes greater they exist in solid form (Table 3-1). When mechanical stress is applied to such low-molecular-weight solids the molecules slide over each other and separate from each other. The solid has very little mechanical strength and is of very little structural value (Figure 3-7). With continuing increase in molecular weight, the molecules become entangled and the attractive forces between them become greater, and mechanical strength begins to appear. Ultimately

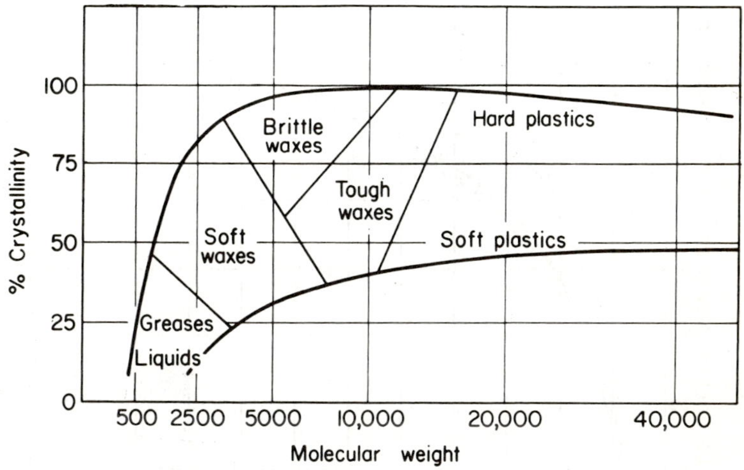

Figure 3-7. Effect of molecular weight and crystallinity upon physical properties of polyethylene (ref 35).

the sum of the entanglements and intermolecular attractive forces becomes greater than the strength of the individual covalent bonds in the backbone of the polymer chain. When mechanical stress is applied to such a high-molecular-weight system it may be easier to break primary covalent bonds in the polymer backbone than to separate one polymer molecule from another. Beyond this point further increase in molecular weight produces no further increase in mechanical strength.

Mechanical properties may best be divided into three types or classes: (1) Reversible Rigidity/Flexibility or modulus which measures the stiffness of the polymer system; (2) Mechanical Failure by yielding (Permanent Deformation) or breaking at low (Conventional Ultimate Strength and Elongation) or high (Impact) rates of test or under a

Multiple Cycling Stress; and (3) Complex Mechanical Properties involving a combination of these simpler processes. These classes are now considered in detail.

Reversible Rigidity/Flexibility

The rigidity, hardness, stiffness, or flexibility of a solid material is generally measured by its modulus, the slope of its stress/strain curve

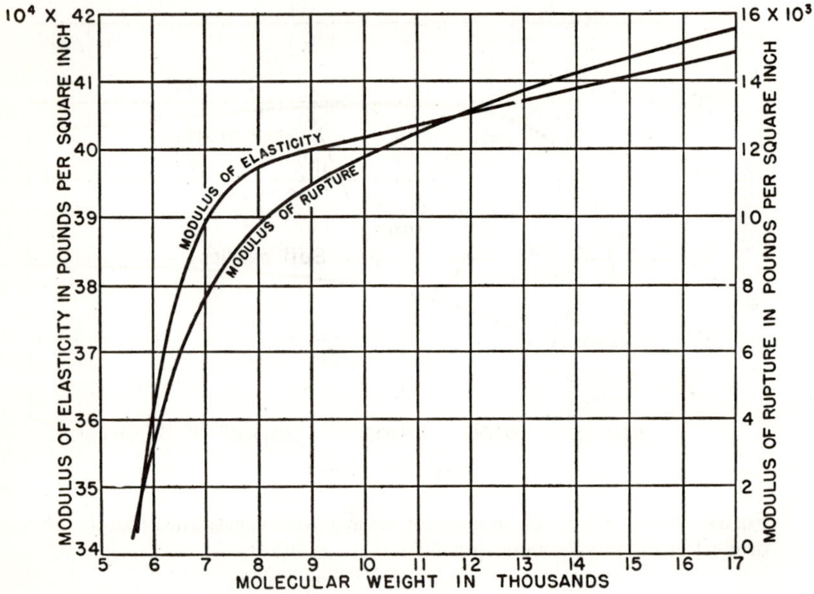

Figure 3-8. Effect of molecular weight upon modulus of vinyl chloride/vinyl acetate copolymers (ref 37).

either at the origin or on the secant to any standard elongation. From one theoretical point of view, such phenomena involve only segments of polymer molecules, and should therefore be independent of molecular weight; some experimental data actually conform to this hypothesis (ref 36), and other published results approach it (Figure 3-8). It is true in general that amorphous polymers in the glassy state below T_g all tend to have tensile moduli in the range of about 500,000 psi (ref 38).

Flexible Polymers

For polymers with a glass transition near or below room temperature, on the other hand, and therefore in a leathery or rubbery state, such properties are definitely dependent upon molecular weight (ref 39) (Figure 3-9). In general, low-molecular-weight polymers have a greater number of chain ends which do not pack as neatly in the solid polymer, have more mobility, and occupy a larger free volume; and this larger free volume permits greater mobility and motion of other segments of the polymer chain, making the material softer (ref 42).

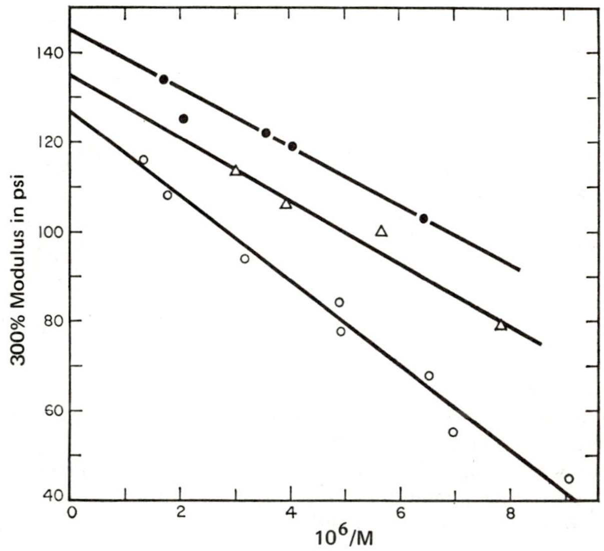

Figure 3-9. Effect of molecular weight upon modulus of butyl rubber. The 3 lines represent 3 different levels of vulcanization (ref 40).

More specifically, at low speeds of testing which permit large intertwined molecules to untangle and separate, increasing molecular weight will have a significant effect upon polymer modulus; whereas at high rates of testing which do not permit sufficient time for intertwined molecules to disentangle, molecular weight will have little or no effect upon modulus (Figure 3-10). Increasing molecular weight produces an increasing degree of entanglement, reducing the ability of individual molecules to move and deform; this produces a broader plateau of rubbery properties (ref 43). Thus the main practical effects of molecular weight upon modulus appear primarily in long-term permanent deformation and in higher-temperature transition to rubbery flow, rather than in reversible modulus itself. These phenomena will be considered later.

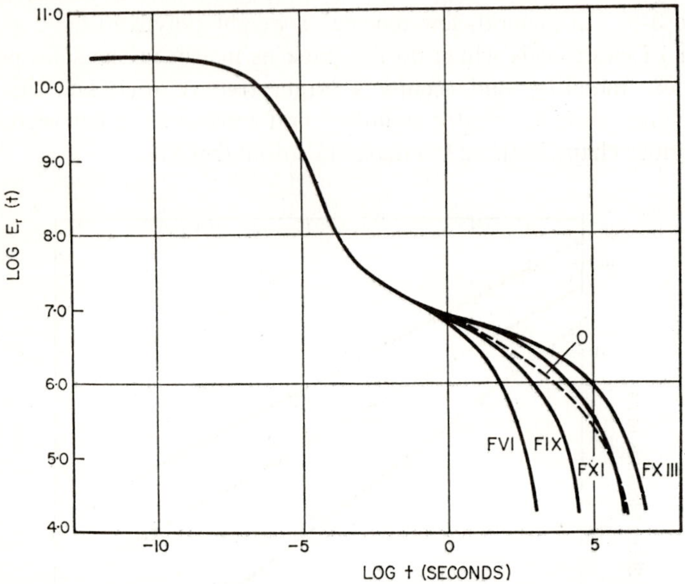

Figure 3-10. Effect of rate of testing upon the relationship between molecular weight and modulus (ref 41).

Master stress relaxation curves reduced to 75°C showing the time dependence of Young's modulus ($\simeq 3 \times$ shear modulus) for polyvinyl acetate samples of various molecular weights.

sample	*M (viscosity average)*
F VI	83,600
F IX	184,000
F XI	420,000
F XIII	784,000
O (unfractionated)	358,000

Plasticizers

Plasticizers are monomeric or low-molecular-weight polymeric liquids which are added to polymers to lower their hardness (Figure 3-11), stiffness (Figure 3-12), and modulus (Figure 3-13), and generally make them more soft and flexible. Plasticizers are most often employed in flexible and elastomeric polyvinyl chloride compositions used in upholstery, clothing, shower curtains, wire insulation, packaging film, sporting goods, and a host of other soft, pliable, and rubbery products; the most common types of plasticizers are phthalic, phosphoric, and adipic esters. In coatings, plasticizers generally increase the flexibility, pliability, and distensibility of the polymeric binder (ref 46); probably most common are phthalic esters. Petroleum hydrocarbon plasticizers or softeners in rubber have similar effects, producing lower

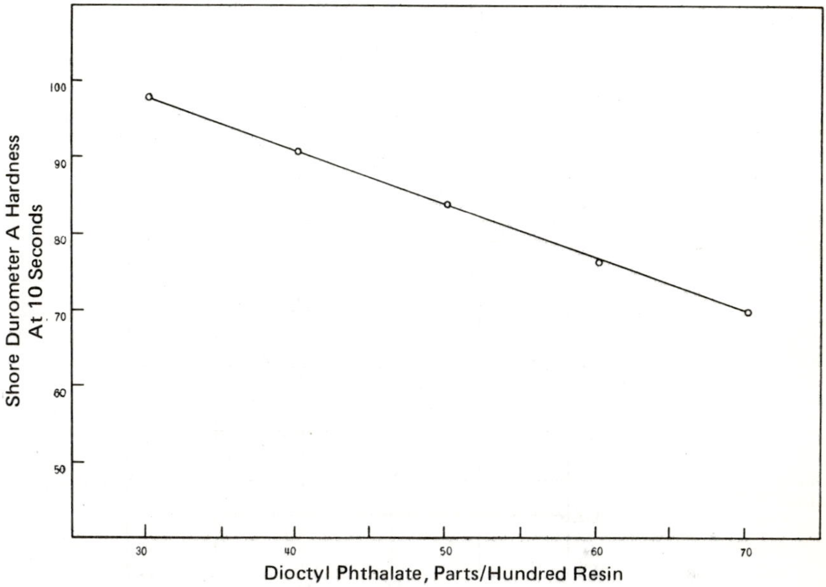

Figure 3-11. Effect of plasticizer upon hardness of polyvinyl chloride (ref 44).

hardness and modulus and higher flexibility, elasticity, and resilience (ref 47,48). Water absorption, especially by polar polymers such as nylons 6 and 66, also produces plasticization resulting in lower modulus (Figure 3-14) and hardness (Figure 3-15) and increased ductility (ref 51,52); but these effects vary with environmental conditions and tend to make such polymers unreliable in their performance.

The mechanism of plasticizer action results from the high mobility of the small plasticizer molecules, which occupy a large free volume, and bring this high free volume and mobility to a high-molecular-weight polymer when they are added to it. This increased free volume permits

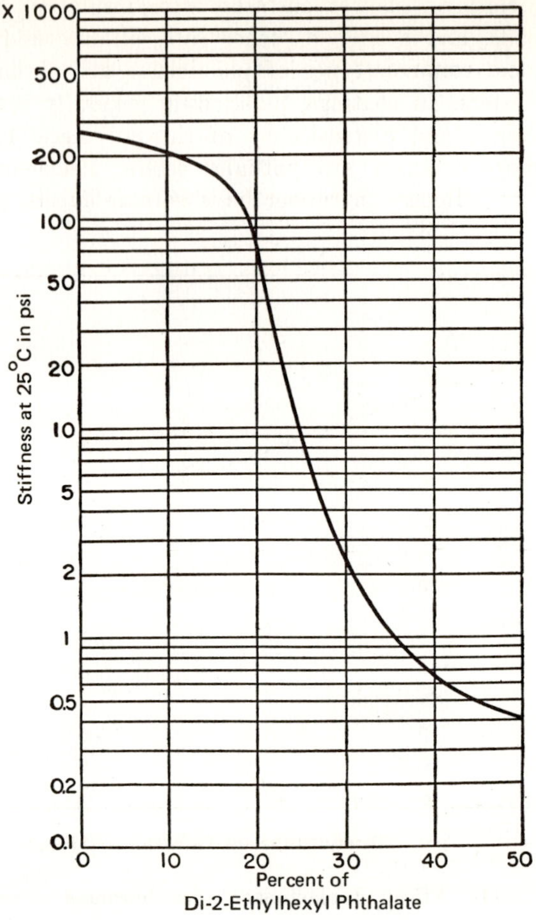

Figure 3-12. Effect of plasticizer upon stiffness of vinyl chloride/vinyl acetate copolymer (ref 45).

greater mobility of larger segments of the polymer molecule and thus produces lower modulus. Theoretically modulus is proportional to $V_p^{\frac{1}{3}}$ where V_p is the volume fraction of polymer in the plasticized system; but experimental data only approximate this semiquantitatively, indicating that the actual mechanism is more complex than the present theory (ref 53,54).

In general the plasticizer should be soluble in the polymer and have a low viscosity and a low temperature coefficient of viscosity (ref 55).

It should have a low glass transition temperature, in order to lower the glass transition temperature of the total plasticized polymeric composition; in this respect plasticizers usually have lower glass transition temperatures than comonomers and therefore are more effective in introducing flexibility without creep (ref 25) (Figure 3-16). Plasticizers also increase the rate at which a polymeric material can respond flexibly, particularly at lower temperatures (Figure 3-17). A poorly compatible

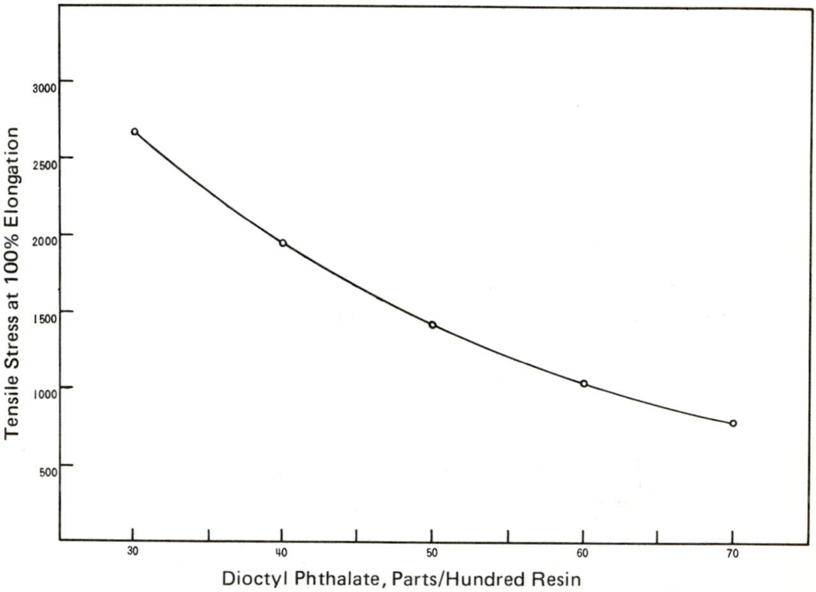

Figure 3-13. Effect of plasticizer upon modulus of polyvinyl chloride (ref 44).

plasticizer will give a much broader and more leathery transition region (ref 55); this effect is even more pronounced in crystalline polymers and polymers having several transitions (ref 57) as shown later.

In all of these ways, adding monomeric liquid plasticizers, as low-molecular-weight ingredients in an otherwise high-molecular-weight polymeric material, lowers and broadens the overall molecular weight by increasing the lowest-molecular-weight portion of the material, and thus increases molecular flexibility and softens the reversible mechanical properties such as rigidity, hardness, stiffness, flexibility, and modulus. In some respects these softening effects of plasticizers are also due to

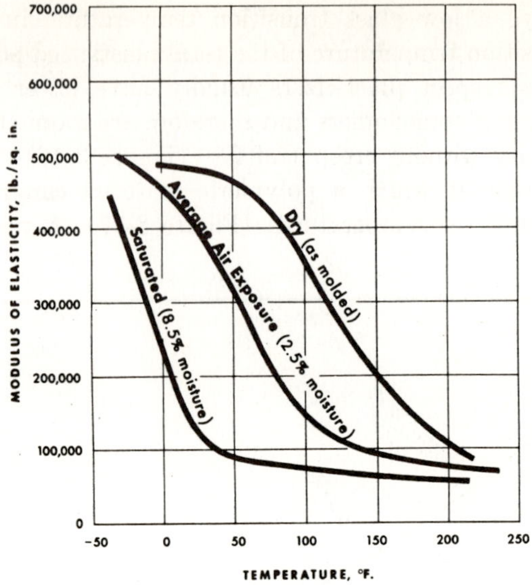

Figure 3-14. Effect of moisture absorption upon modulus of nylon 66 (ref 49).

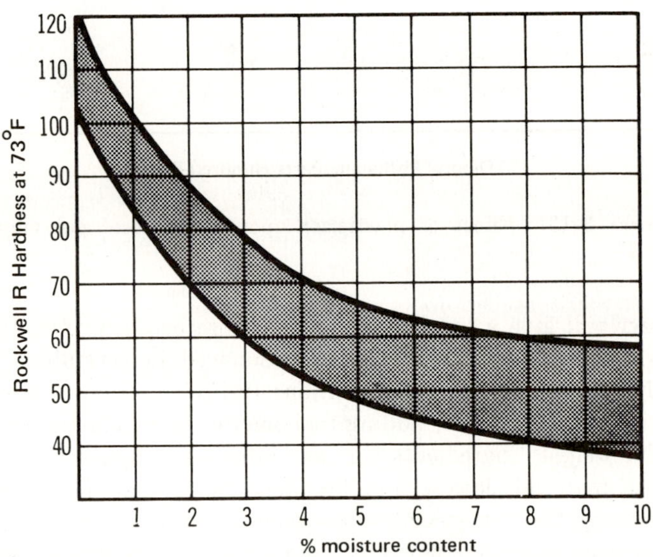

Figure 3-15. Effect of moisture absorption upon hardness of nylon 6 (ref 50).

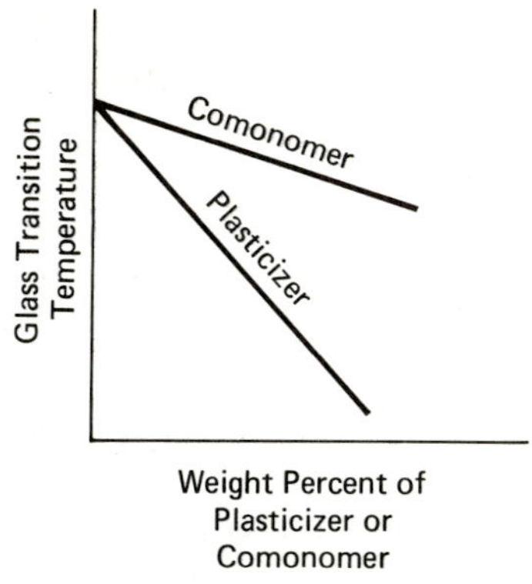

Figure 3-16. Effect of plasticizer and comonomer upon glass transition temperature (ref 24).

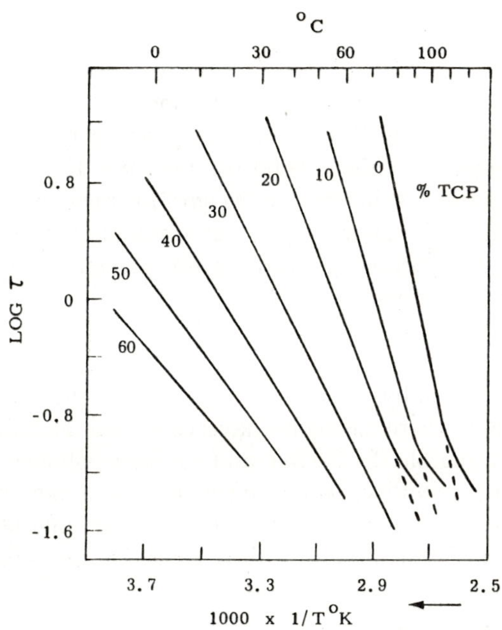

Figure 3-17. Effect of plasticizer on rate of response (ref 56). Log of the relaxation time, τ, versus reciprocal of the absolute temperature; polyvinyl chloride-tricresyl phosphate system.

their ability to reduce intermolecular bonding between polymer molecules; these mechanisms will be discussed later.

Mechanical Failure

When polymeric materials are subjected to excessive mechanical stress, they will deform permanently or break catastrophically. At lower molecular weights the polymer molecules will simply disentangle, slide over each other, and separate. Beyond the critical molecular weight for entanglement, these forces of intermolecular entanglement and attraction are greater than the strength of the covalent bonds within the polymer molecule itself, and excessive mechanical stress will produce complete rupture of these primary valence bonds within the polymer molecule. In either case such failure is permanent, and no longer reversible when the stress is removed. In practical measurements, mechanical failure may be observed as: (1) permanent deformation; (2) ultimate strength and elongation in low-speed testing; (3) impact strength in high-speed testing; and (4) multiple cycling stress which results in fatigue failure.

Permanent Deformation

Permanent deformation is generally measured as yield strength, creep, or stress-relaxation. Rigid glassy polymers generally tend to have tensile yield strengths of about 8000 psi (ref 38). Under sustained mechanical stress, creep is inverse to weight average molecular weight; and beyond a critical minimum $\bar{M}_w$ intermolecular entanglements act like cross-links in restricting, retarding, or preventing creep (ref 58). Stress-decay or relaxation in rubber similarly depends on molecular weight (ref 59).

In general, permanent deformation under sustained stress results from molecules gradually disentangling and slipping past each other. In the glassy state, where molecules have very little mobility, permanent deformation is negligible. In flexible and rubbery polymers, where molecules have greater mobility, increasing molecular weight increases their entanglement and decreases the rate at which linear amorphous polymers can disentangle, and thus decreases their rate of permanent deformation.

Ultimate Strength and Elongation

Catastrophic failure under gradually increasing mechanical stress is generally measured in terms of ultimate strength (stress) and ultimate elongation (strain) at time of break. Whether such gradual mechanical stress is tensile, flexural, compressive, torsional, or shear, the fundamental molecular mechanism is the same. In low-molecular-weight materials, the bonding forces within the molecules are high but the attraction between molecules is small, and failure occurs at low stress by separation between molecules. As polymer molecular weight increases, the sum of the attractive forces between adjacent polymer molecules increases correspondingly, and so does the mechanical stress required to separate

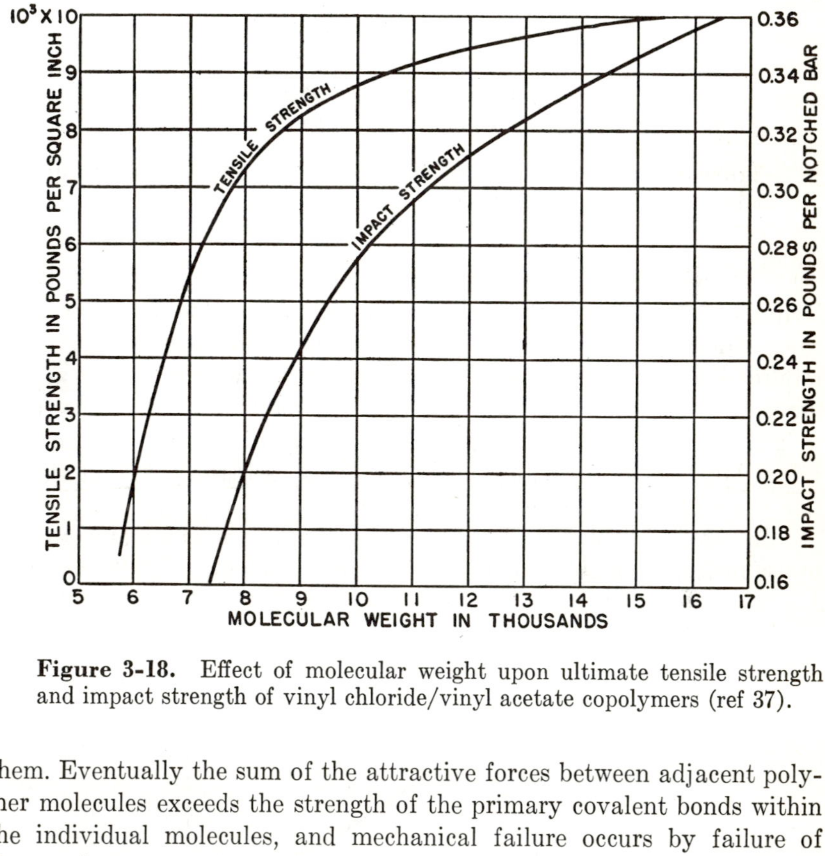

Figure 3-18. Effect of molecular weight upon ultimate tensile strength and impact strength of vinyl chloride/vinyl acetate copolymers (ref 37).

them. Eventually the sum of the attractive forces between adjacent polymer molecules exceeds the strength of the primary covalent bonds within the individual molecules, and mechanical failure occurs by failure of these primary covalent bonds; beyond this point, further increase in molecular weight cannot contribute any more to mechanical strength. Thus ultimate strength at break is generally observed to depend on molecular weight (ref 60) (Figures 3-18, 3-19), and this relationship is important in plastics, in fibers (ref 62,145), and even in coatings (ref 30).

Ultimate elongation, while more complex and less reliable, also ap-

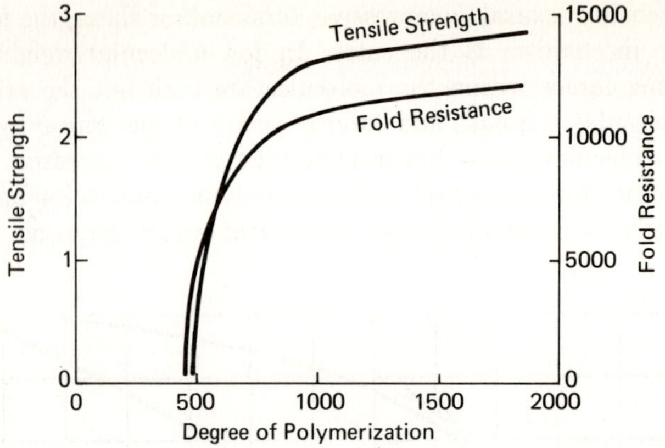

Figure 3-19. Effect of molecular weight upon ultimate tensile strength and fold resistance of cotton fibers (ref 61).

pears to depend similarly upon molecular weight (Figures 3-20, 3-21). Here elongation probably represents the extent to which polymer molecules slide past each other before separating completely at catastrophic failure. Thus elongation increases with molecular weight so long as failure occurs by separation between polymer molecules. When molecular weight becomes so high that failure occurs by breaking of primary covalent bonds within the polymer molecules, it is reasonable that elongation should level off as well.

In most plastic and elastomeric applications it is usually possible to use molecular weights high enough so that ultimate strength properties

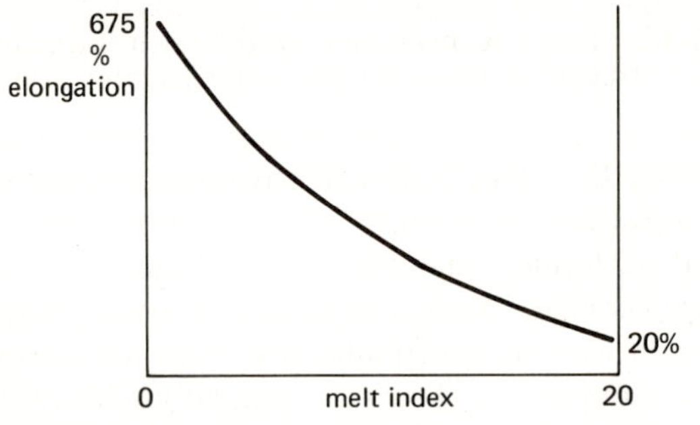

Figure 3-20. Effect of molecular weight upon ultimate elongation of polyethylene (ref 63). Note that increasing melt index indicates decreasing molecular weight.

are already in the optimum plateau region. When further increase in molecular weight produces continuing significant improvement in strength, at a sacrifice of melt processability, these higher molecular weights are used only in products which require the extra strength, such as extruded polyethylene pipe, wire coating, and monofilaments, and especially the new ultra-high-molecular-weight mill shapes which can only be machined into end-products such as machine parts (ref 8). In applications requiring higher fluidity during processing, however, such

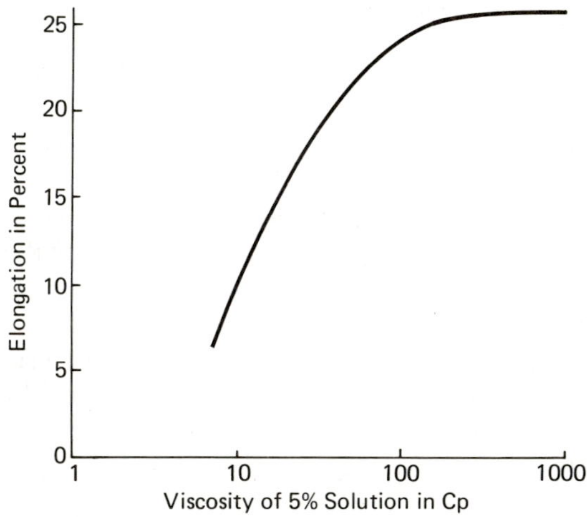

Figure 3-21. Effect of molecular weight upon ultimate elongation of ethyl cellulose (ref 64). Note that solution viscosity increases with molecular weight.

as coatings, adhesives, and solution-spinning of fibers, the need for low molecular weight during processing may conflict seriously with the desire for high-molecular-weight to yield optimum ultimate strength in the final product, and the practical compromise may necessarily be somewhat less than completely satisfactory.

Molecular Weight Distribution In many studies, ultimate tensile strength appears to correlate well with $\bar{M}_n$, thus tensile strength $= a - (b/\bar{M}_n)$ (ref 65).

This is believed to be due to the high concentration of chain ends resulting from a low-molecular-weight fraction and producing imperfections, high free volume, and mobility in the polymer structure. Other studies have observed that ultimate tensile strength, ultimate elongation, and stress-strain properties in general all are similarly related to $\bar{M}_n$ (ref 66). More specifically it has been suggested that minimum degrees of polymerization to produce good tensile strengths should be at least 40–80 for plastics and 150 for fibers, and further that both ultimate tensile and flexural strengths increase rapidly with number average degrees of polymerization up to 200–250 and finally level off at 600–700 (ref 67).

On the other hand, other studies have indicated that ultimate tensile strength and ultimate elongation may actually depend on average molecular weights between $\bar{M}_n$ and $\bar{M}_w$ or even between $\bar{M}_w$ and $\bar{M}_z$ (ref 10,65). In a particularly detailed study of the relationship between ultimate elongation and molecular weight in acrylic films, it was observed that ultimate elongation correlated most reproducibly with $\bar{M}_w$, whereas it gave a family of curves with $\bar{M}_n$ depending upon the type of molecular weight distribution of the different samples (ref 7) (Figure 3-3).

Thus there is general agreement that increasing molecular weight produces increasing entanglement and attraction between polymer molecules, which in turn produce increasing resistance to deformation and breakage, and that ultimately failure may occur by breakage of primary valence bonds in the polymer molecules themselves; but further study is needed to determine whether such ultimate mechanical strength properties correlate best with number, weight, or other type of average molecular weights.

PLASTICIZATION. Addition of monomeric liquid plasticizers to high-molecular-weight polymers introduces a large low-molecular-weight fraction which reduces the entanglement and bonding forces between the polymer molecules, and increases their free volume and mobility. This generally lowers ultimate tensile, compressive, and tear strengths and increases ultimate elongation (Table 3-2, Figures 3-22 to 3-25), whether

Table 3-2. Plasticization of Polyvinyl Chloride by Tricresyl Phosphate (ref 68)

Tricresyl phosphate wt %	Tensile strength psi	Elongation at break %
0	5,690	2
30	3,840	170
60	1,000	500

it occurs in plastics (ref 53,68,69), rubber (ref 48), or coatings (ref 70). In most cases plasticizer is added to plastics to increase flexibility or impact strength, to rubber to increase the rate of elastic response, and to coatings and adhesives to increase flow during processing; and the sacrifice in ultimate strength is an unwanted accompaniment of the major desired effect. Absorption of moisture produces similar plasticization in both plastics and coatings (Figures 3-26, 3-27) (ref 51,70), providing a somewhat unreliable flexibilizing effect and a sacrifice of strength.

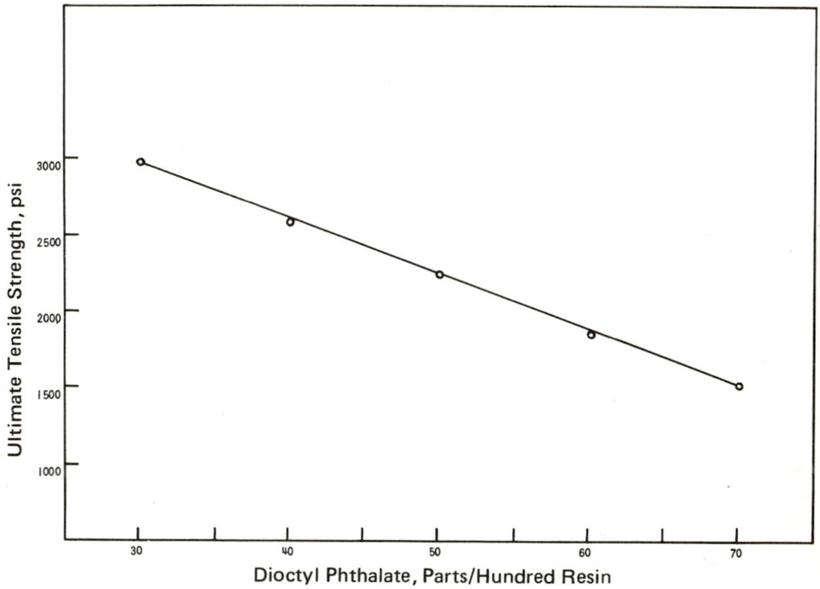

Figure 3-22. Effect of plasticizer upon ultimate tensile strength of polyvinyl chloride (ref 44).

In all of these cases the increased low-molecular-weight fraction of plasticizer is responsible for the lower entangling and attractive forces between the larger polymer molecules and the resultant decrease in ultimate strength properties.

Impact Strength

Another important type of catastrophic mechanical failure is brittle breakage under high-speed impact. Resistance to brittle failure generally

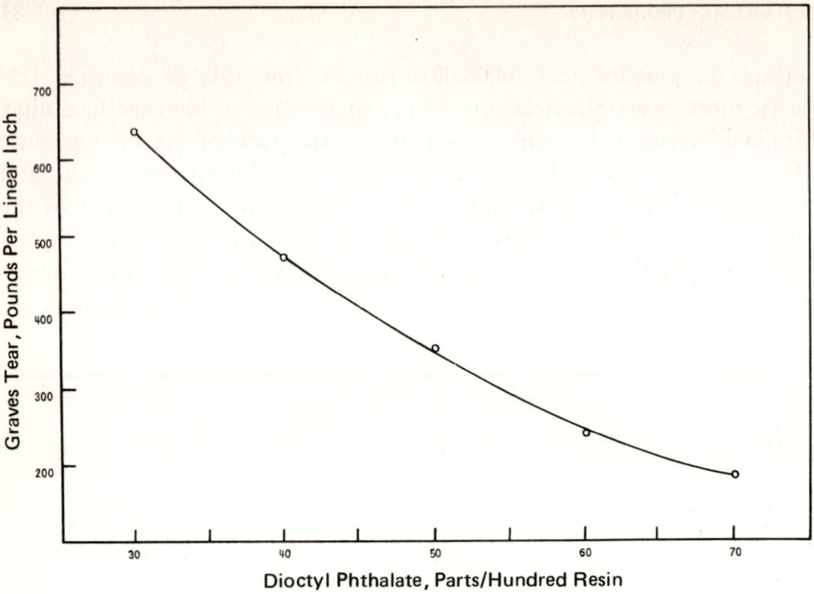

Figure 3-23. Effect of plasticizer upon tear strength of polyvinyl chloride (ref 44).

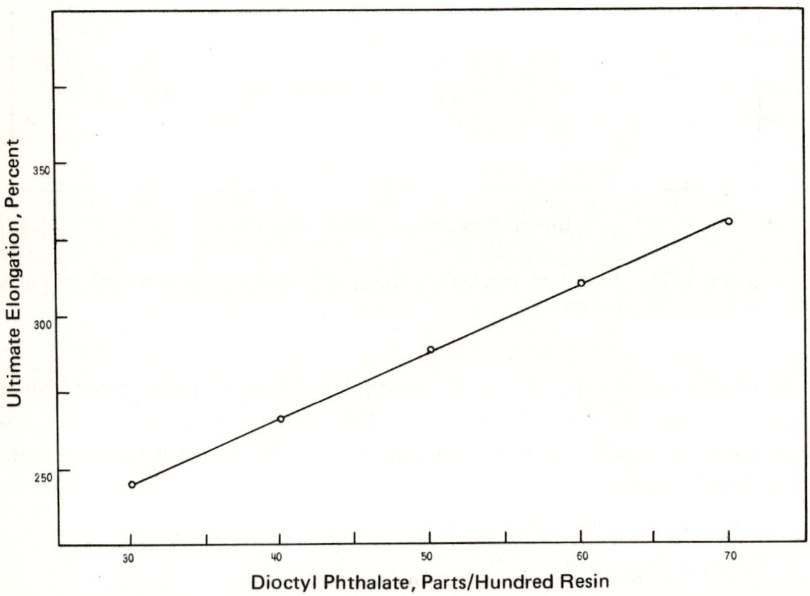

Figure 3-24. Effect of plasticizer upon ultimate elongation of polyvinyl chloride (ref 44).

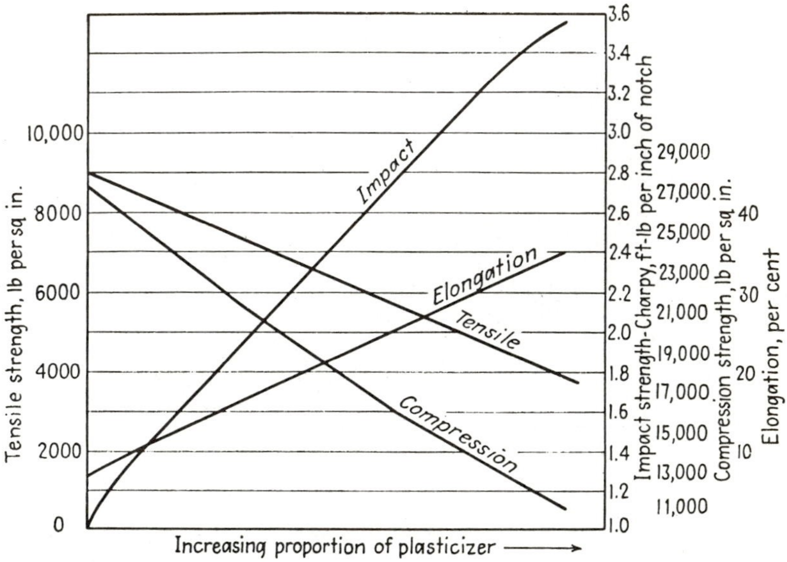

Figure 3-25. Effect of plasticizer upon mechanical properties of cellulose acetate (ref 69).

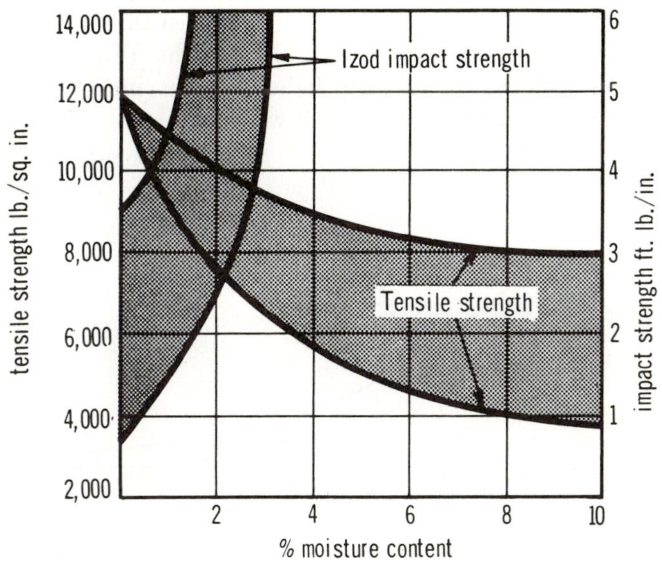

Figure 3-26. Effect of moisture absorption upon tensile and impact strengths of nylon 6 (ref 50).

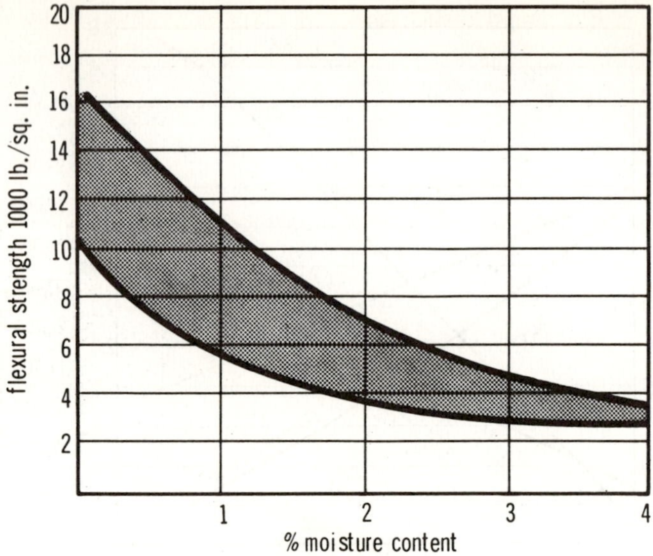

Figure 3-27. Effect of moisture absorption upon flexural strength of nylon 6 (ref 50).

improves with increasing molecular weight of the polymer (ref 9) (Figures 3-18, 3-28, 3-29). Some studies have indicated that impact strength increases with number average molecular weight (Figure 3-30) (ref 67,73,74), while other studies have indicated that impact strength increases with weight average molecular weight (ref 75). Tensile impact

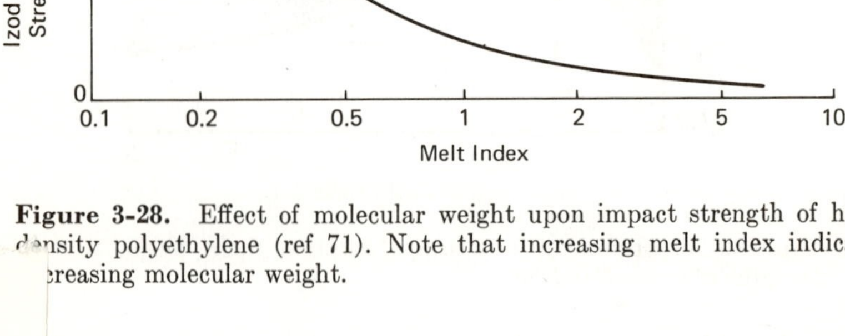

Figure 3-28. Effect of molecular weight upon impact strength of high-density polyethylene (ref 71). Note that increasing melt index indicates decreasing molecular weight.

strength also appears to be primarily a function of weight average molecular weight (ref 10).

On the other hand, addition of plasticizers increases the lowest-molecular-weight portion of the composition and therefore its mobility, flexibility, and impact strength (ref 68); the use of phthalic ester plasticizers in cellulose ester plastics is practiced largely for this reason (Figure 3-25). In coatings, addition of plastitizer improves not only impact strength but also related qualities such as resistance to checking and

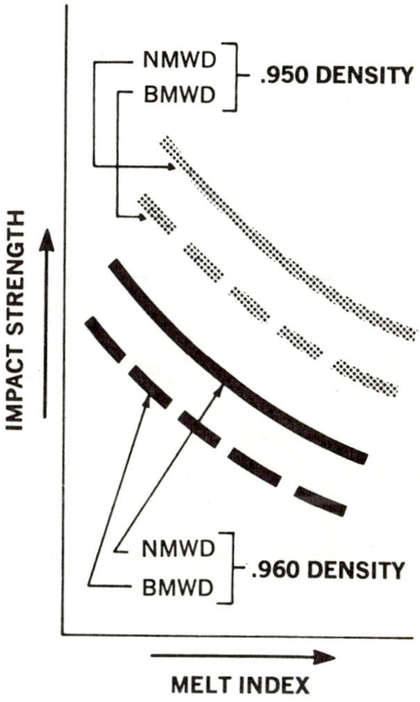

Figure 3-29. Effect of molecular weight and distribution upon impact strength of high-density polyethylenes (ref 72).

chipping (ref 46). Here again, absorption of moisture in polar polymers like nylon produces similar plasticizing effects (Figure 3-26).

Multiple Cycling Stress

Another type of failure results from repeated cyclic mechanical stress below the level that would produce failure in a single application. Resistance to such failure, often called fatigue resistance, is generally measured in terms of flex life at a particular amplitude and speed of stress. Such flex life correlates directly with molecular weight (ref 76).

Complex Mechanical Properties

Many important mechanical properties are more complex, and seem to involve combinations of a number of simple types of stress. For example:

Stress cracking of rigid plastics, when immersed in certain reagents under mechanical strain, is a major problem in the use of plastics both

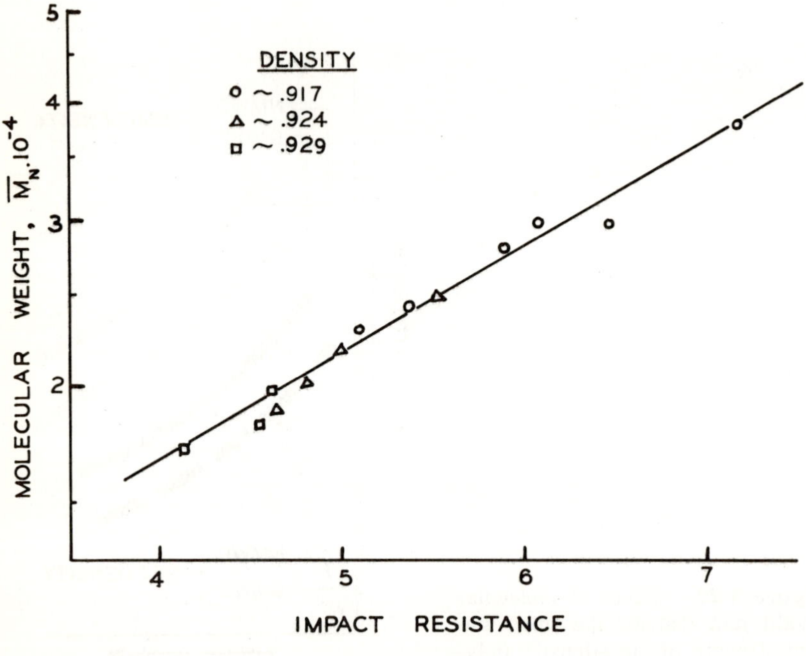

Figure 3-30. Effect of number average molecular weight upon impact strength of low-density polyethylene (ref 73).

in the home and in industry. Experimental studies indicate that it is inverse to molecular weight (Figure 3-31), specifically number average molecular weight (Figure 3-32) and appears to be due primarily to extraction or leaching of the low-molecular-weight fraction (ref 79).

Adhesion of high-molecular-weight polymeric coatings on smooth metal surfaces may be relatively poor, but can be greatly improved by addition of low-molecular-weight plasticizers and resins (ref 80). Here the improvement may be primarily in process fluidity, which permits the adhesive to wet the surface more thoroughly.

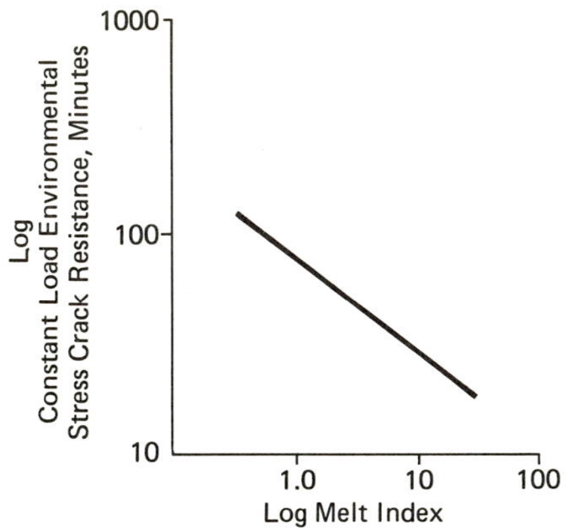

Figure 3-31. Effect of molecular weight upon stress-cracking of high-density polyethylene (ref 77). Note that increasing melt index indicates decreasing molecular weight.

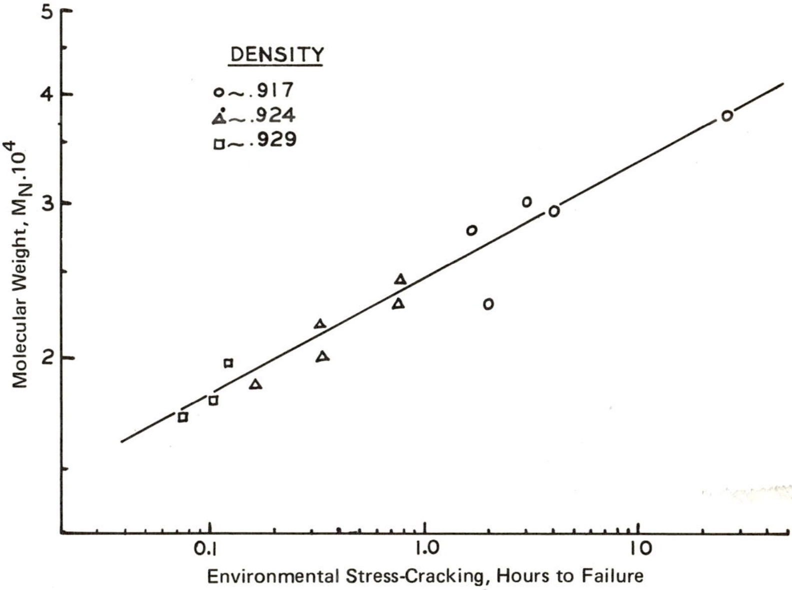

Figure 3-32. Effect of number average molecular weight upon stress-cracking of low-density polyethylene (ref 78).

Friction is generally believed to depend upon several properties including adhesion, and appears to be reduced when molecular weight becomes very high, as in ultra-high-molecular-weight polyethylene machine bearings and ski bottoms (ref 17).

Abrasive attack and wear are due to a combination of modulus, stiffness, resilience, adhesion, friction, and tearing (ref 81); and abrasion-resistance also improves at extremely high molecular weight, as in the use of ultra-high-molecular-weight polyethylene for rough-duty machinery (ref 17).

To the extent that such properties involve mechanical stress, high-molecular-weight appears to be an important factor in these complex properties as well.

3-3 THERMAL PROPERTIES

The effects of molecular weight on thermal properties are most evident in three areas: (1) Thermal Mechanical Properties; (2) Thermodynamic Properties; and (3) Thermal Stability.

Thermal Mechanical Properties

The effect of molecular weight on thermal mechanical properties is best understood by examining its effects on the relationship between modulus and temperature (Figure 3-33). In a linear amorphous polymer at low temperature, the atoms in the polymer molecule have little thermal energy of vibration, and the heavily intertwined polymer molecules are quite immobile and unable to move past each other, even under mechanical stress; in this region the polymer is rigid and glassy. As the temperature increases, the atoms and groups in the polymer molecule acquire an increasing amount of thermal energy, causing them to vibrate and move about; the higher the temperature the greater the thermal motion and the larger the sequences of atoms in the polymer chain that become mobile in this way. At some critical temperature called the glass transition temperature T_g, or more properly in some critical narrow temperature range, this mobility begins to involve polymer chain segments large enough to permit substantial motion within the polymer mass. Beyond this point, when mechanical stress is applied, motion of such large polymer chain segments is sufficient to permit motion of the entire polymer mass; in this transition region the polymer's mechanical behavior is stiffly flexible or leathery.

At still higher temperatures the mobility of large polymer segments is sufficiently high so that the polymer can yield easily under mechanical stress and its total behavior is flexible and rubbery; this flexible rubbery plateau is the temperature range in which the polymer yields easily to mechanical stress, and recovers readily when the stress is released. At still higher temperatures the thermal energy and mobility of the atoms and larger segments of the polymer molecule become so high

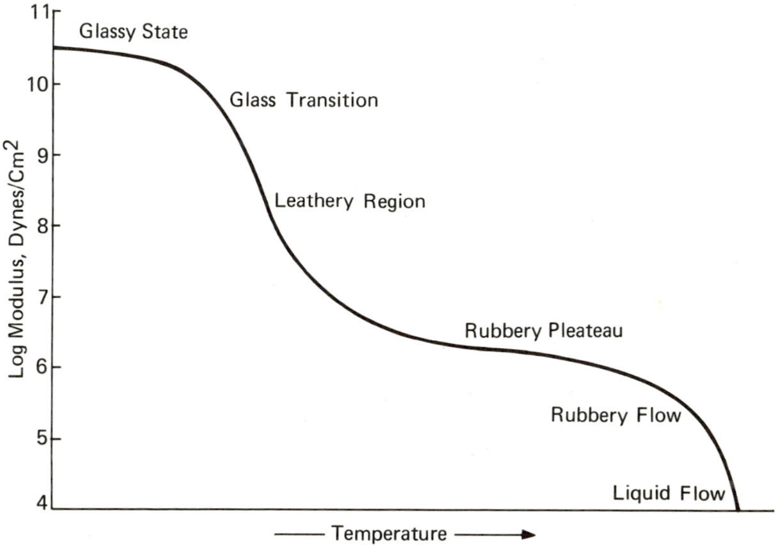

Figure 3-33. Effect of temperature upon modulus of a linear amorphous polymer (ref 82–84).

that, when mechanical stress is applied, entire polymer molecules tend to disentangle and flow past each other; in this second transition region polymer behavior is described as rubbery flow, meaning that it will be mildly elastic under low mechanical stress but permit permanent deformation and flow under higher mechanical stress. Finally at still higher temperatures the thermal energy and mobility of the polymer molecules become so high that they flow easily and permanently under mechanical stress and are describable as truly liquid in character. The effects of molecular weight on thermal mechanical properties are best understood by examining their effects on each of these regions in turn.

Rigidity/Flexibility and the Glass Transition Temperature

Although some studies have found that molecular weight has no effect on modulus in the glassy and leathery regions (ref 36), and that in certain polymers such as polyoxymethylene and polyoxypropylene, T_g is independent of molecular weight (ref 85), most studies have indicated that molecular weight has a distinctive effect upon the glass transition region (ref 86,87). This effect is most evident at molecular weights up to 20,000 (ref 88), and becomes negligible at very high molecular weights (ref 89) (Figure 3-34).

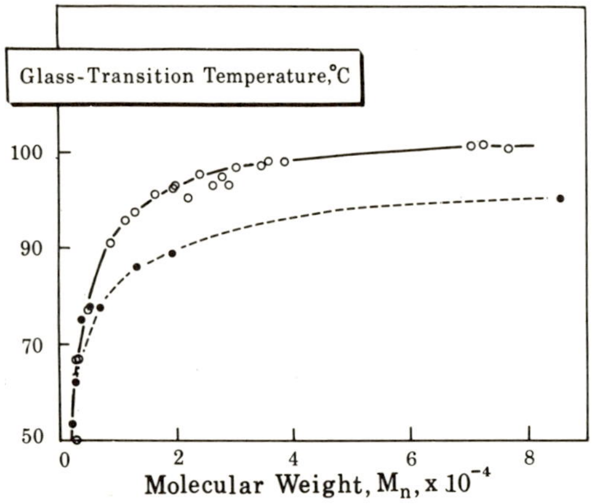

Figure 3-34. Effect of molecular weight upon glass transition temperature (ref 85). ○ Polymethyl methacrylate; ● Polystyrene.

Mechanism The effect of molecular weight on the glass transition region is due primarily to the ends of the polymer molecules (ref 10,89–92). Whereas the mobility of segments within the polymer molecule is limited by their attachment at both ends, the mobility of segments at the ends of the polymer molecule is limited by their attachment only at one end. Thus the chain ends are much more flexible and mobile, and generate a larger free volume into the space that they occupy. This free volume in turn provides empty space in which all the segments of the polymer molecule have some freedom to move, and this produces a greater flexibility in the polymer molecule as a whole. Since chain ends are measured primarily by number average molecular weight, T_g

relates best with M_n. This relationship may be simply expressed as $T_g = a + b/M$ (ref 2). With greater precision this has been written as

$$T_g = T_\infty - \frac{B}{\bar{M}_n} \qquad \text{or} \qquad \frac{1}{T_g} = \frac{1}{T_\infty} + \frac{K}{M_n}$$

where T_∞ is the glass transition temperature at infinite molecular weight (ref 92,93) (Figure 3-35). Thus the mechanism of the relationship between molecular weight and glass transition region is both well understood and well described with mathematical precision.

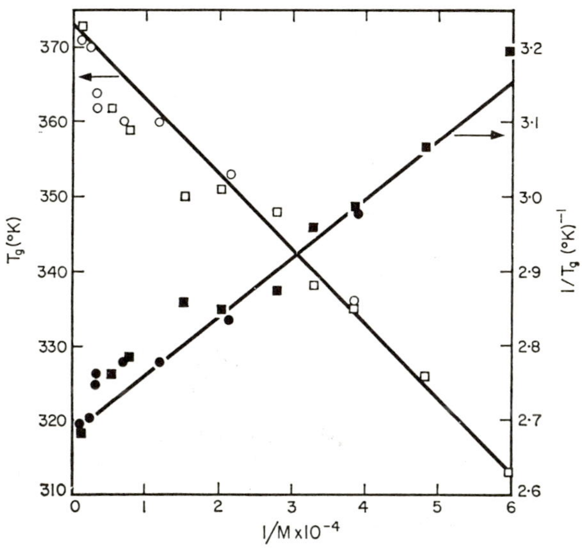

Figure 3-35. Relationships between molecular weight and glass transition temperature of polystyrene (ref 93).

PRACTICAL EFFECTS Glass transition temperature alone is a fairly abstract quantity, measured primarily in theoretical studies; but it is often assumed that it should correlate fairly well with a number of practical properties and phenomena, which are similar in mechanism and would be expected to occur in or near the glass transition region. For example, when a tough rigid polymer is cooled to lower and lower temperatures, its impact strength tends to drop, and this decrease is often most marked near the glass transition temperature. When a flexible or rubbery polymer is cooled, it reaches a temperature at which it becomes inflexible and often brittle. On the other hand the basic mechanism of glass transition may not be quite the same as the mechanism of some types of practical mechanical failure, and the correlation may at times be poor but still informative. For example, in studies on polyisobutylene (Figure 3-36) it has been observed that at low molecular weight the brittle temperature is considerably higher than the glass transition temperature, because T_g measures the mobility of much smaller molecular segments than are involved in actual embrittlement. With increasing molecular weight T_g rises as predicted mathematically, but brittle temperature actually decreases because the increasing length of the molecules

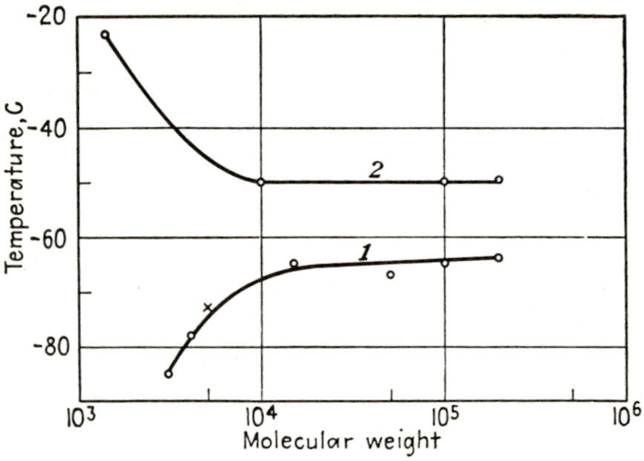

Figure 3-36. Effect of molecular weight upon low-temperature flexibility of polyisobutylene (ref 94).
Curve 1: glass transition temperature
Curve 2: brittle temperature

provides greater mechanical strength as noted earlier. Finally at higher molecular weights T_g and brittle temperature do finally approach each other (ref 95,96). A similar mechanism probably explains the inverse effect of molecular weight on brittle temperature of polyethylene (Figure 3-37).

Moving in the other direction, when a rigid polymer is heated it reaches a temperature at which it begins to soften and become more flexible, and this is generally considered the maximum temperature at which it should be used. It is often measured by heat distortion temperature, deflection temperature, or Vicat softening point, and these have generally been observed to correlate with glass transition temperature (ref 97) and with molecular weight (ref 1,76). The softening point of polyethylene, for example, increases regularly with molecular weight (ref 98). Related to this is the rate and degree of distortion above the glass transition temperature, such as is measured in creep tests; and here the creep of an amorphous linear polymer is restrained only by the entanglement of the larger molecules and therefore proves to be inverse to weight average molecular weight (ref 99).

PLASTICIZERS *Mechanism* Plasticizers are monomeric or low-molecular-weight polymeric liquids which are added to high-molecular-weight

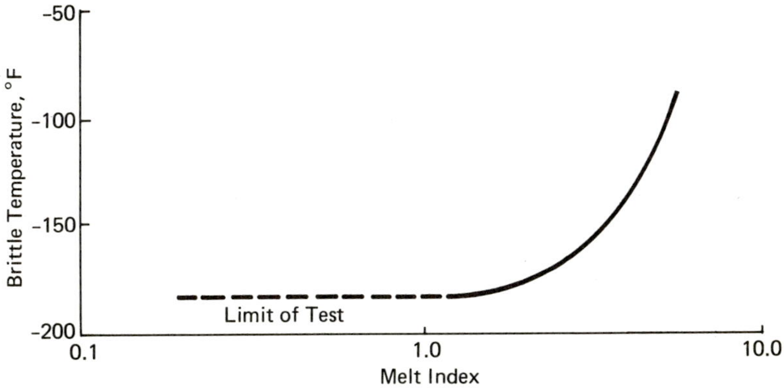

Figure 3-37. Effect of molecular weight upon low-temperature brittleness of high-density polyethylene (ref 71). Note that increasing melt index indicates decreasing molecular weight.

polymers to reduce their rigidity or improve their flexibility. With respect to thermal properties, plastiticzers are used primarily to improve the low-temperature-impact strength of rigid plastics and to lower the stiffening and brittle temperatures of flexible plastics and rubbers. Plasticizers accomplish these effects by lowering the glass transition temperature (Figure 3-38).

An effective plasticizer must be compatible with the polymer to which it is added, and is therefore useful only in amorphous polymers or in the amorphous regions of partly crystalline polymers (ref 100,101). It should be a mobile low-viscosity liquid (ref 101) with a large free volume which it can contribute to raising the free volume of the total plasticized polymer (ref 54). In addition to having a low viscosity it should have a low temperature coefficient of viscosity (ref 101). And of utmost importance, the plasticizer itself should have a low glass transition temperature (Figure 3-39) (ref 87,101).

As a low-molecular-weight fraction with all of these low-temperature properties, plasticizers can be highly efficient in lowering the glass transition temperature (ref 55,89,100,103–105). In addition to lowering the glass transition temperature, a plasticizer may also broaden the transition region (ref 55).

In addition to acting as a low-molecular-weight fraction, plasticizers may also reduce intermolecular forces between polymer molecules, and thus lower glass transition temperatures in this way (ref 100,106). This mechanism will be considered later.

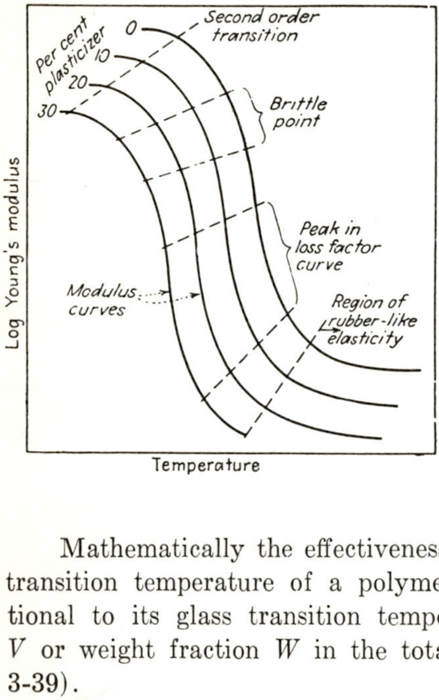

Figure 3-38. General effect of plasticizer upon thermal mechanical properties of a polymer (ref 69).

Mathematically the effectiveness of plasticizer in lowering the glass transition temperature of a polymeric composition is directly proportional to its glass transition temperature and to its volume fraction V or weight fraction W in the total composition (ref 19,102) (Figure 3-39).

$$T_g = V_1 T_{g1} + V_2 T_{g2} \qquad \frac{1}{T_g} = \frac{W_1}{T_{g1}} + \frac{W_2}{T_{g2}}$$

More complicated equations can even show the relationship to free volume (ref 54).

Practical Effects The use of plasticizers to lower the glass transition temperature is most widely practiced in plasticized flexible polyvinyl

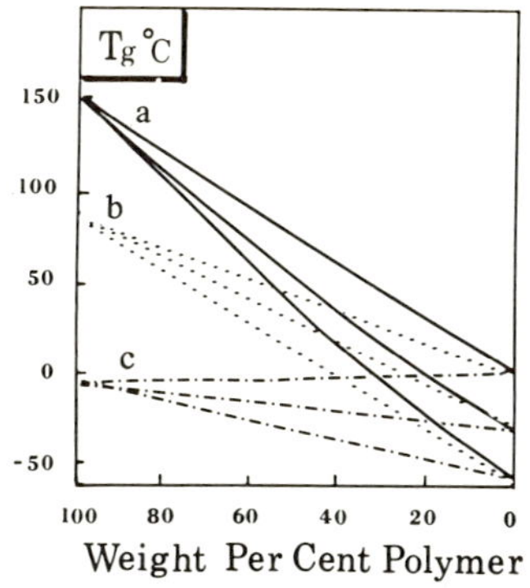

Figure 3-39. Effect of three different plasticizers on the glass-transition temperature of three polymers (ref 102).

Glass-Transition Temperatures of Polymers and Plasticizers Used in a Study of Plasticizer Action

Polymer	T_{gp} (°C)	Plasticizer	T_{gl} (°C)
Polystyrene-acrylonitrile copolymer (a)	151	Dimethylcyclohexyl #1	+6
Poly(vinyl acetate) (b)	84	Tricresyl phosphate #2	−30
Polychloroprene (c)	−9	Di-*n*-hexyl phthalate #3	−57

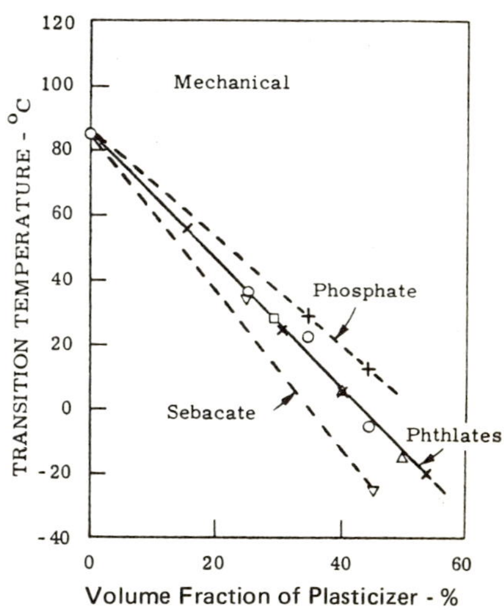

Figure 3-40. Effect of plasticizer structure and concentration upon its efficiency in lowering the transition temperature (ref 107).

chloride, particularly for applications like automotive upholstery, refrigerator gasketing, food packaging, and winter outerwear such as coats, boots, and gloves.

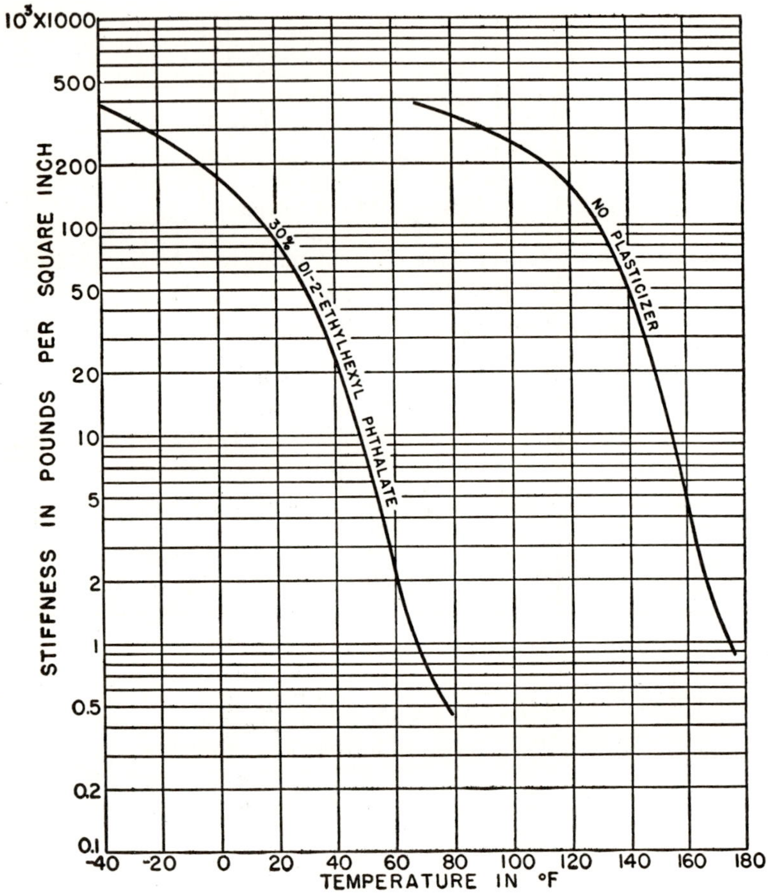

Figure 3-41. Effect of 30% di-2-ethylhexyl phthalate upon stiffness versus temperature of vinyl chloride/vinyl acetate 90/10 copolymer (ref 108).

The efficiency of different plasticizers in lowering the transition temperature depends both upon their individual structure and properties and upon the amount of plasticizer added to the composition (Figure 3-40). Since plasticizers have lower glass transition temperatures than comonomer units in a copolymer, plasticizers are more efficient in lowering the glass transition temperature of the total composition (ref 25) (Figures 3-16).

The effect of plasticizers on glass transition temperature is in

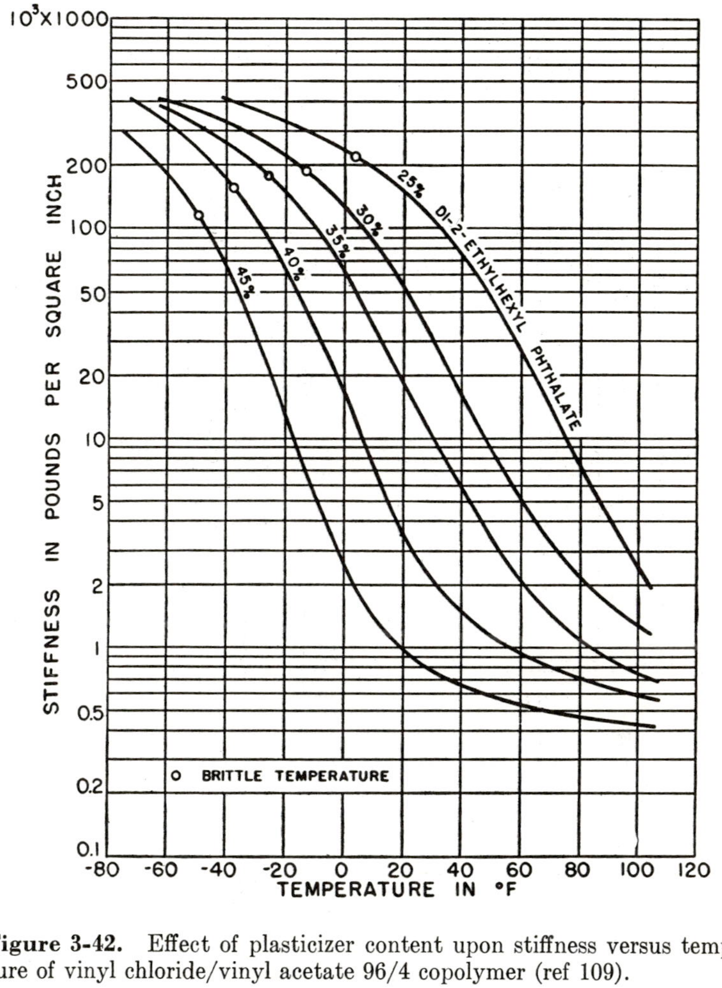

Figure 3-42. Effect of plasticizer content upon stiffness versus temperature of vinyl chloride/vinyl acetate 96/4 copolymer (ref 109).

itself a rather abstract and theoretical observation. It has been correlated frequently with practical properties, such as low-temperature impact strength of rigid plastics, and low-temperature stiffness (Figures 3-41 to 3-43) and brittle temperature (Figure 3-44) of flexible plastics and rubbers. The addition of softeners to rubber formulations is quite effective in lowering the freezing point of the final rubber composition (ref 112). On the other hand, addition of plasticizers to rigid plastics is apt to lower the heat distortion temperature as well, and

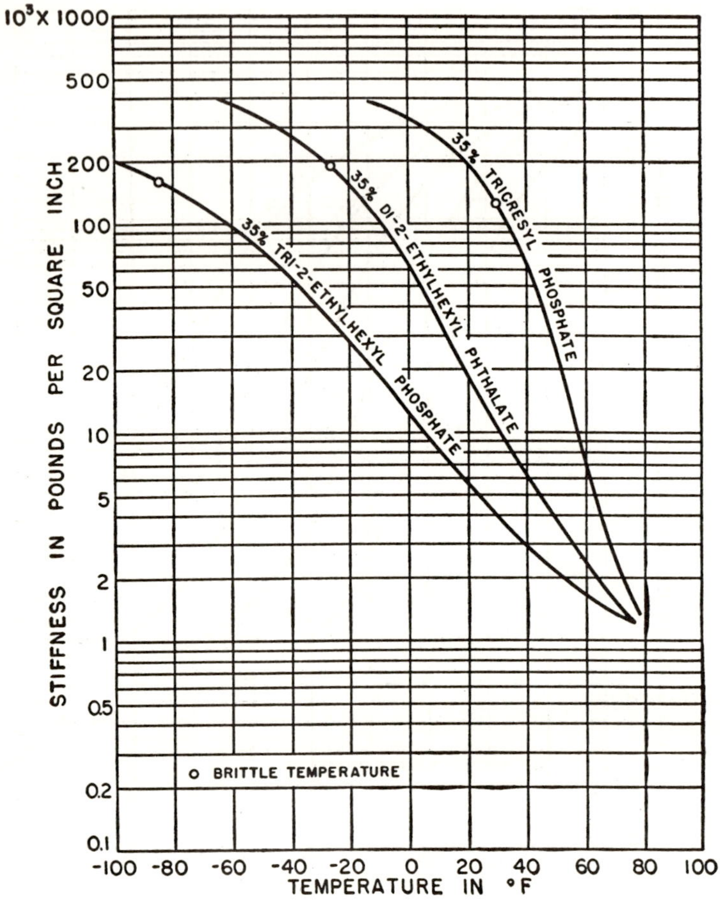

Figure 3-43. Effect of plasticizer structure upon stiffness versus temperature of vinyl chloride/vinyl acetate 96/4 copolymer (ref 110).

to increase the rate and extent of creep above the glass transition temperature (ref 113). In addition to plasticizers which are added intentionally, residual monomers and residual solvents which remain accidentally after processing have often been observed to exert similar effects (ref 113,114).

Leathery Region

As the temperature rises above the glass transition region, the thermal energy and mobility of the atoms and groups within the polymer molecule continue to increase, and with them the size of the segments of the polymer molecule that are able to flex and move, and the frequency with which they undergo such motion. This makes the polymer stiffly flexible or leathery in nature. As the temperature increases and the mobility increases, stiffness decreases and the leatheriness becomes soft. These

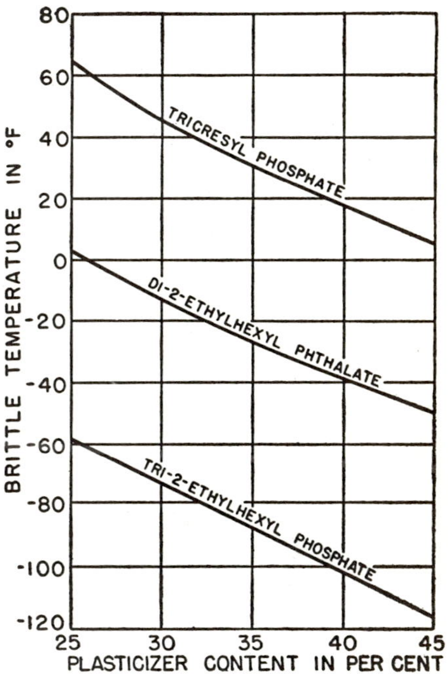

Figure 3-44. Effect of plasticizer structure and concentration upon brittle temperature of vinyl chloride/vinyl acetate 96/4 copolymer (ref 111).

changes occur continuously so that this is generally regarded as a leathery transition region.

The presence of plasticizers tends to smooth and moderate the effect and spread it over a longer temperature range, thus broadening the leathery transition zone. This is particularly true when the plasticizer is not entirely compatible with the polymer (Figure 3-45), or when the polymer is partially crystalline and has several intermediate transition areas (ref 55,116).

Rubbery Plateau

At still higher temperatures the mobility of large segments of the polymer molecule is sufficient to give fast, soft, flexible, and rubbery responses to mechanical stress. The breadth of this rubbery plateau corresponds approximately to the useful temperature range of such soft,

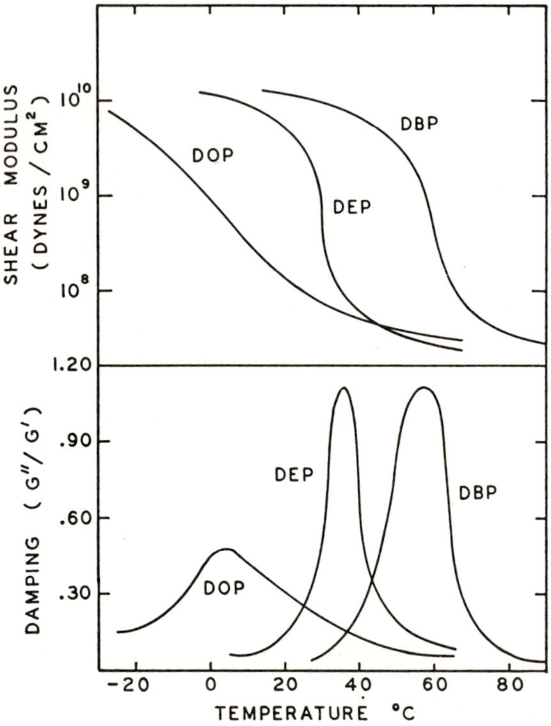

Figure 3-45. Effect of plasticizer structure upon leathery transition of polyvinyl chloride (ref 115).

Dynamic mechanical properties of plasticized polyvinyl chloride. Diethyl phthalate (DEP, 0.254 volume fraction), dibutyl phthalate (DBP, 0.158 volume fraction), *n*-dioctyl phthalate (DOP, 0.402 volume fraction).

flexible plastics and elastomers, and is particularly important whenever they must be used over a wide temperature range. The higher the molecular weight of the polymer, the greater the number of entanglements, and the broader the rubbery plateau will be. These entanglements simulate true cross-links and retard the molecules from sliding permanently past each other into permanent creep, thus providing reversible elasticity (ref 99), and extending it to higher temperatures.

Addition of monomeric liquid plasticizer introduces a mobile low-molecular-weight fraction of high free volume, which facilitates the motion of chain segments into different configurations and shifts rubber-like properties to lower temperatures (ref 53) (Figure 3-38). It also increases the rate of flexible or elastomeric response to mechanical stress, and thus decreases the relaxation time of the plasticized polymer (Figure 3-17). In these respects plasticizer is more effective than comonomer because it has a lower glass transition temperature and greater ability to provide flexibility without creep (ref 25) (Figure 3-16). This is what makes plasticized polyvinyl chloride useful in such a wide range of flexible applications.

If the polymer contains true stable cross-links or if it contains a low degree of crystallinity, holding certain segments of the polymer molecules together in fixed position, these effects are more permanent than simple amorphous entanglements and persist up to higher temperatures, thus extending the breadth of the rubbery plateau over a much greater and higher range. This will be discussed later (Chapters 5, 6).

Crystalline Fusion: The Melting Point Transition

The modulus-temperature curve of linear amorphous polymers is completely described by the five successive regions of Rigidity, Leathery Transition, Rubbery Plateau, Rubbery Flow, and Liquid Flow. Crystalline polymers have an additional and very characteristic transition at the melting point of the crystalline phase. This would generally appear between the regions of Rubbery Plateau and Rubbery Flow, and is therefore best considered at this point.

Although some studies have reported that the polymer melting point is not dependent on molecular weight (ref 117), it is fairly generally accepted that the melting point increases with molecular weight (Table 3-1, Figure 3-46). Whereas pure monomeric crystalline solids have sharp, constant, and reproducible melting points, this is not quite so true of polymers. Polymers are only partially crystalline at best, and contain many imperfections which decrease the regularity and perfection of crystalline structure and melting. One of the most common types of imperfection in polymer structure is the ends of the polymer molecule, which are sufficiently different from the rest of the molecule to create serious imperfections in the crystal structure, and cause a consequent reduction in crystallinity. The ends of the polymer molecules are also fairly mobile and flexible, creating looser packing and lower crystallinity for this reason as well. Consequently it is not surprising that the melting point of a crystalline polymer is a precise function of the number average molecular weight. Thus

$$\frac{1}{T_m} - \frac{1}{T_m^\circ} = \frac{2RM_0}{\Delta H_u \bar{M}_n}$$

or approximately

$$T_m = \frac{T_m^\circ \left(\dfrac{\bar{M}_n}{M_0}\right)}{B + \left(\dfrac{\bar{M}_n}{M_0}\right)}$$

where temperatures are given in °K, T_m° is the melting point of pure polymer at infinite molecular weight, M_0 is monomer molecular weight,

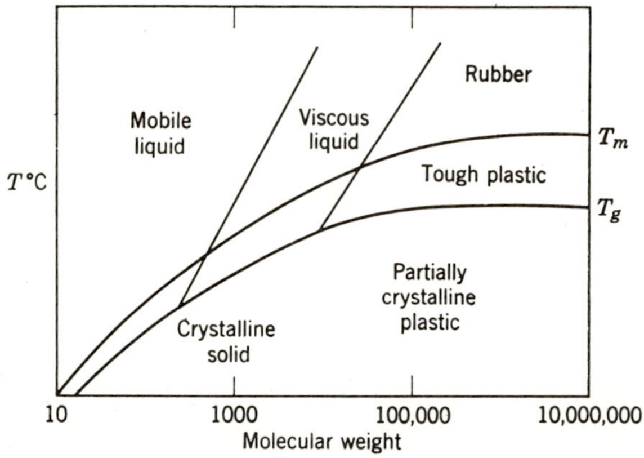

Figure 3-46. Effect of molecular weight upon transition temperatures and properties of polymers (ref 23).

R is the gas constant, ΔH_u is molar heat of fusion of the repeat unit in the crystalline polymer, and B is an empirical constant (ref 118).

It has been noted that both T_m and T_g increase with molecular weight (Figure 3-46). The ratio between these two temperatures is frequently a fairly characteristic one (Figure 3-47): T_g °K = 1/2 or 2/3 T_m °K (ref 117). These temperature effects on properties are also quite characteristic. Below T_g increasing molecular weight produces a change from crystalline solids to partly crystalline plastics. Between T_g and T_m increasing molecular weight produces increasingly tough plastics. And above T_m increasing molecular weight produces a change, already described, from fluid to viscous melts to very-high-molecular-weight rubbery solids.

Addition of monomeric plasticizers, solvents, or other low-molecular-weight materials to a higher-molecular-weight polymer, lowers its crystalline melting point in proportion to the molar volume V_1 of the low-molecular-weight liquid added: $1/T_m - 1/T_m^\circ = R/\Delta H_u \cdot V_u/V_1(v_1 - x_1 v_1^2)$, where V_u is the molar volume of the repeat unit in the polymer,

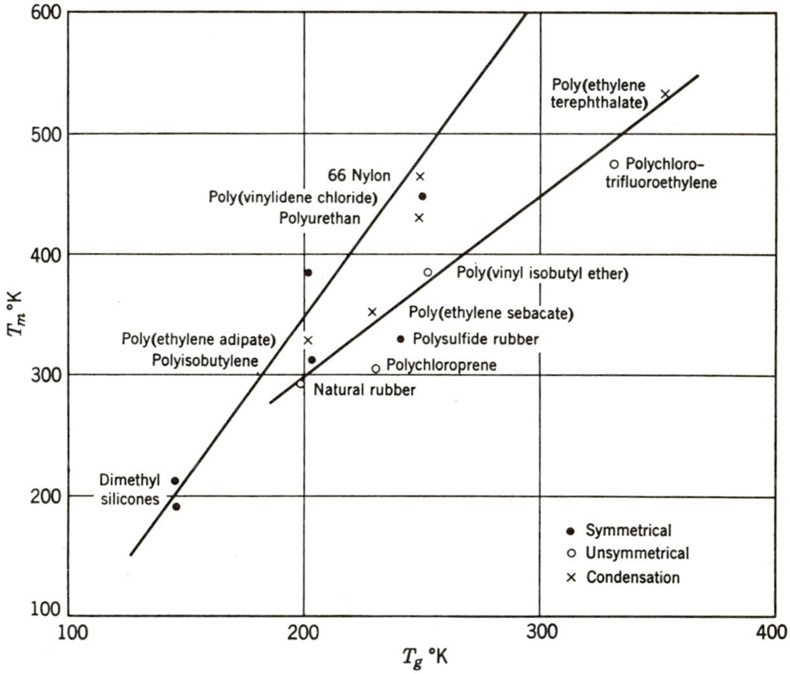

Figure 3-47. Relationship between melting and glass transition temperatures of various polymers (ref 90).

v_1 is the volume fraction of low-molecular-weight liquid and x_1 is an interaction parameter (ref 119). While either an added liquid monomeric plasticizer or a copolymerized second monomer can be effective in lowering the regularity of crystallinity and melting point of a polymer, a comonomer occupies less volume than a plasticizer and is therefore more effective in this respect (Figure 3-5). Both techniques have been practiced commercially to adapt nylons as hot-melt adhesives.

Rubbery and Liquid Flow

When linear amorphous polymers are heated to a temperature beyond the rubbery plateau, or when crystalline polymers are heated to a temperature above the melting point, the thermal energy and motion of large segments and whole polymer molecules are sufficient to permit these molecules to disentangle, and even occasionally to flow past each other under high mechanical stress. In the borderline transition region from the rubbery plateau up to true liquid flow, this transition region is characterized by rubbery liquid flow in which the soft polymer remains elastic under low mechanical stress or short time intervals but shows true permanent plastic flow under higher mechanical stress or longer time intervals. Naturally, increasing molecular weight produces an increasing degree of intermolecular entanglements, requiring a greater amount of stress, time, and/or thermal energy to produce plastic flow, and thus driving the rubbery flow transition region to higher temperatures. This is particularly important when high-molecular-weight causes rippling and even melt fracture in high-speed extrusion; and it becomes most pronounced when extremely-high-molecular-weight polymers, such as ultra-high-molecular-weight polyethylene, become completely unprocessable by conventional melt processing techniques.

At still higher temperatures, indicating still greater thermal energy and mobility of the segments of the polymer molecules, moderate mechanical stress is sufficient to produce true liquid flow even in short periods of time. This is the region of normal melt processing, and the liquid flow required in each type of process is controlled by balancing molecular weight against temperature, pressure, and time. Most of the major polymers are produced in a number of molecular weight grades specifically to meet the process temperature and viscosity requirements for each type of processing technique.

In both the rubbery flow and the true liquid flow regions, modulus or viscosity is a direct function of molecular weight (ref 84). At low molecular weight the polymer melts to a mobile liquid; at higher molecular weight to a more and more viscous liquid; and at extremely high molecular weights, even at high temperatures, the flow remains primarily rubbery and elastic in nature (Figure 3-46).

Thermodynamic Properties

Fundamental thermodynamic properties of polymers have not yet been thoroughly explored. There has been a start toward the understanding of the relationship between molecular weight and two fundamental thermodynamic properties: (1) Thermal Conductivity and (2) Coefficient of Thermal Expansion.

Thermal Conductivity

Thermal conductivity of polymers is generally much lower than that of most other materials, and in fact many uses of polymers are based on their efficiency as thermal insulators. Some of the deficiencies of polymers also result from their inability to conduct heat. Thermal conductivity does not vary greatly with conditions. It is somewhat dependent upon molecular weight, because the atomic vibrations of thermal energy are transmitted more directly along the covalent bonds in the polymer molecule than across the weakly-bonded regions between polymer molecules. At molecular weights up to 100,000, thermal conductivity is proportional to the square root of weight average molecular weight. At molecular weights above 100,000, thermal conductivity becomes less sensitive to molecular weight and considerably more constant (ref 120).

Coefficient of Thermal Expansion

Polymers have a much higher coefficient of thermal expansion than do most other materials, particularly metals, and this is a serious problem in highly precise design of components. The coefficient of thermal expansion depends upon molecular weight, primarily because the end groups of the polymer molecule are more flexible and mobile and occupy a higher free volume. Thus it is not surprising that coefficient of thermal expansion is inversely related to number average molecular weight. For example, for polystyrene in the rubbery state, $\alpha_l = 5.46 \times 10^{-4} + (3.45 \times 10^{-4})/P_n$; and in the glassy state, $\alpha_g = 1.81 \times 10^{-4} + 2.76 \times 10^{-4}/P_n$, where P_n is number average degree of polymerization and α is given in $cm^3\ g^{-1}\ deg^{-1}$ (ref 121).

Thermal Stability

While most commercial polymers are reasonably stable at temperatures used with final products and up to normal processing temperatures, they are generally less stable than metals and ceramics when heated to very high temperatures. Since thermal degradation, oxidation, and hydrolysis are characteristic of most organic materials at higher temperatures, it is not surprising that similar reactions should occur in organic polymers. Since there is a great need and demand for thermal stability in many of the potential new applications for polymers, however, thermal stability remains a critical concern to polymer scientists and has been the basis for many fundamental studies.

In many practical studies, thermal stability appears to be directly related to molecular weight. This is due to the fact that while the polymer itself has a fairly stable structure, much of the instability of polymers arises from the different nature of the end groups on these polymer molecules. For example, the initiators used in most addition polymerization reactions often leave initiator fragments attached to the ends of the polymer molecules. When vinyl polymerization reactions terminate by disproportionation, the resulting unsaturated C═C end groups are a point of instability in the finished polymer molecule. When vinyl polymerization reactions terminate by chain transfer with initiator, solvent, or purposeful chain transfer agent, these end groups are frequently less stable than the rest of the polymer molecule and form a point of thermal instability. In condensation polymers the end groups are frequently polar reactive functional groups of much higher chemical reactivity than the rest of the polymer molecule. Thus in most polymers the end groups form points of particular reactivity and thermal instability. Consequently it is not surprising that thermal stability is directly related to molecular weight.

Conversely, when polymers are heated to the point at which further reactions occur, either during processing or during final use, these reactions may cause either degradation to lower molecular weight or cross-linking to higher, sometimes almost infinite, molecular weight. Thus in the relationships between molecular weight and thermal stability, molecular weight may be cause and/or effect.

3-4 ELECTRICAL PROPERTIES

While electrical properties are generally independent of the molecular weight of the polymer itself, they are very strongly affected by the addition of monomeric liquid plasticizer to form a mobile low-molecular-weight fraction of the total composition. The mechanism of the effects on (1) Resistance, and (2) Dielectric Properties is quite different, and they are best considered separately.

Resistance

Most polymers are excellent electrical insulators, and display high electrical resistance and very little conduction of an electric current. The small conductivity which is observed is probably due primarily to electrolytic motion of ionic impurities in the polymer. Thus a decrease in electrical resistivity is generally the result of both (1) ionic impurities and (2) mobility. While ionic impurities may be difficult to determine,

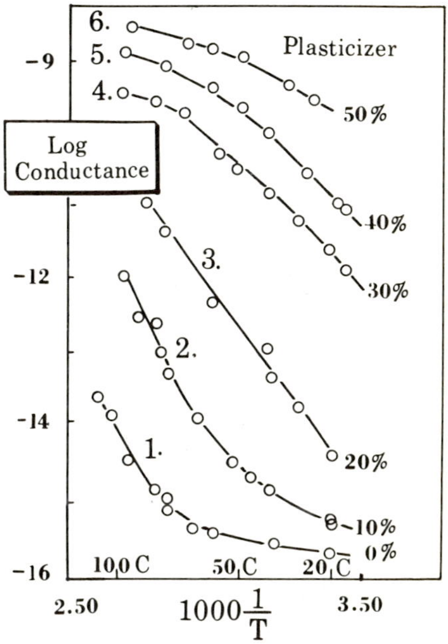

Figure 3-48. Effect of plasticizer content upon electrical conductance of polyvinyl chloride (ref 123).

mobility is easily measured by mechanical and thermal properties, and is particularly affected by the addition of plasticizers. Thus conductance is directly related to plasticizer content (ref 122) (Figure 3-48). The resistivity of plasticized polyvinyl chloride decreases regularly with increasing plasticizer content (Figure 3-49). The relationship between mobility in general and electrical conduction is clearly evident in the twin effects of plasticizer upon mechanical modulus and volume resistivity of plasticized polyvinyl chloride (Table 3-3).

While these effects are quite large, the electrical resistivity of plasticized polyvinyl chloride is still high enough for wide use in electrical

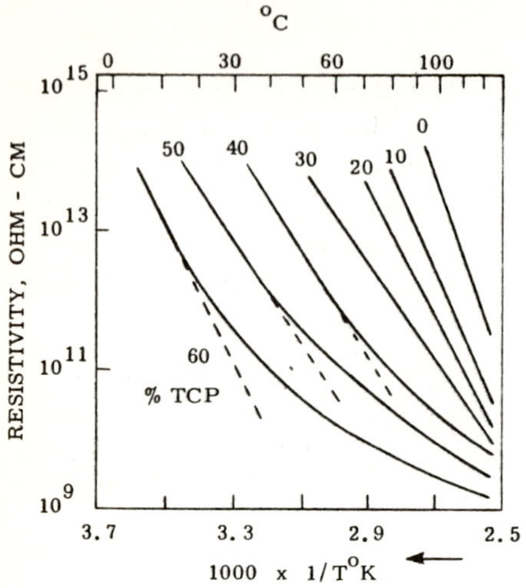

Figure 3-49. Effect of plasticizer content upon electrical resistivity (ref 124). Tricresyl phosphate in polyvinyl chloride.

wire and cable insulation, largely because of its combination of easy processing, flexibility, color range, weather-resistance, and low cost.

Dielectric Properties

When a polar polymer is placed in an electric field, the polar groups in the polymer will tend to orient in that field. If the polymer is very flexible, or at least if the polar groups are very flexibly attached to

Table 3-3. Effect of Plasticizer upon Mechanical Modulus, Hardness, and Volume Resistivity of Polyvinyl Chloride (ref 125)

Plasticizer			Properties		
Dioctyl phthalate	Polyester plasticizer	Nitrile rubber	100% modulus	Shore durometer D hardness	Volume resistivity
—	—	—	—	75	1500 × 10^{13} ohm-cm
—	—	25 PHR	—	75	810
25 PHR	—	—	3700 psi	67	270
—	—	50	2700	64	140
—	—	75	1700	44	30
50	—	—	1600	34	13
—	—	100	1200	31	8.7
75	—	—	800	18	1.2
—	100 PHR	—	760	20	0.4
100	—	—	515	14	0.18

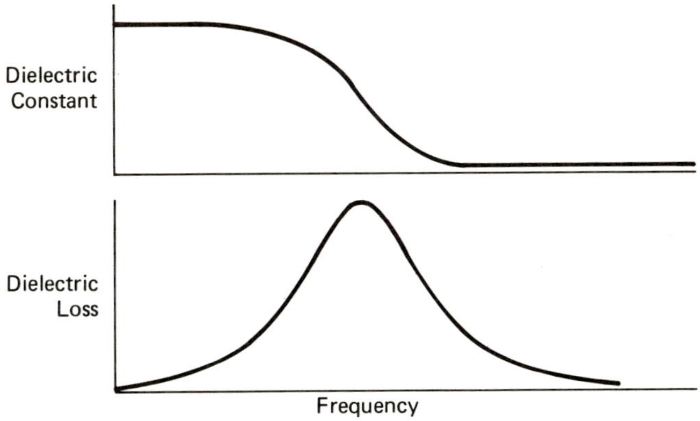

Figure 3-50. Dielectric response of a polar polymer in an alternating electrical field (ref 126).

the polymer, they will orient easily and quickly. If the polymer is very rigid and the polar groups are rigidly attached to it, they will orient only slowly and with great difficulty. Thus in an alternating electric field, the polar groups in a polymer will orient and give high dielectric constant only when the frequency of alternation is low enough to permit motion and orientation of these polar groups (Figure 3-50). With increasing frequency in the alternation of the electric field, the polar groups will become less and less able to orient rapidly enough to affect dielectric constant; and at very high frequencies they will barely be able to orient

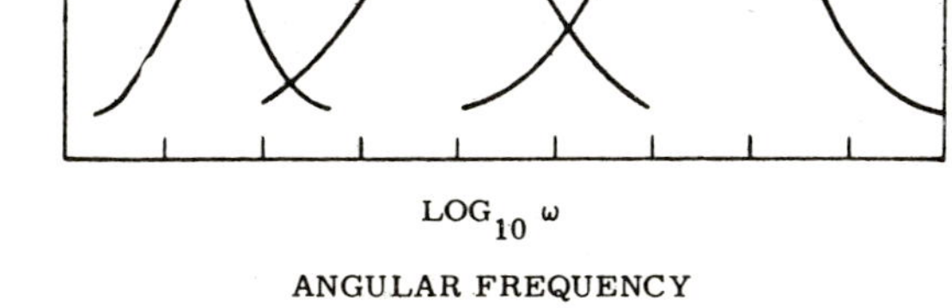

Figure 3-51. Effect of plasticizer upon dielectric loss index of a polar polymer (ref 127).

at all, and the polymer will exhibit only a low dielectric constant. Loss of electrical energy by conversion to thermal energy will be minor at low frequencies when the polar groups orient freely, and also at high frequencies when the polar groups are unable to orient at all; but it

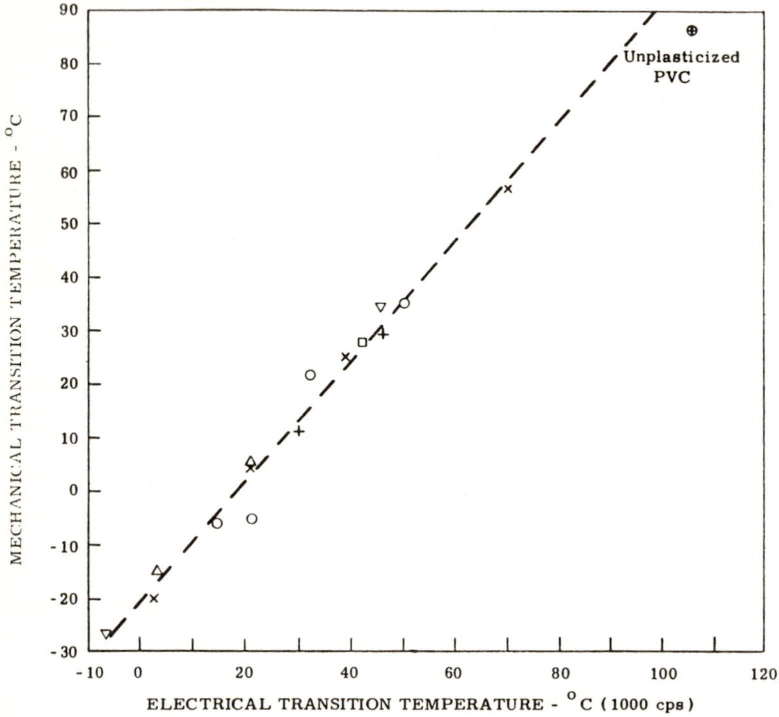

□—Dimethyl phthalate
○—Diethyl phthalate
×—Dibutyl phthalate
△—*n*-dioctyl phthalate
+—Tricresyl phosphate
▽—Dibutyl sebacate

Figure 3-52. Correlation between mechanical and electrical transition temperatures of plasticized polyvinyl chlorides (ref 129).

will become a major factor at the intermediate frequencies in the transition region where the polar groups are partially able to orient at the rate and frequency of alternation of the electric field.

Addition of monomeric liquid plasticizer to polyvinyl chloride increases the flexibility and mobility of the polymer molecule and the ability of the C—Cl bond attached to the polymer molecule to rotate and orient in an electric field (Figure 3-51). With increasing plasticizer content, and increasing mobility of the polymer and of the polar C—Cl bonds attached to the polymer, the transition region increases to higher and higher frequencies; at the same time the amount of loss also increases. Polar polymers which can absorb moisture are plasticized by the moisture and similarly exhibit a higher dielectric constant and dissipation factor because of this plasticization (ref 128).

It has already been observed that the mobility of the segments in the polymer molecule, and the mechanical flexibility of the polymer as a whole, increase with increasing temperature and thermal energy

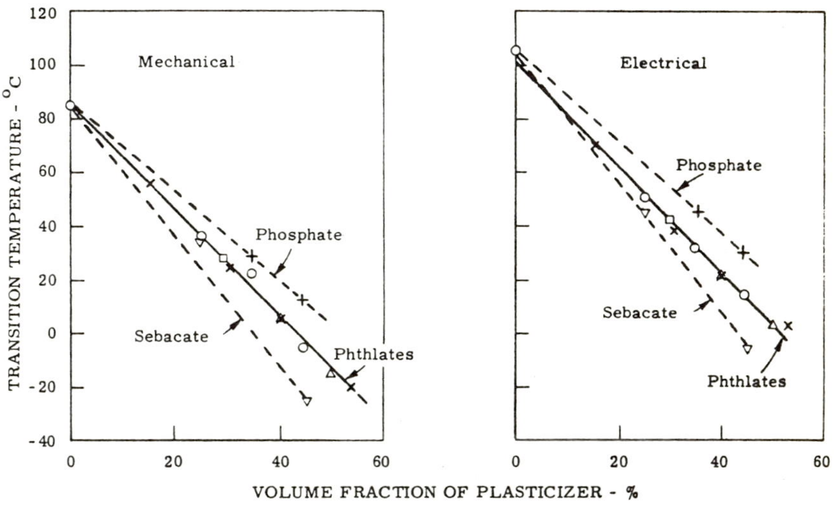

Figure 3-53. Effect of plasticizer structure and concentration upon mechanical and electrical transition temperatures (ref 107).

content until they reach a characteristic point at which mobility is high enough so that the polymer changes from rigid to flexible in nature at the glass transition temperature; and it has also been observed that the addition of monomeric liquid plasticizer to the polymer increases mobility and thus decreases this characteristic glass transition temperature. Dielectric properties similarly depend upon mobility and similarly show a characteristic transition temperature. Addition of monomeric liquid plasticizer to polyvinyl chloride affects both the mechanical and electrical transition temperatures directly (Figure 3-52). The effect of different plasticizers on the electrical transition temperature similarly depends upon the efficiency of the individual plasticizer, just as does their effect on mechanical transition temperature (Figure 3-53).

Practical use of plastics in dielectric applications generally requires reasonably constant dielectric properties over a broad range of temperature, frequency, and/or humidity, and therefore generally avoids both plasticized and moisture-sensitive polymers for such applications.

3-5 OPTICAL PROPERTIES

While it has been observed that optical properties do not generally depend upon molecular weight (ref 130), it seems reasonable to expect

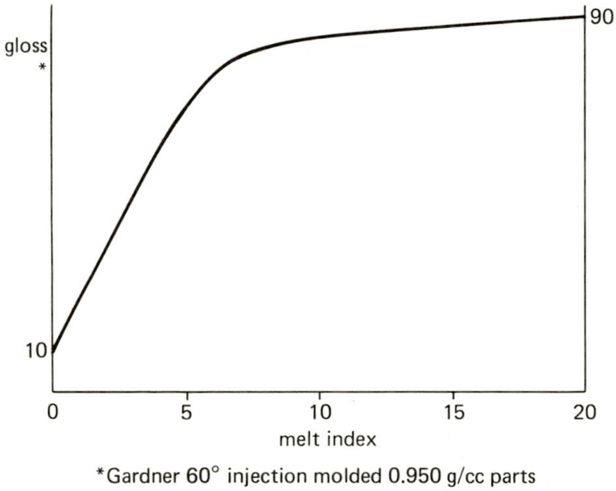

Figure 3-54. Effect of molecular weight upon gloss of extruded high-density polyethylene film (ref 63). Note that increasing melt index indicates decreasing molecular weight.

that in low-molecular-weight polymers, the different nature of the end group on most polymer molecules would produce some changes in absorption and refraction of light.

It has already been explained how molecular weight and distribution affect processability of polyethylene into film form. This in turn affects the optical properties of the resulting film. Thus low molecular weight favors high clarity and gloss (Figure 3-54). Narrow molecular weight distribution favors high clarity, while broad distribution favors high gloss (ref 9,131). Similar relationships undoubtedly apply to other polymers as well.

3-6 CHEMICAL PROPERTIES

The molecular weight of the polymer molecule has a number of significant effects upon its chemical properties. While there is not universal agreement on these relationships, they are important enough to deserve careful note. They are best divided into several groups of related properties: (1) Solubility, Extraction, and Stress Cracking; (2) Swelling, Absorption, and Permeability; and (3) Reactivity and Aging.

Solubility, Extraction and Stress Cracking

Solubility

It is generally accepted that solubility of a polymer in solvents is inverse to its molecular weight (ref 132–4). One of the most striking and classic examples was Ostromislensky's early pioneering study of the relationship between molecular weight and solubility of polyvinyl chloride (ref 135–8). Such considerations are particularly important in the formulation of solution coatings such as paints and lacquers, where the solvent balance is often very critical. Compatibility of polymers with plasticizers may similarly be inverse to the molecular weight (ref 134), which would thus affect the latitude available to formulators of plasticized polymers.

Extraction

It follows then that in a polymer of broad molecular weight distribution, even though the majority of high-molecular-weight polymer is solvent resistant, the low-molecular-weight fraction may provide a significant basis of solubility and extractability which must be considered in practical applications. Particularly when the high-molecular-weight polymer is cross-linked during processing, the low-molecular-weight fraction is statistically least likely to cross-link, and therefore remains as a highly soluble and extractable fraction in the final polymer.

As an extreme case, addition of plasticizers and other monomeric and low-molecular-weight ingredients, to an otherwise solvent-resistant polymer during compounding, introduces a low-molecular-weight fraction of high solubility and extractability and thus lowers the overall chemical resistance of the total composition (ref 139).

Stress Cracking

Stress-cracking appears to be a specialized case of the extractability of low-molecular-weight fractions in a polymer (Table 3-4), which probably explains why it is inversely related to number average molecular weight (ref 73,79) (Figures 3-31, 3-32, 3-55). This effect has been most

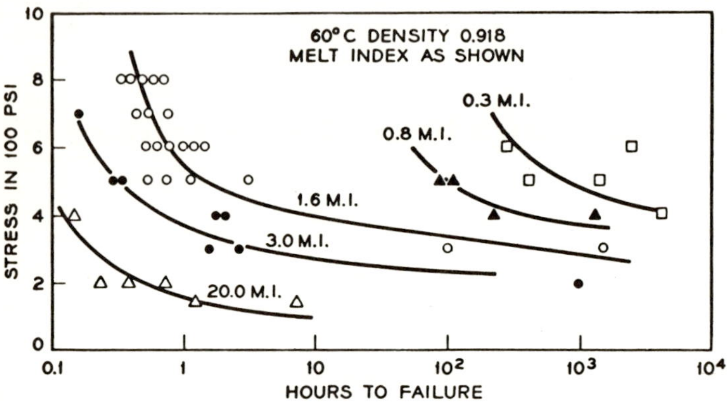

Figure 3-55. Effect of molecular weight upon stress-crack-resistance of low-density polyethylene (ref 141). Note that increasing melt index indicates decreasing molecular weight.

thoroughly studied in polyethylene, but undoubtedly is of major importance in many other polymers as well.

Table 3-4. Effect of Chloroform-extractable Fraction upon Stress-Crack-Life of Low-density Polyethylene (ref 140)

Chloroform-extractable fraction	Stress-cracking time in Igepal at 50°C
11.8%	4 hr.
10.6	25
9.7	>1000
9.3	>1000
7.9	>1000

Swelling, Absorption, and Permeability

There is less agreement on the effects of molecular weight on swelling, absorption, diffusion, and permeability through a polymer. While it has been stated that molecular weight does not affect diffusion and permeation by liquids and gases (ref 142), it has also been reported that molecular weight has inverse effects on swelling, absorption, and permeability (ref 134). In polyethylene, increasing molecular weight produces increased grease-proofness and slightly lower gas permeability (ref 98). A more detailed examination of absorption in glassy polymers suggests that when such a glassy polymer absorbs solvent, this results in plasticization which lower the glass transition temperature, increases mobility, and therefore produces even higher absorption (ref 143).

Reactivity and Aging

Reactivity

The end group on a polymer molecule is frequently the most reactive point in the entire molecule. Vinyl polymerizations which terminate by disproportionation leave a C═C double bond on the end of the polymer molecule, and it is still quite reactive. Vinyl polymerizations which terminate by transfer to solvent, initiator, monomer, or chain-transfer agent, all attach reactive groups to the ends of the polymer molecules. Most condensation polymerization reactions leave a few unreacted polymerizable groups at the ends of the polymer molecules, and these remain quite as reactive as any of those which had entered the polymerization reaction. Thus in most polymer systems, the end group on the polymer molecule is one of the most reactive positions in the entire molecule. Since the number of end groups is exactly inverse to number average molecular weight, it is to be expected that polymer reactivity would be inverse to molecular weight.

It has long been observed that the lower mobility of high-molecular-weight polymers makes them much less reactive to normal chemical reactions than similar functional groups would be in conventional low-molecular-weight monomeric structures. To this extent, increasing molecular weight would be expected to decrease reactivity quite rapidly and then taper off at higher molecular weights.

Aging

Reactivity with oxygen, ozone, and moisture, and ultraviolet light sensitization of such reactions, all become important during long-term aging of polymers in use, particularly at high temperature or in outdoor weathering. Resistance to deterioration should be directly related to molecular weight for three distinct reasons: (1) Reactive end groups are inverse to number average molecular weight; (2) Reactivity of high-molecular-weight species of low mobility is much less than for low-molecular-weight species of high mobility; and (3) Degradation of very high-molecular-weight polymers during aging may still only degrade them to medium-molecular-weight polymers of almost equivalent properties, whereas degradation of medium-molecular-weight polymers during aging may degrade them to such small molecules that a significant loss in properties results (Figure 3-56).

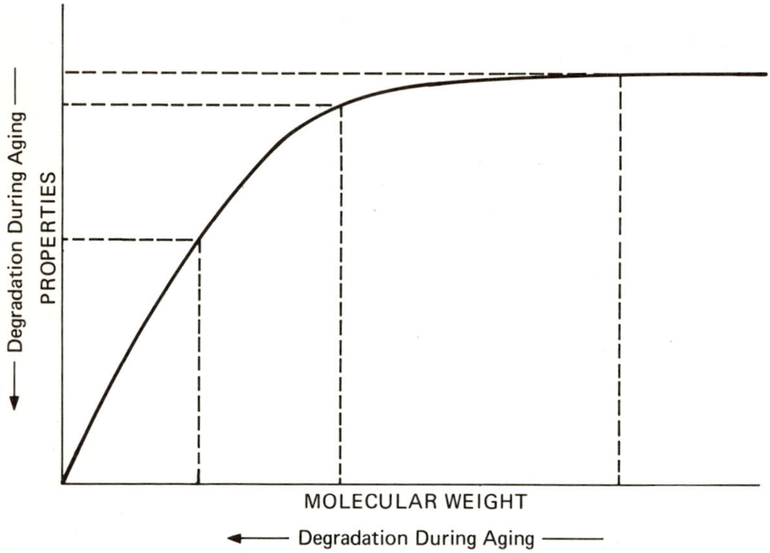

Figure 3-56. Effect of molecular weight upon degradation of properties during aging.

Not only does molecular weight affect aging, but aging also affects molecular weight. Aging may either lower the molecular weight of the polymer by cleavage or may produce cross-linking up to infinite molecular weight. Thus molecular weight and aging may each be cause and/or effect.

3-7 PRICE

The molecular weight of a commercial polymer may have a number of significant effects upon its price. Most often, high-molecular-weight polymers are more expensive to produce and process for a number of reasons. In general, they require the use of more highly purified monomer, which is costly to produce. In many addition polymerization reactions, higher molecular weight requires lower polymerization temperature, often involving costly refrigeration and longer polymerization time to carry the polymerization to completion. In conventional molten bulk condensation polymerization reactions, higher-molecular-weight polymers require longer polymerization time, higher temperatures often approaching the decomposition point, higher vacuums to remove the low-molecular-weight by-products and to help drive the equilibrium reaction to completion, and higher power to stir the polymerization reaction at higher melt viscosity. Finally, high-molecular-weight polymers of high viscosity require more power and time during compounding and processing. For all of these reasons the highest-molecular-weight commercial polymers are apt to be more expensive to produce than a normal molecular weight would be.

Occasionally it is the lower-molecular-weight polymers which are more costly to produce. This may happen for a variety of specific reasons. For example, in coordination polymerization of olefins, it is often very easy to reach extremely high molecular weights, and much more difficult to stop at controlled lower molecular weight. In the production of rigid polyvinyl chloride the lower-molecular-weight required for processability is best obtained by polymerization at higher temperatures, which create higher vinyl chloride monomer pressures, often exceeding the design ratings of conventional polymerization equipment, and thus requiring higher capital investment in heavier-duty polymerization equipment capable of being used at higher temperatures and pressures. Many of the B-stage thermosetting resins and liquid prepolymers are so reactive that they have at best a limited shelf life during room-temperature storage, and thus require careful control during production and stabilization during storage, and still often involve losses of material due to premature over-cure. For all these reasons low-molecular-weight polymers may sometimes be more expensive to produce than polymers of conventional molecular weight.

When a specific molecular weight distribution is required, this may also involve additional manufacturing costs. In particular, the production of very narrow-molecular-weight monodisperse systems can be accomplished either through specialized polymerization techniques or through highly selective fractionation after polymerization. Both of these methods can be extremely expensive. In many cases it would appear that the best balance of properties is obtained at a rather broad molecular weight distribution, so that the cost of producing such narrow molecular weight monodisperse systems may be quite impractical. In polyethylenes, on the other hand, molecular weight distribution is most conveniently controlled by degree of long-chain branching.

3-8 BALANCE OF PROPERTIES

Up to this point we have been considering primarily the effects of molecular weight upon individual properties. At this point it is appropriate to note that in any specific application a number of different properties are involved, and the effects of molecular weight upon these may conflict and thus complicate our choice of an optimum molecular weight for any such application.

Processability vs Final Properties

This conflict is most often observed in the balance between processability and final properties. Thus, in general, low molecular weight gives low melt viscosity which is desirable for melt processability, whereas high molecular weight gives the best final mechanical properties (ref 144). In the production of synthetic fibers in particular, low-molecular-weight polymers give weak brittle fibers, higher-molecular-weight polymers give high melt viscosity and difficult processing, and it is only at an intermediate molecular weight that easy processing and good final properties of fibers can be combined (ref 145). In the production of coatings, high-molecular-weight polymers give the highest strength in the final coating, but give very viscous solutions during application, requiring high dilution to give low viscosity. This in turn produces thinner coats and thus requires multiple coating, which is more expensive, in order to produce the desired final thickness. This balance between viscosity during coating and strength after coating also requires choice of a compromise intermediate molecular weight (ref 30). Similarly adhesives need low molecular weight for easy flow onto an irregular substrate surface, and especially for wetting and mixing with the surface; but they also need high molecular weight for cohesion and bond strength in the final bond.

Optimum Molecular Weight Distribution

In some cases it has been suggested that the balance of properties may actually be achieved without serious compromise or sacrifice, by choice of an optimum molecular weight distribution. Thus for example low weight-average-molecular-weight may be required for easy melt processing while high number-average-molecular-weight is desirable for high tensile strength and impact strength; in such a situation a narrow molecular weight distribution would give the best balance between processability and properties (ref 74). In the production of polyethylene bottles, wire coating, and monofilaments, broad molecular weight distribution produces the best combination of melt flow at high shear rates plus melt strength at low shear rates during processing, and also provides high load-bearing strength during use (ref 8).

Complex Relationships

On more detailed consideration it has been observed that certain properties are best at high molecular weight, others at low molecular weight; some at narrow distribution, others at broad distribution; and still others depend upon the combination of high-molecular-weight polymer with low-molecular-weight plasticizer. Thus the choice of optimum molecular weight for any specific application may depend upon a complex balance of such considerations.

Processability

Processability in particular requires an optimum molecular weight for each type of process technique being used. Stretching, orientation, and thermoforming involve the ability of a polymer film or fiber to withstand high mechanical stress and permit considerable elongation without failure in the solid or semisolid state, and are thus best practiced at quite high molecular weights. Extrusion of polymer in the form of continuous profile requires sufficient melt strength so that the molten product coming from the hot die in the unsupported form will be strong enough to retain its shape until it is cooled to the point of substantial strength. In injection molding the molten final product is well supported on all sides by the cool mold and does not require such high melt strength, so somewhat lower molecular weights are optimum for maximum melt flow into the mold. Most adhesive applications require sufficient fluidity to penetrate all of the tiny irregularities in a substrate surface, and optimally to mix with the substrate itself; but they still require sufficient viscous strength to provide a quick bond, and are thus best designed at a fairly low molecular weight. Coating applications require all of the fluidity of adhesives but place fewer demands upon quick strength, and are therefore generally best carried out with a still lower average molecular weight, most often obtained by the addition of large amounts of monomeric solvents. Monomer casting operations require sufficient fluidity to pour the liquid monomer into a mold of any shape in which it will fill the mold completely under the minimal force of gravity, and these are therefore carried out at quite high fluidity, generally using monomeric or at most low-molecular-weight polymeric species. Thus each type of process technique involves its own optimum molecular weight range for optimum processability (Table 3-5).

Table 3-5. Optimum Average Molecular Weight Ranges for Polymer Processing

Average Molecular Weight Range	Type of Application or Process Technique
Highest	Stretching, orientation, thermoforming
	Extrusion, blow molding
	Injection molding
	Adhesives
	Coatings
Lowest	Monomer casting

Optimum Balance of Properties

Similar considerations may well be applied to the different types of mechanical, thermal, electrical, optical and chemical properties, depending upon the specific application in which they will be used. Ultimately each type of application will require a balance of certain specific properties, each of which depends upon molecular weight in their own specific way, and require much more detailed consideration and study to optimize them in such final applications.

REFERENCES

1. Miller, M. L., *The Structure of Polymers,* Reinhold, 1966, p 3.
2. *Ibid.,* p 98.
3. Billmeyer, F. W., Jr., *Textbook of Polymer Science,* Interscience, 1964, p 7.
4. *Ibid.,* p 56.
5. *Ibid.,* p 66.
6. *Ibid.,* p 76.

7. Miller, *op. cit.*, p 11.
8. Modern Plastics Encyclopedia 1968, McGraw-Hill, 1967, pp 206–8.
9. *Ibid.*, p 545.
10. Miller, *op. cit.*, p 99.
11. Meares, Patrick, *Polymers: Structure and Bulk Properties,* Van Nostrand, 1965, p 351.
12. Nielsen, L. E., *Mechanical Properties of Polymers,* Reinhold, 1962, pp 56–57.
13. *Ibid.*, p 60.
14. Modern Plastics Encyclopedia, *op. cit.*, p 207.
15. *Ibid.*, p 22.
16. Schmidt, A. X., and Marlies, C. A., *Principles of High Polymer Theory & Practice,* McGraw-Hill, 1948, p. 66.
17. Hercules, Inc., Technical Bulletins on Hi-Fax 1900.
18. Golding, Brage, *Polymers and Resins,* Van Nostrand, 1959, p 559.
19. Nielsen, *op. cit.*, p 27.
20. Morton, Maurice, *Introduction to Rubber Technology,* Reinhold, 1959, pp 40–41.
21. *Ibid.*, p 153.
22. Skeist, Irving, *Handbook of Adhesives,* Reinhold, 1962, pp 452–59.
23. Billmeyer, *op. cit.*, p 233.
24. Nielsen, *op. cit.*, p 37.
25. *Ibid.*, pp 32–37.
26. *Ibid.*, p 39.
27. *Ibid.*, p 41.
28. *Ibid.*, pp 254–55.
29. Miller, *op. cit.*, p 214.
30. Payne, H. F., *Organic Coating Technology,* Vol. I, Wiley, 1954, p 7.
31. *Ibid.*, p 37.
32. Alfrey, Turner, and Gurnee, E. F., *Organic Polymers,* Prentice-Hall, 1967, p 115.
33. Miller, *op. cit.*, p 100.
34. *Ibid.*, pp 99–100.
35. Alfrey, *op. cit.*, p 51.
36. Tobolsky, A. V., *Properties and Structure of Polymers,* Wiley, 1960, p 75.
37. Wakeman, R. L., *The Chemistry of Commercial Plastics,* Reinhold, 1947, p 314.
38. Brydson, J. A., *Plastics Materials,* Iliffe, 1966, p 58.
39. Nielsen, *op. cit.*, pp 165–68.
40. Meares, *op. cit.*, p 221.
41. *Ibid.*, p 305.
42. *Ibid.*, pp 263–66.
43. Nielsen, *op. cit.*, pp 165–66.
44. General Tire & Rubber Company, Technical Bulletin on *Vygen.*
45. Wakeman, *op. cit.*, p 340.

46. Golding, *op. cit.,* p 560.
47. Morton, *op. cit.,* p 41.
48. *Ibid.,* p 155.
49. DuPont, Technical Bulletin, *Designing with Zytel Nylon Resin.*
50. Gulf Oil Corporation, Technical Bulletin, *Effects of Moisture on Physical Properties of Gulf Nylon.*
51. Golding, *op. cit.,* p 573.
52. Nielsen, *op. cit.,* p 111.
53. Meares, *op. cit.,* p 233.
54. *Ibid.,* pp 266–67.
55. Nielsen, *op. cit.,* pp 168–70.
56. Baer, Eric, *Engineering Design for Plastics,* Reinhold, 1964, p 555.
57. Nielsen, *op. cit.,* pp 110–11.
58. *Ibid.,* pp 59–60.
59. *Ibid.,* pp 83–86.
60. Miller, *op. cit.,* p 539.
61. Wakeman, *op. cit.,* p 638.
62. Morton, W. E., & Hearle, J. W. S., *Physical Properties of Textile Fibers,* Butterworths, 1962, p 410.
63. Celanese Polymer Company, Technical Bulletin P1A on *Fortiflex Polyethylene Properties.*
64. Wakeman, *op. cit.,* p 691.
65. Billmeyer, *op. cit.,* p 210.
66. Nielsen, *op. cit.,* pp 115–16.
67. Golding, *op. cit.,* pp 572–73.
68. Schmidt, *op. cit.,* p 84.
69. *Ibid.,* p 83.
70. Payne, *op. cit.,* pp 17–18.
71. Phillips Chemical Company, Technical Bulletin 4 on *Marlex Types.*
72. DuPont, Technical Bulletin on *Key Molecular Parameters and Their Effects on Properties: DuPont Alathon for Injection Molding.*
73. Wild, Leslie, Woldering, J. F., and Guliana, R. T., SPE Annual Technical Conference, *13,* 94, May 1967.
74. Billmeyer, *op. cit.,* p 211.
75. *Ibid.,* p 222.
76. *Ibid.,* pp 221–22.
77. DuPont, *op. cit.*
78. Wild, *op. cit.,* p 95.
79. Baer, *op. cit.,* pp 748–52.
80. Payne, *op. cit.,* p 17.
81. Nielsen, *op. cit.,* pp 228–29.
82. Modern Plastics Encyclopedia 1968–9, McGraw-Hill, 1968, p 31.
83. Nielsen, *op. cit.,* p 90.
84. Tobolsky, *op. cit.,* p 73.
85. Miller, *op. cit.,* p 292.

86. Baer, *op. cit.*, p 402.
87. Miller, *op. cit.*, p 291.
88. Golding, *op. cit.*, p 554.
89. Brydson, *op. cit.*, p 49.
90. Billmeyer, *op. cit.*, p 213.
91. Meares, *op. cit.*, pp 263–64.
92. Miller, *op. cit.*, pp 291–92.
93. Meares, *op. cit.*, p 264.
94. Schmidt, *op. cit.*, p 193.
95. *Ibid.*, pp 191–93.
96. Golding, *op. cit.*, pp 553–58.
97. Baer, *op. cit.*, p 404.
98. U.S. Industrial Chemicals Co., Technical Bulletin, "Polyethylene Processing Tips."
99. Nielsen, *op. cit.*, pp 55–60.
100. Billmeyer, *op. cit.*, p 225.
101. Nielsen, *op. cit.*, p 168.
102. Miller, *op. cit.*, pp 290–91.
103. Golding, *op. cit.*, p 556.
104. Miller, *op. cit.*, pp 287–91.
105. Billmeyer, *op. cit.*, p 220.
106. *Ibid.*, pp 215–20.
107. Baer, *op. cit.*, p 558.
108. Wakeman, *op. cit.*, p 339.
109. *Ibid.*, p 342.
110. *Ibid.*, p 341.
111. *Ibid.*, p 343.
112. Morton, *op. cit.*, pp 40–42.
113. Nielsen, *op. cit.*, pp 58–59.
114. Miller, *op. cit.*, p 287.
115. Nielsen, *op. cit.*, p 170.
116. *Ibid.*, p 110.
117. Billmeyer, *op. cit.*, p 212.
118. Nielsen, *op. cit.*, p 31.
119. *Ibid.*, p 32.
120. Miller, *op. cit.*, p 563.
121. *Ibid.*, p 296.
122. *Ibid.*, pp 671–74.
123. *Ibid.*, p 673.
124. Baer, *op. cit.*, p 508.
125. Deanin, R. D., Skowronski, B. J., and Wixon, F. R., *Advances in Chemistry, 48,* 140 (1965).
126. Schmidt, *op. cit.*, p 440.
127. Baer, *op. cit.*, p 542.
128. *Ibid.*, pp 565–69.

129. *Ibid.*, p 557.
130. Billmeyer, *op. cit.*, p 227.
131. Modern Plastics Encyclopedia 1968, *op cit.*, p 19.
132. Billmeyer, *op. cit.*, p 25.
133. *Ibid.*, Chapter 2.
134. *Ibid.*, p 226.
135. I. I. Ostromislensky, *Chem. Zentr.*, *I*, 1980, 1983 (1912).
136. I. I. Ostromislensky, U.S. Patents 1,721,034; 1,791,009.
137. I. I. Ostromislensky, *J. Russ. Phys. Chem. Soc.*, *44*, 204, (1912).
138. I. I. Ostromislensky, *Chem. Ztg.*, *36*, 199 (1912).
139. Baer, *op. cit.*, pp 643–44.
140. *Ibid.*, p 751.
141. *Ibid.*, p 749.
142. *Ibid.*, p 661.
143. *Ibid.*, p 648.
144. Billmeyer, *op. cit.*, pp 232–33.
145. McFarlane, S. B., *Technology of Synthetic Fibers*, Fairchild Publications, 1953, pp 85a–87.
146. Power, G. E., Meyers, K. D., & Fremont, R. E., U.S. Patent 3,286,008 (1966).

Chapter Four

Molecular Flexibility

The flexibility of the individual polymer molecule is the most general and fundamental factor in its mechanical properties, and is also very important in many of its other properties. In addition to its direct effect upon certain properties, it also interacts with the other structural features of the polymer, and thus has further indirect effects on many other properties. It is best analyzed by first examining the effects of molecular structure upon molecular flexibility, and then considering the effects of this molecular flexibility on other properties and applications.

4-1 MOLECULAR STRUCTURE → MOLECULAR FLEXIBILITY

The relationship between molecular structure and molecular flexibility is best understood by first examining the general fundamental concept of a completely flexible chain molecule, as derived from mathematics and physics, and then considering the ways in which chemical structure introduces restriction of rotation, and thus limits the flexibility of actual polymer molecules.

General Fundamental Concepts

The fundamental flexibility of a generalized polymer molecule is most simply analyzed by basic concepts of mathematics and physics.

Tetrahedral Carbon → Random Coil

This begins by assuming that when two carbon atoms are joined by a single pair of electrons in a single covalent bond, the two carbon atoms are free to rotate about this bond. The carbon atom has four such equal bonds equidistant from each other in space, and therefore tetrahedral in geometry, producing a 109° angle between any two bonds (ref 1,2). Thus a chain of carbon atoms joined by such bonds is not

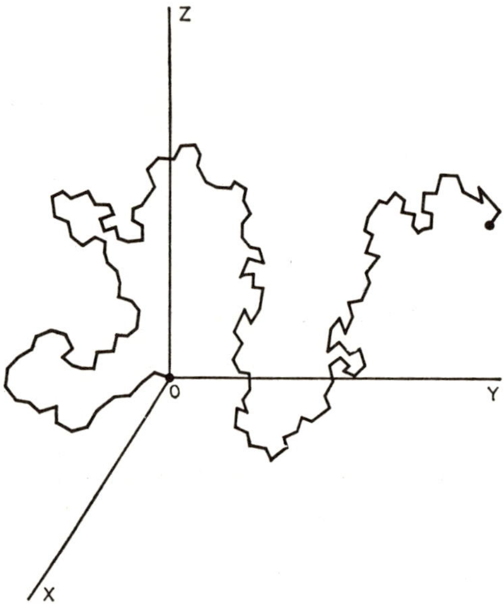

Figure 4-1. Random zig-zag coil. A freely jointed chain in three dimensions (ref 7).

really a simple straight line, as shown in most two-dimensional diagrams, but actually a very random zig-zag coil (ref 3–6) (Figure 4-1). Free rotation about each C—C bond (Figure 4-2) produces a great many equivalent conformations (Figure 4-3), a state of disorder producing high entropy and stability for such a randomly-coiled condition.

Thermal Motion → Free Volume → Segment Mobility

Thermodynamically, temperature means that these atoms are in constant motion, and that the polymer molecule is not frozen into a single random coil conformation, but is constantly fluctuating between

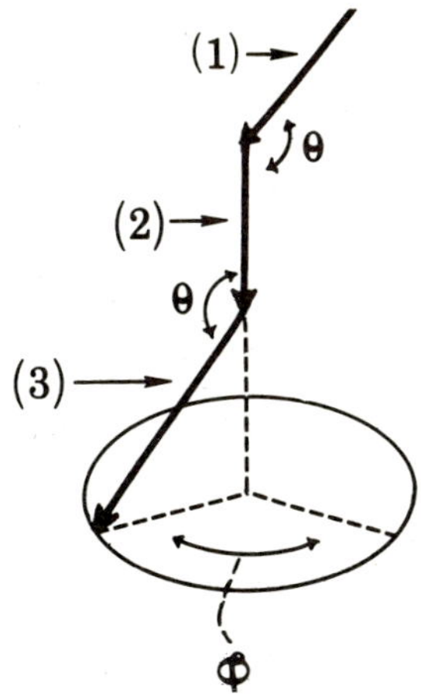

Figure 4-2. Free rotation about a C—C bond (ref 8).

Diagrammatic representation of the rotation of bond #3 in a portion of a carbon-carbon chain containing three bonds, #1, 2 and 3, in which the bond angle, θ, is fixed. The angle of rotation of the bond is ϕ

many such equivalent states. At low temperature this motion is restricted to the vibration of individual atoms by stretching, bending, and rotation about individual bonds, but they are too closely packed for any greater motion than this.

With increasing temperature the vibrations increase, and the atoms and molecules move somewhat further apart, creating a certain amount of free volume between them. When an individual vibrating atom finds

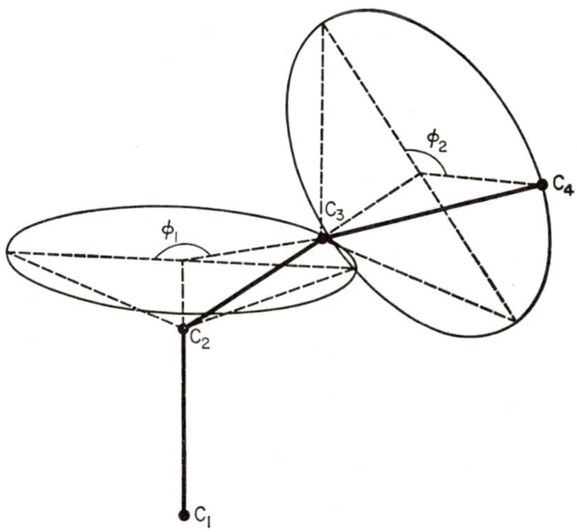

Figure 4-3. Free rotation about two successive C—C bonds (ref 9). Rotational freedom in a chain of four carbon atoms.

itself adjacent to such an element of free volume, it can jump into it, thus producing a small change in the overall random-coil conformation. The extent to which an individual atom can move, however, is distinctly limited by its attachment to adjacent atoms on either side in the polymer chain. For any major amount of motion to occur, an entire group of such atoms adjacent to each other would have to move all at once or in sequence (Figure 4-4). At low temperatures it is unlikely that such a group of adjacent atoms would all find elements of free volume adjacent to them at the same time.

With continuing increase in temperature and in the vibrations of the atoms, the molecules continue to move further apart, leaving more and more free volume between them. Eventually they reach a temperature at which the critical free volume is about 2.5 percent, generally called the glass transition temperature T_g. At this point, chains of five

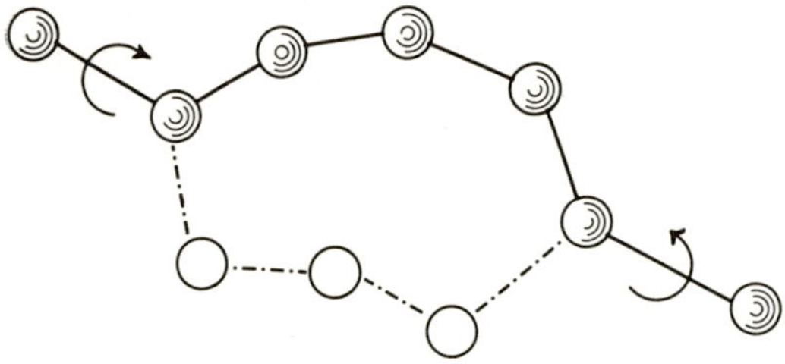

Figure 4-4. Motion of a segment in a polymer chain (ref 8).

or more adjacent atoms are all likely to find elements of free volume adjacent to them, and be able to move as a segment into a new conformation. Thus the polymer, which was previously rigid and glassy in nature, now becomes stiffly flexible or leathery.

With continued increase in temperature, atomic vibration, and free volume, chain segments of 10–30 carbon atoms are able to move simultaneously and rapidly, and the polymer becomes soft and flexible in nature (ref 10–14).

In the absence of any directional stress, all of this segmental motion is simply Brownian in nature, and does not result in any significant permanent displacement—it goes nowhere. Under directional stress, however, such segmental motion will occur more frequently in the direction which will release this potential energy and reduce the stress, disentangling adjacent molecules and thus permitting flexibility and extensibility in the total material (ref 15). A major sustained stress may actually stretch the molecules into a much more linear and less randomly-coiled state, which will be a much less probable conformation of lower entropy; when such a stressed molecule is released, it is very likely to recoil into its more probable randomly-coiled conformation of higher entropy (ref 6,16–18). At still higher temperatures, whole linear molecules under stress may tend to slip past each other, producing a rubbery type of liquid flow; and at even higher temperatures whole molecules may flow readily past each other, producing fairly true liquid flow (ref 13).

Freedom of Rotation

Even in the simplest carbon chain, the bonds on each carbon atom which are not used in forming the chain must be attached to some other atoms.

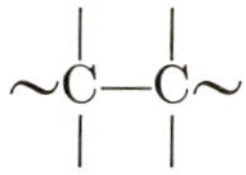

In the simplest and the most general case they are attached to hydrogens.

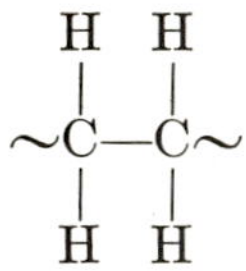

While hydrogen is the smallest known atom, the size of its electronic orbit is still sufficient to have a significant effect upon the flexibility of the chain. Even in the simple molecule like ethane, the conflict between adjacent hydrogen atoms is sufficient to make the ethane molecule stablest when the two methyl groups are 60° out of phase (ref 19) (Figure 4-5); and the energy barrier to rotation is about 2 kilocalories per mol (ref 20). The significance of this repulsion between adjacent hydrogen atoms is best seen by comparison with methylacetylene

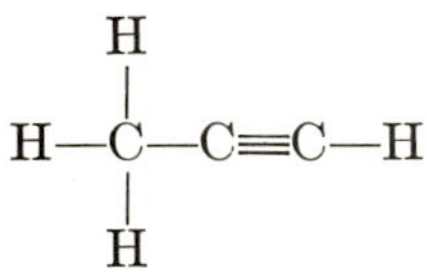

where the rotation energy barrier is only 0.5, and with neopentane

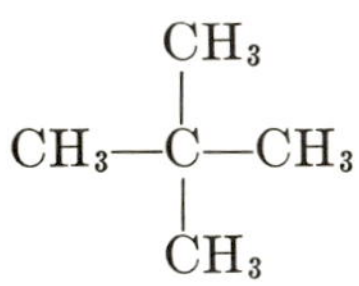

where it is 4.2 (Figure 4-6).

The same effects are observed in polyethylene and other polymers. In polyethylene the trans conformation is 0.5 kilocalories stabler than the gauche (Figure 4-7). At low temperatures polyethylene crystallizes and has low kinetic energy, and the conformation is primarily trans; but at higher temperatures and in solution, this energy barrier is low enough to permit rotation, and the molecule becomes partially gauche, permitting coiling (ref 22,23). Many isotactic polymers have a regular sequential arrangement of trans and gauche (ref 23).

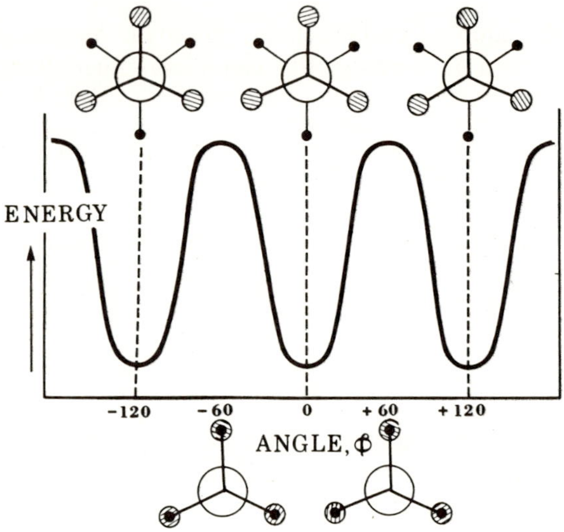

Figure 4-5. Steric hindrance in ethane (ref 19).
Diagrammatic representation of the change in potential energy of a molecule of ethane as the bond between the methyl groups rotates through the angle ϕ.

In general, bond rotation energy barriers are about 1–5 kilocalories per mol, which is comparable with the forces of molecular cohesion; so molecular flexibility correlates with glass transition temperature and also with melting point (ref 24). Low energies of rotation permit easy motion of large segments and produce low glass transition temperatures; restriction of rotation creates large potential energy barriers and high glass transition temperatures (ref 25). Internal friction (viscosity) in rigid amorphous solid polymers below the glass transition temperature is 10^{12-13}; in rubber above the glass transition temperature it is about 10^6 poises (ref 26,27).

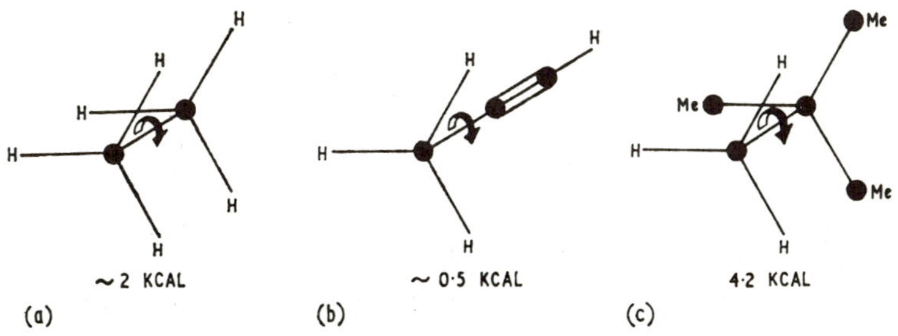

Figure 4-6. Steric hindrance in hydrocarbons (ref 20).

Rotational-energy barriers as a function of substitution. The small barrier (~2 kcal) in ethane (a) is lowered even further (~0·5 kcal) if three bonds are "tied back" by replacing three hydrogen atoms of a methyl group by a triply-bonded carbon, as in methylacetylene (b). The barrier is raised (4.2 kcal) when methyl groups replace the smaller hydrogen atoms, as in neopentane (c).

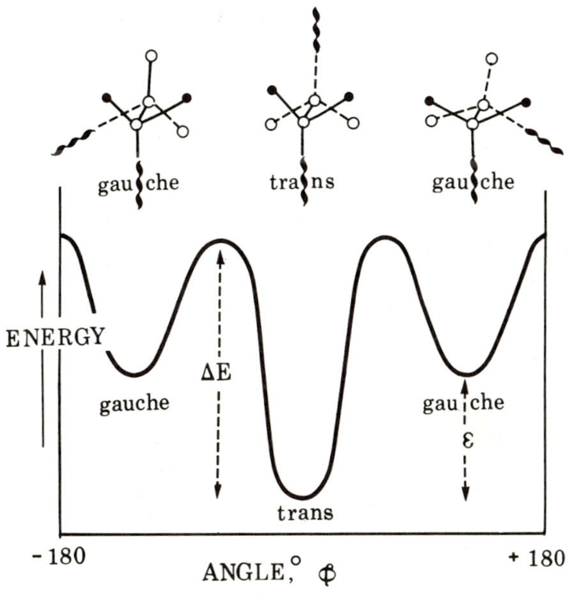

Figure 4-7. Steric hindrance in polyethylene (ref 21).

Diagrammatic representation of the change in potential energy of a bond between two carbon atoms in a chain of polyethylene as the bond rotates through the angle ϕ.

Restriction of Rotation

As has just been seen, polyethylene provides the best fundamental model for a mathematical and physical analysis of molecular flexibility. It is now appropriate to examine the ways in which the structure of other polymer molecules affects the fundamental flexibility of the simple polyethylene backbone. Most substituents and variations in structure will tend to restrict rotation of a simple polyethylene molecule. The structural factors which restrict rotation may be divided into: (1) those found in the main chain or backbone of the polymer molecule; and (2) those found on side chains attached to the main chain.

Main-Chain Structure

The main chain of a polymer molecule may contain two types of structural features which affect its flexibility: (1) Number of bonds an atom shares in the chain, and (2) Resonance.

Number of Bonds an Atom Shares in the Chain Obviously carbon or any other atom in the main chain of the polymer molecule must share at least two covalent bonds in order to form this chain.

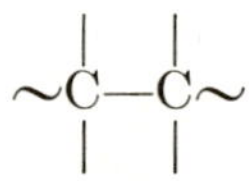

It is generally assumed that the atoms enjoy free rotation about such a single covalent bond.

Double Bonds If two adjacent atoms share another pair of electrons to form a covalent double bond such as the C═C bond, it is generally believed that such a double bond cannot rotate freely and therefore introduces a point of stiffness in the main chain of the polymer molecule (ref 28).

The effect on the flexibility of the entire molecule, however, depends upon the interaction of the C═C bond with adjacent structural features. Paradoxically an individual double bond reduces the number of bonding directions on its carbon atoms to 3 instead of 4,

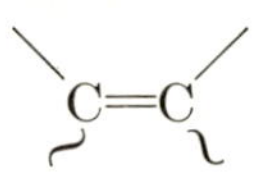

reducing the conflict between adjacent side groups even as small as hydrogen, and thus increases the freedom of rotation about the adjacent C—C single bonds in the main-chain (ref 29,30). Thus the overall effect, as seen in the diene rubbers, generally is to retain the freedom of rotation, softness, flexibility, and elasticity.

Small Rings An atom may also share more than two bonds in forming the main-chain, if it is part of a small ring structure.

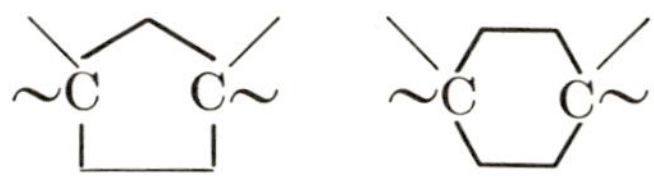

Five-membered rings are generally quite rigid, six-membered rings only slightly flexible. The stiffening effects of such rings in the main-chain are commonly seen in polyvinyl acetals, cellulosic polymers, and cyclohexane dimethylene terephthalate (ref 31,32). These effects are most easily seen in glass transition temperatures and melting points.

RESONANCE → PLANARITY When several conjugated alternating double and single bonds in a functional group are stabilized by resonance between them,

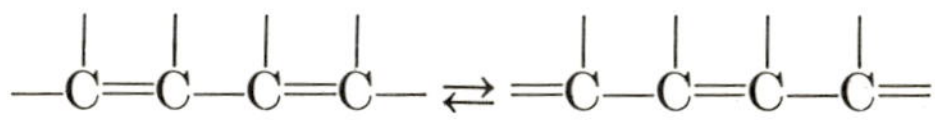

this imposes a flat planar structure upon such a functional group; and this flat planar structure constitutes a significant reduction in rotation and flexibility, and a significant increase in the stiffness of the polymer molecule (ref 30).

Resonance in the benzene ring

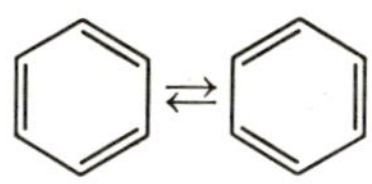

produces a flat rigid structure, which raises both glass transition temperature and melting point (ref 28,33–37). Typical commercial polymers include poly-*p*-xylylene and phenolic resins. Similar effects are observed when resonance-stabilized unsaturated heterocyclic rings are incorporated into the main-chain of the polymer molecule.

When resonance stabilization and planarity extend beyond the ring structure to include adjacent functional groups, the size of the rigid planar unit increases and the stiffening effect becomes even greater. Polyethylene terephthalate

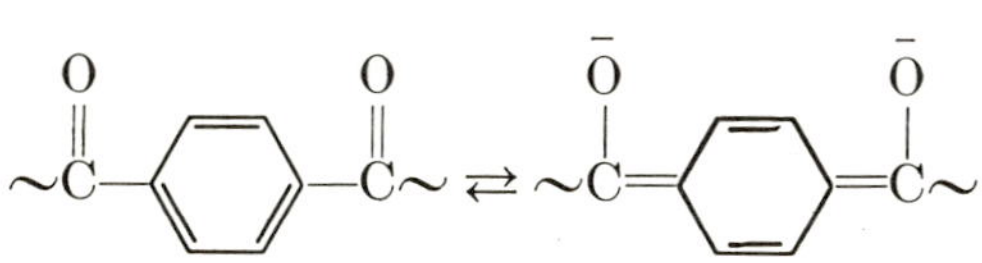

is typical (ref 38); and other commercial polymers such as phenoxy, epoxy, polyphenylene oxide, polysulfone, polycarbonate, polyesters, diallyl phthalate, alkyd, polyurethane, and melamine resins probably owe part of their rigidity to such extended resonance planarity.

Still further, polycyclic structures all resonating in a single plane, such as diphenyl and naphthyl groups,

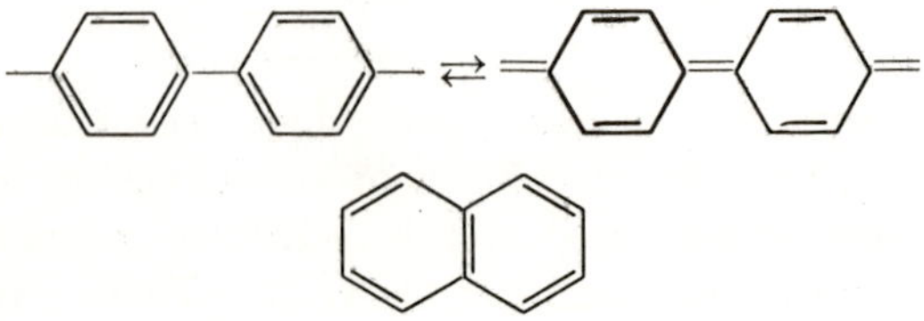

produce even more extensive stiffening (ref 38). Further yet, polyheterocyclic structures produce even greater flat planar areas of resonance stabilization, and greater stiffening and temperature resistance, for example in polyimides and polybenzimidazoles. The ultimate stiffening effects of such large resonating planar structures may be observed in the new heterocyclic ladder polymers, which contain a continuous resonance-stabilized planarity throughout the entire polymer molecule (Figure 4-8).

Side-Chain Structure

Substituents attached to the atoms in the main-chain of a polymer molecule have profound and often complex effects upon the flexibility of the molecule as a whole. These effects may be due to: (1) Size and Shape of the side chain, or to its (2) Polarity.

SIZE AND SHAPE It is not surprising that the size and shape of side-chains attached to the main-chain of the polymer molecule will determine the extent of steric hindrance, restriction of rotation, and stiffening of the polymer molecule (ref 29,35).

Hydrogen Steric hindrance by hydrogen atoms is slight but significant, as can be seen by eliminating them. The C═C double bond in the main-chain of a polymer molecule has only half as many hydrogen atoms attached to it, and produces less steric hindrance and greater freedom of rotation and flexibility in the C—C single bonds adjacent to it, particularly in diene elastomers (ref 30).

Introduction of main-chain atoms with a valence of 2, particularly

Polyquinoxalines

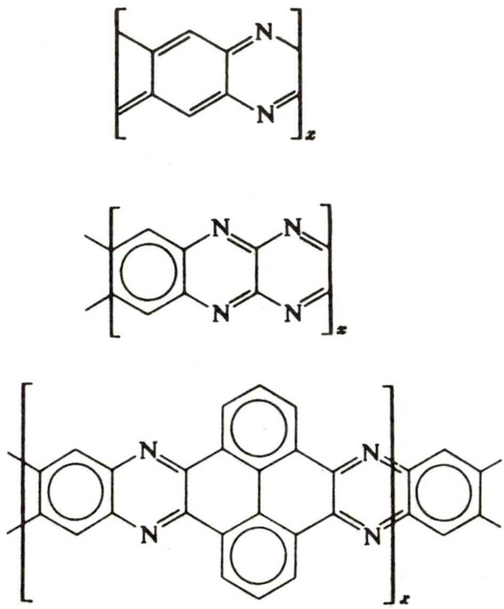

Poly(imidazopyrrolones)

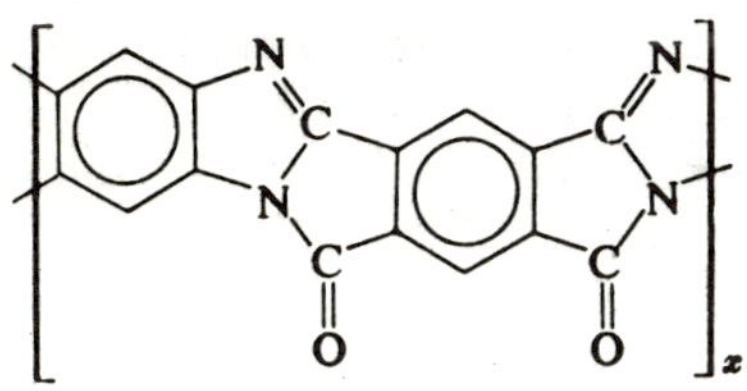

Poly(benzimidazo-benzophenanthrolines)

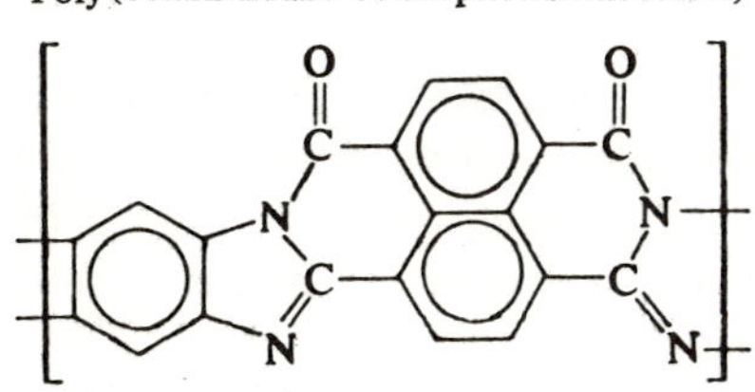

Figure 4-8. Heterocyclic ladder polymers (ref 39).

oxygen and sulphur, produces links which do not need any side groups, not even as small as hydrogen:

$$\mathrm{C{-}\underset{\cdot\cdot}{\overset{\cdot\cdot}{O}}{-}C} \qquad \mathrm{C{-}\underset{\cdot\cdot}{\overset{\cdot\cdot}{S}}{-}C}$$

and these therefore tend to permit freer rotation and produce greater molecular flexibility. Thus the C—O—C bond in aliphatic polyethers and polyesters, and the C—S—C bond in aliphatic sulfides, all produce greater flexibility and lower melting points (ref 30,40,41). Similarly the C—S—S—C bond in aliphatic polysulfides, and especially the Si—O—Si bond in silicones, are so flexible that they actually produce excellent elastomeric properties. These molecules are all more flexible because they lack even the minor steric hindrance of the hydrogen atom as a side-chain.

In comparison with most other structural features in polymers, however, the hydrogen atoms on the methylene group offer relatively little steric hindrance, and the presence of a polymethylene $—(CH_2)_n—$ group in the polymer chain generally has a flexibilizing effect, as has been noted particularly in polyesters and polyamides.

Methyl The methyl side-chains in polypropylene produce considerable steric hindrance, forcing the molecule into the shape of a helical coil even in its liquid form (ref 42), and making the amorphous polymer turn glassy near room temperature. Similar stiffening effects are observed in poly (α-methyl styrene) and polymethyl methacrylate.

In natural rubber and synthetic polyisoprene, where the methyl group is attached to a C═C double bond at a 124° angle (ref 43), the steric hindrance of the methyl group is less important and the molecule remains flexible and rubbery to quite low temperatures (ref 30). In silicone rubber, the large silicon atoms in the main chain, and the unsubstituted oxygens alternating with them, space the methyl groups so far apart that there is little or no steric hindrance between them, and the chain remains highly flexible down to very low temperatures. On the other hand, the flexibility of two methyl groups on polyisobutylene and butyl rubber is remarkable and requires further study and explanation; present theories based on symmetry (ref 44,45), polarity (ref 22), entropy, and the low energy barrier between equally strained conformations (ref 46) offer some guidance in this direction.

Longer Alkyl Groups Lengthening the *n*-alkyl side-chain does not produce any further steric hindrance (Figure 4-9A). On the other hand, it does push the main-chains of the polymer molecules further apart, creating poorer packing and higher free volume (ref 47) (Figure 4-9B). This results in lower modulus, greater flexibility, lower glass transition temperature, and lower melting point (ref 30,35,48–53) (Table 4-1) (Figure 4-10).

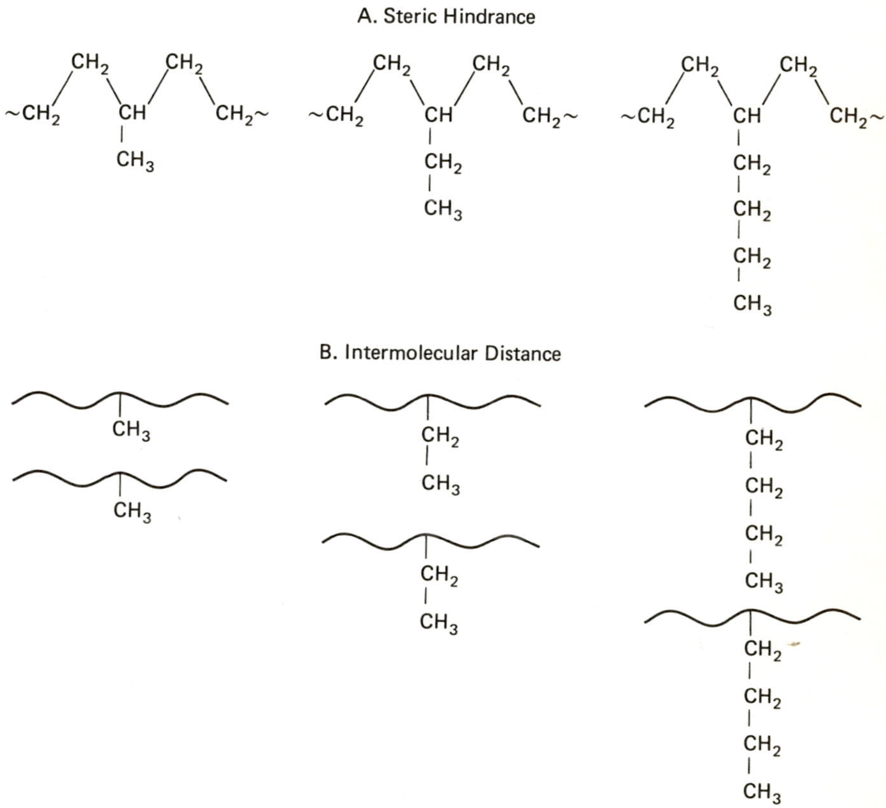

Figure 4-9. Effect of *n*-alkyl side-chain length on molecular flexibility.

In the polyalkyl acrylate series, for example, polymethyl acrylate is stiffly flexible, polyethyl acrylate flexible, polybutyl acrylate rubbery and polyoctyl acrylate tacky and adhesive at room temperature. Similarly in the polyvinyl acetals, polyvinyl butyral is generally softer and more flexible than polyvinyl formal.

Table 4-1. Effect of n-Alkyl Side-Chain Length on Glass Transition Temperatures and Melting Points of Polyolefins (ref 52)

Olefin	T_g °C	T_m °C
Ethylene	−122	137
Propylene	−19	176
1-Butene	−24	120
1-Pentene	−47	70
1-Hexene	−50	−55
1-Heptene		−40
1-Octene	−60	−38
1-Dodecene		45
1-Octadecene		70

When these n-alkyl side-chains grow beyond a certain optimum length (4–12 carbon atoms in different polymer systems), however, they begin to resemble the linear aliphatic hydrocarbon molecules in crystalline paraffin waxes. At this point the n-alkyl side-chains begin to cluster into small microcrystalline aggregates (Figure 4-11). Such side-chain

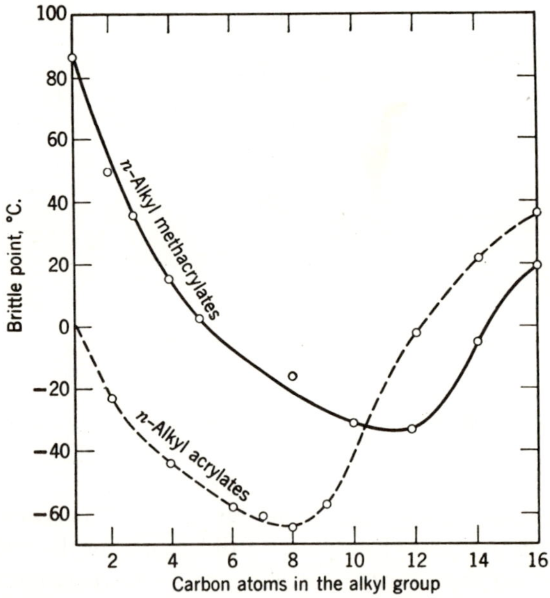

Figure 4-10. Brittle points of polymeric n-alkyl acrylates and methacrylates (ref 163).

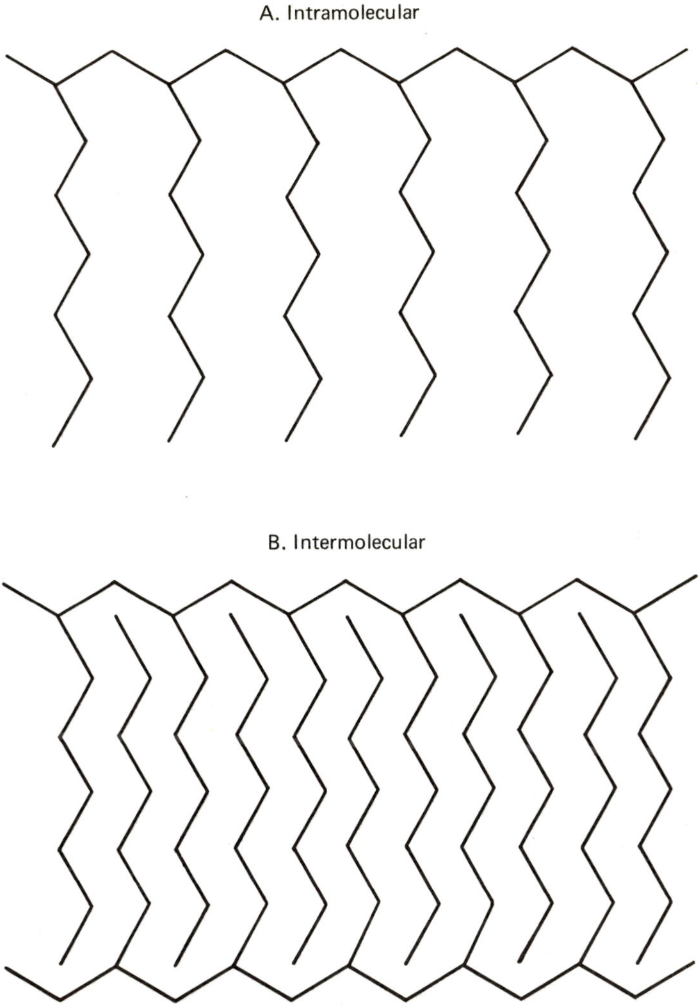

Figure 4-11. Side-chain crystallization.

crystallization reverses the effects of internal plasticization and produces a progressive increase in stiffness and melting point (ref 30,51,53–56). It also gives polymers such as polyoctadecene, polyvinyl stearate, and polyoctadecyl acrylate a very waxy quality which can be useful in a variety of applications (ref 57,58).

Branched Side Chains Branching of the aliphatic side chain makes it much bulkier, producing greater steric hindrance (Figure 4-12), less flexibility, and higher glass transition temperatures and melting points (ref 30,52,59–61). These effects are strikingly obvious in commercial poly(4-methylpentene-1). Very heavy branching crowds the main-chain and increases the glass transition temperature (ref 62). When the side groups are extremely bulky they may produce so much steric hindrance

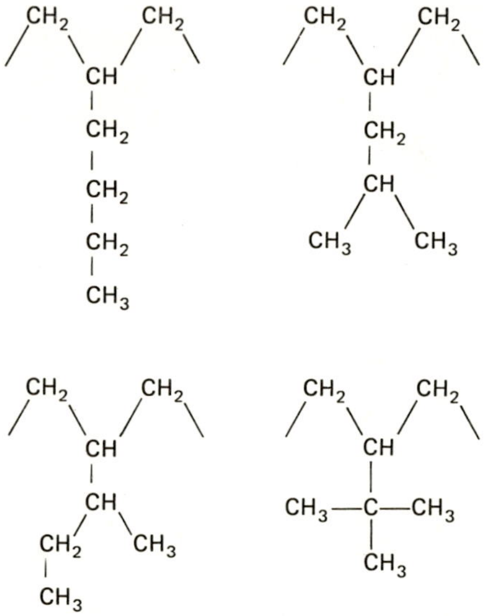

Figure 4-12. Side-chain branching → restriction of rotation.

and strain that they even lower the ceiling temperature for polymerization ⇌ depolymerization (ref 63,64).

Cyclic Side Groups Cyclic substituents in the side-chains make them very bulky, and their restriction of rotation increases as they come closer to the main-chain (ref 28) (Table 4-2). The benzene rings on the main-chain of polystyrene produce a very rigid brittle polymer. In o-substituted styrenes, steric hindrance becomes still greater (ref 65,66) (Table 4-3); and in polyvinyl cyclohexane and fused ring systems the effects of steric hindrance become even more obvious (ref 28,68,69). This stiffening of the molecule is generally most readily seen in its effect on glass transition temperature and melting point.

In impact styrene and ABS, the sterically-hindered bulky polystyrene and styrene/acrylonitrile polymer chains are actually large polymeric side-chains attached to rubbery backbones; but they are spaced far enough apart so that they do not destroy the flexibility of the rubbery backbone, and still permit it to contribute high impact resistance in the otherwise rigid polymer.

POLARITY When the side-chain substituents are polar in nature they tend to repel each other, restricting rotation and stiffening the polymer chain further (ref 29,51,70). Thus in polytetrafluoroethylene the fluorine

Table 4-2. Steric Hindrance by Phenalkyl Side-Chains (ref 28,52)

Side-Chain	T_g °C	T_m °C
 CH_2CH_2—	10	160
 CH_2—	60	208
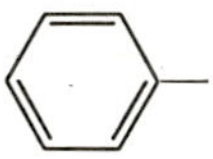	83	240

atoms are not only larger in size than hydrogen, but their high electronegativity also makes them repel each other and stiffen the polymer molecule into a rod-like structure which is very inflexible and very high-melting. Chlorine atoms are also quite electronegative and larger in size, making polyvinyl chloride a fairly rigid molecule; addition of liquid ester plasticizers neutralizes some of this polarity and repulsion, and thus helps to increase the flexibility of the polymer molecule. In polyacrylonitrile, too, the —C≡N side-groups provide both steric hindrance and dipole repulsion, stiffening the molecule into a rigid rod (ref 71).

The most extreme effects are seen in studies of polyelectrolyte solutions. Highly-ionized polyacrylate, carboxymethyl cellulose, and polyquaternary ammonium ions contain so many like electrical repulsions that the normally random coil expands into a much stiffer, more rod-like structure (ref 72,73), (Figures 4-13, 4-14). Such expanded structures

Table 4-3. Glass-Transition Temperature of Substituted Polystyrenes (ref 67)

Polymer	$T_g(°C)$
Polystyrene 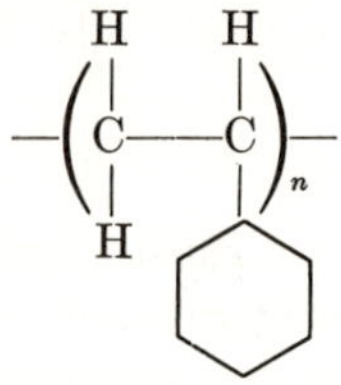	100, 105
o-Methyl	115, 125
m-Methyl	72, 82
p-Methyl	101
p-Ethyl	27, 78
p-*n*-Butyl	6
p-*n*-Hexyl	−27
p-*n*-Octyl	−45
p-*n*-Nonyl	−53
p-*n*-Decyl	−65
p-*n*-Dodecyl	−52 (?)
p-*n*-$C_{14}H_{29}$	−36 (?)
p-*n*-$C_{16}H_{33}$	+4.5 (?)
p-*n*-$C_{19}H_{39}$	+32 (?)
2,4-Dimethyl	119, 129
2,5-Dimethyl	122
3,4-Dimethyl	102, 83
p-*tert.*-Butyl	118, 131
p-Methoxy	<90
p-Phenyl	138
p-Phenoxy	100
2,5-Difluoro	101
p-Chloro	128
2,5-Dichloro	130, 115
3,4-Dichloro	138, 103
2,6-Dichloro	167, 132
α-Vinyl naphthalene	162
α-Methyl	180, 192, 150

increase solution viscosity so markedly that they are very useful as thickeners and gelling agents in foods, cosmetics, and latex paints (ref 75).

Additional Miscellaneous Considerations

In addition to main-chain and side-chain structures, there are two other structural features of frequent importance in molecular flexibility: (1) Copolymerization, and (2) Plasticizer Structure.

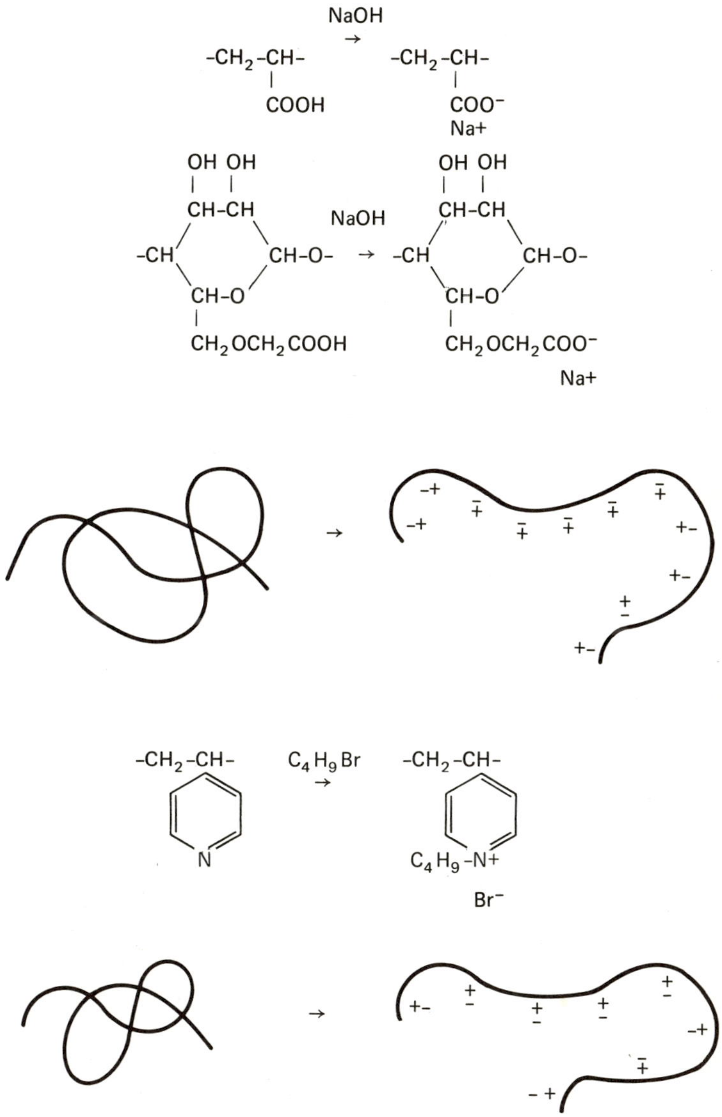

Figure 4-13. Electrical repulsion and molecular stiffening in polyelectrolytes.

Copolymerization

Copolymerization can insert occasional flexible groups into an otherwise rigid polymer molecule (ref 76). In general, random copolymers will have properties more or less average and intermediate between those of the individual homopolymers (ref 77–79) (Figures 4-15, 4-16). Homogeneous copolymers will have fairly sharp glass transition temperatures which are useful in many types of processability and for lowering crystallinity; while heterogeneous copolymers will have broad glass transition temperatures which are best for temperature-insensitivity of properties (ref 82) (Figure 4-17). Random copolymers give weight average properties such as glass transition temperature

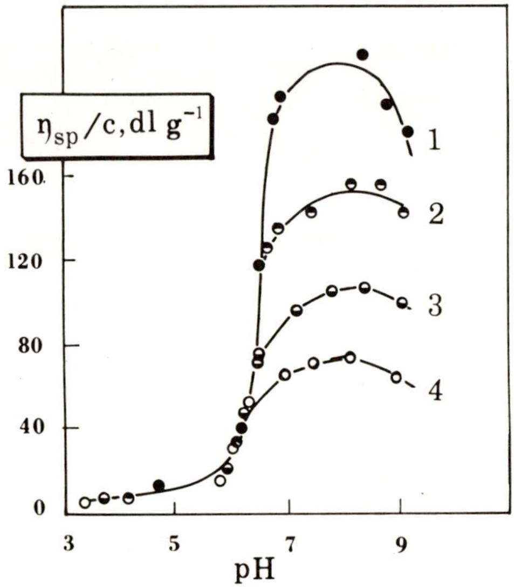

Figure 4-14. Expansion of poly(methacrylic acid) coil in aqueous alkali (ref 74).

Reduced specific viscosity of poly(methacrylic acid) (molecular weight 216,000) in aqueous solution when neutralized to various extents, concentrations of polymer shown on the curves.

$$\frac{1}{T_g} = \frac{1}{w_1 + Bw_2}\left[\frac{w_1}{T_{g1}} + \frac{Bw_2}{T_{g2}}\right]$$

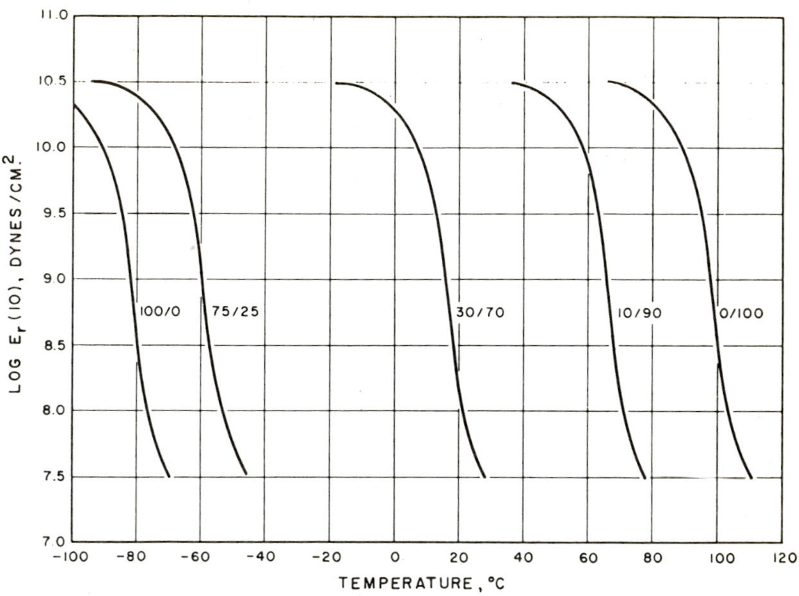

Figure 4-15. Modulus versus temperature curves for random copolymers (ref 80). $E_r(10)$ versus temperature for butadiene-styrene copolymers.

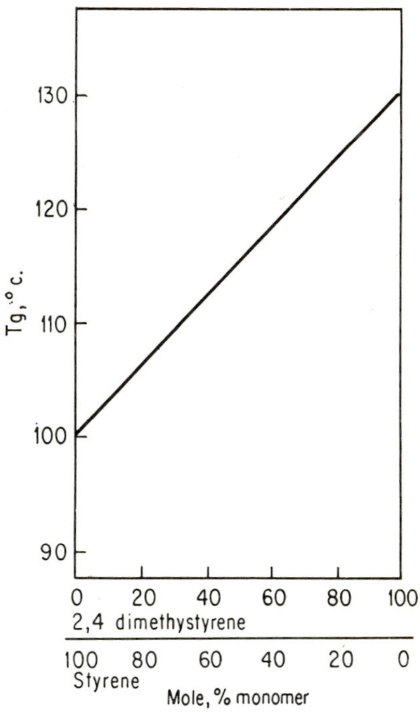

Figure 4-16. Glass transition temperatures for amorphous copolymers (ref 81). Reprinted by permission of Modern Plastics Encyclopedia, McGraw-Hill, Inc.

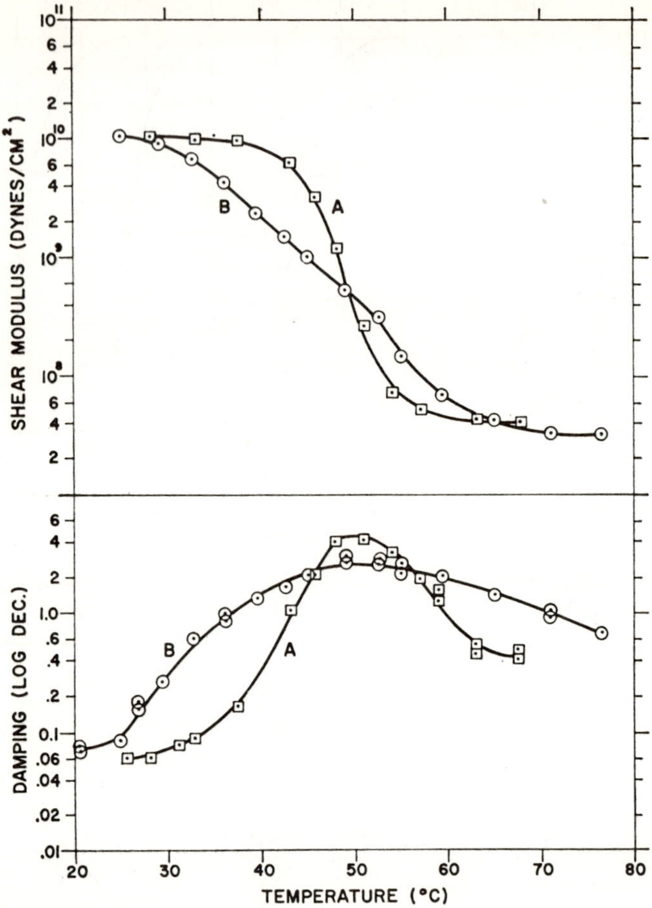

Figure 4-17. Modulus versus temperature curves for homogeneous and heterogeneous copolymers (ref 83).

Dynamic mechanical properties of vinyl chloride-methyl acrylate copolymers. A. Homogeneous; B. Heterogeneous.

where B is a constant approaching ≈ 1, simplifying to

$$\frac{1}{T_g} = \frac{w_1}{T_{g1}} + \frac{w_2}{T_{g2}} \qquad \text{(ref 84)}$$

(Figure 4-18). Block and graft copolymers may actually show two individual glass transition temperatures (ref 84).

The use of copolymerization to modify flexibility is most commonly practiced commercially in the use of butadiene to improve the film-forming ability of polystyrene latex paints and paper coatings; vinyl acetate to reduce melt viscosity of polyvinyl chloride during processing; methyl methacrylate, styrene, or vinyl chloride to toughen polyethyl acrylate latex paints; acrylic or maleic esters to improve the film-forming ability of polyvinyl acetate latex paints; acrylic esters to lower the melt viscosity of polymethyl methacrylate molding powders; and fatty esters to

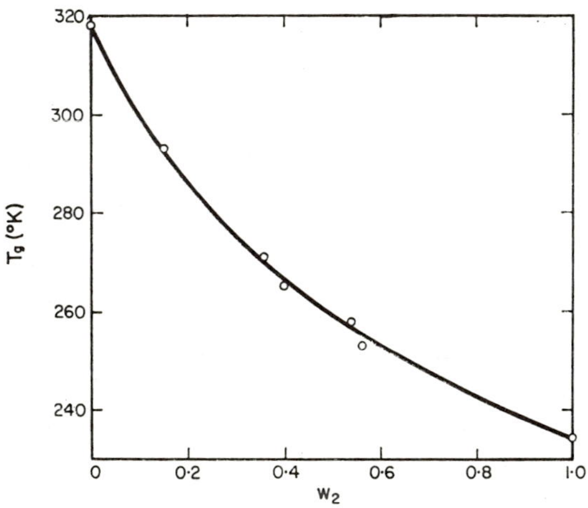

Figure 4-18. Glass transition temperatures of random copolymers (ref 85). Glass temperatures T_g of copolymers of polychlorotrifluoroethylene with polyvinylidene fluoride plotted against the weight fraction w_2 of the latter. The curve is the theoretical equation with $B = 1.75$.

improve the flexibility of alkyd coatings. The technique is particularly readily applied to produce detailed continuous ranges of copolymer properties for different specialty applications, such as coatings, adhesives, polishes, textile assistants, and similar types of formulations.

Plasticizer Structure

Molecular flexibility is important not only in the polymer molecule but also in plasticizer molecules as well (ref 86–89) (Figures 4-19, 4-20)

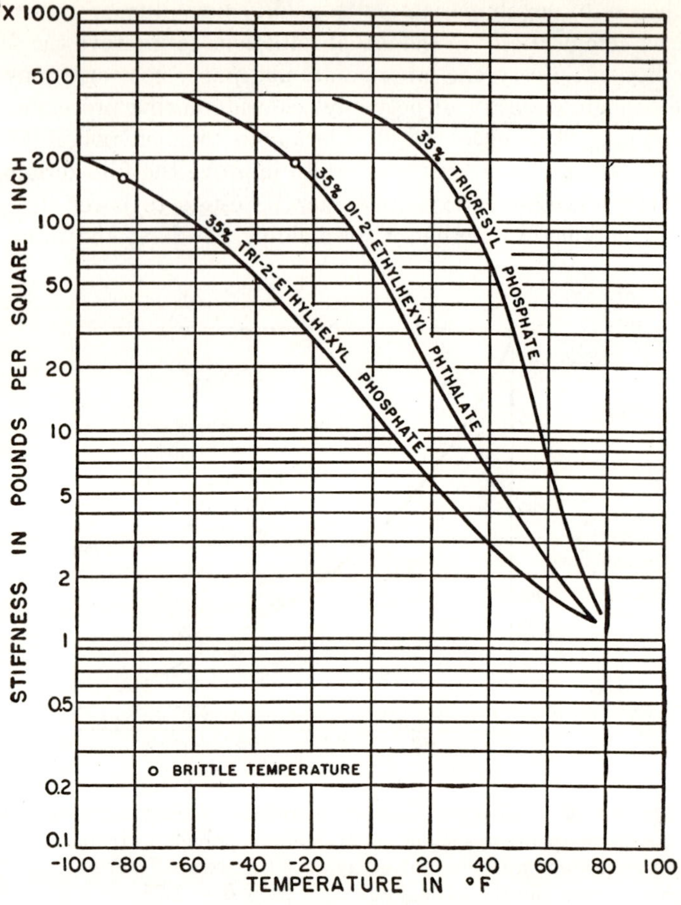

Figure 4-19. Molecular flexibility of plasticizer → flexibility of polymer (ref 90). Stiffness versus temperature of "Vinylite" VYNW plasticized with 35 percent of three different plasticizers.

(Table 4-4). In general, linear plasticizer molecules are themselves more flexible than cyclic ones, and provide more efficient plasticization when they are added to polymer systems. Commercial applications are primarily in soft flexible polvinyl chloride upholstery, wall-covering, outerwear clothing, household furnishing, electrical wire insulation, packaging film, toys, garden hose; in polyvinyl acetate latex paints and adhesives; in polyvinyl butyral safety glass; and in cellulose ester film, sheet, tape, and coatings.

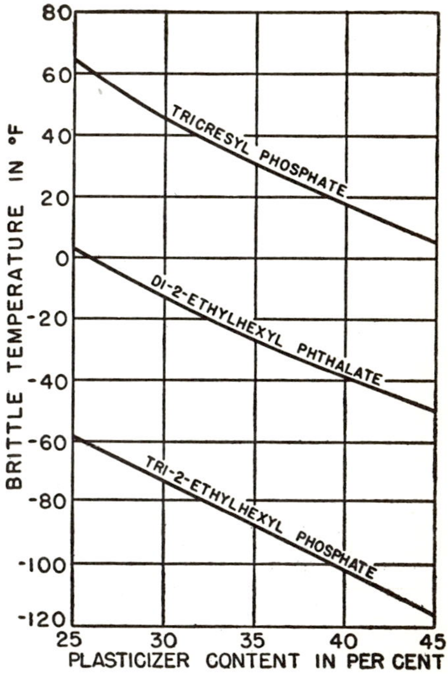

Figure 4-20. Molecular flexibility of plasticizer → brittle temperature of polymer (ref 91). Effect of plasticizer concentration on brittle temperature of "Vinylite" VYNW. (ASTM Method D-746-43T.)

Many of these effects may be summarized by a tabulation of some typical polymer structures, and the ways in which their structural features interact to determine their glass transition temperatures (Table 4-5). In all of these ways, the structure of the individual polymer molecule determines the flexibility of the individual molecule. This in turn contributes largely to the practical properties of the total polymeric composition.

Table 4-4. Effect of Plasticizer Structure on Plasticizing Efficiency: 67 PHR Plasticizer in 40-mil Polyvinyl Chloride Sheet (ref 89)

Plasticizer	Modulus at 100% elongation	Ultimate tensile strength	T_F at which modulus = 45,000 psi
Dioctyl adipate	630 psi	1710 psi	−66°C
Diphenyl octyl phosphate	700	1960	−39
Dioctyl phthalate	830	1960	−39
Butyl benzyl phthalate	830	2240	−24
Tricresyl phosphate	1000	2490	−10

Table 4-5. Glass Transition Temperature T_g of Selected Polymers (ref 92)

Polymer	Structure	T_g(°C)	Feature described in text
Poly(dimethyl siloxane)	$—O·Si(CH_3)_2—$	−123	Siloxane bonds very flexible
Polybutadiene	$—CH_2·CH{=}CH·CH_2—$	−85	C—C bonds flexible
*Cis*polyisoprene	$—CH_2·C(CH_3){=}CH·CH_2—$	−70	Stiffening effect of methyl group
Poly(*n*-heptyl butadiene)	$—CH_2·C·(C_7H_{15}){=}CH·CH_2—$	−83	Chain separation balanced by entanglement
Poly(*n*-decyl butadiene)	$—CH_2·C·(C_{10}H_{21}){=}CH·CH_2$	−53	Side-chain entanglement and crystallisation
Butadiene-styrene copolymer	75% Bu–25% St.	−55	Copolymerisation
Polypropylene	$—CH_2·CH(CH_3)—$	−27	Stiffening effect of methyl group
Polyisobutylene	$—CH_2·C(CH_3)_2—$	−65	Increased flexibility through reduction of dipole moment
Poly(3-methyl butene-1)	$—CH_2·CH·[CH(CH_3)_2]—$	+50	Stiffening effect of isopropyl group
Polyoxymethylene	$—CH_2—O—$	−73	Flexibility of C—O bonds
Poly(ethylene adipate)	$—(CH_2)_2OOC·(CH_2)_4COO—$	−70	Flexibility of C—O and C—C bonds
Poly(ethylene terephthalate)	$—(CH_2)_2OOC·C_6H_4COO—$	+67	Stiffening effect of phenylene group in backbone
Polycarbonate of (bisphenol A)	$—C_6H_4·C(CH_3)_2·C_6H_4·O·CO·O$	+149	Stiffening effect of phenylene groups
Poly(vinyl chloride)	$—CH_2·CH(Cl)—$	+80	Dipole attraction of chlorine atoms
Poly(vinylidene chloride)	$—CH_2·C·(Cl)_2—$	−17	Reduction of dipole moment
Polystyrene	$—CH_2·C(C_6H_5)—$	+100	Stiffening effect of attached benzene ring

4-2 MOLECULAR FLEXIBILITY → PROPERTIES → APPLICATIONS

Having outlined in detail the ways in which molecular structure controls and determines molecular flexibility, it is now appropriate to examine the ways in which the flexibility of the individual polymer molecule affects the overall properties and applications of the polymeric material. While molecular flexibility is of great fundamental importance, in actual polymer systems it is often complicated or even masked by the effects of intermolecular and supermolecular structure, so its practical effects are often difficult to discern clearly. In most cases its practical significance must be considered as one interacting component of the total structure → properties relationship.

Thermal properties are so fundamentally involved that they will be considered first, after which each of the other types of properties will be considered in its usual sequence.

Thermal Properties

As has already been explained, thermal motion of atoms and groups within the polymer molecule determines its flexibility, its mobility, and its ability to respond to mechanical stress.

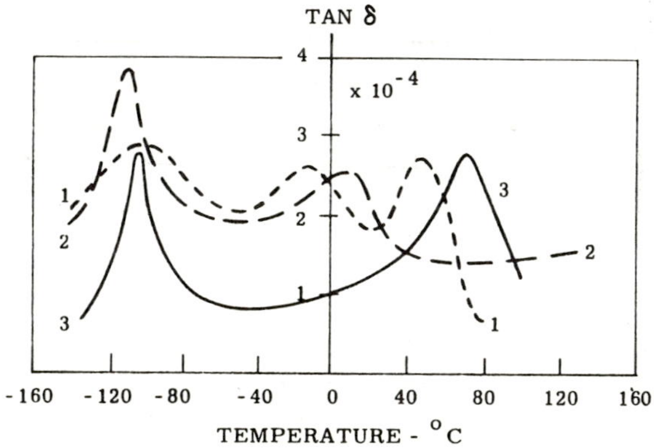

Figure 4-21. Effect of secondary transitions on dielectric properties (ref 93).

Dissipation factor versus temperature at 1000 cps; different types of polyethylene.

1—High-pressure polyethylene
2—"Quenched" low-pressure polyethylene
3—Normal low-pressure polyethylene

Low-Temperature Rigidity

At very low temperatures, thermal motions of the individual atoms are so small that the polymer has no significant mobility at all and is highly rigid, producing modulus measurements of about 10^{10} dynes/cm^2.

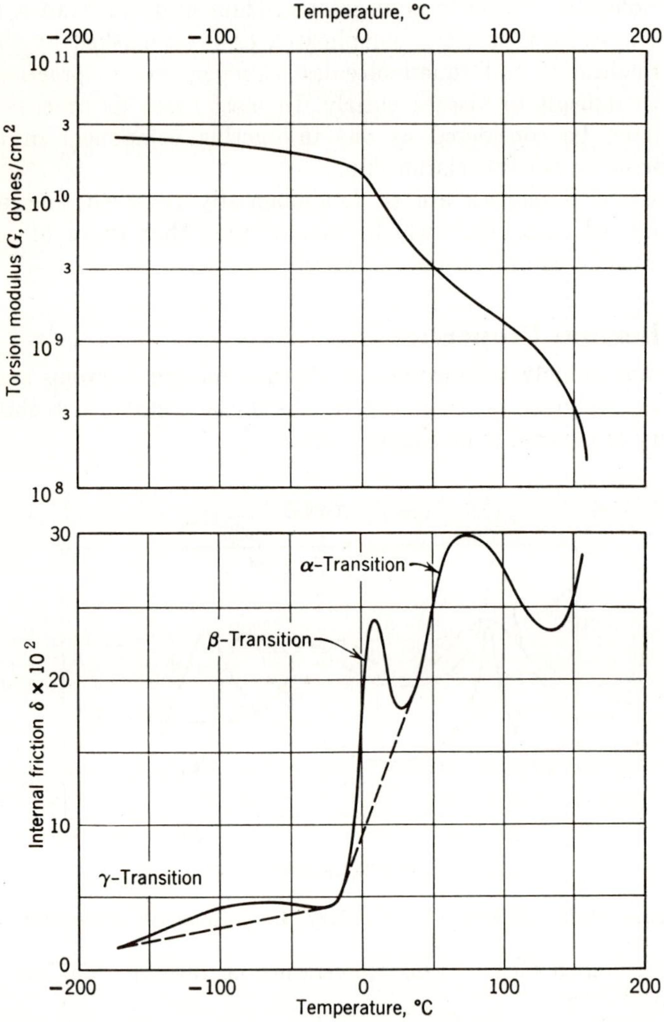

Figure 4-22. Effect of secondary transitions on modulus of polypropylene (ref 94).

Secondary Transition

With increasing temperature the motion of individual atoms and small groups becomes possible. If these groups have any polarity, their increased mobility may first be seen by a change in dielectric properties (Figure 4-21). Motion of small side-groups, or 2–5 atom segments in the main chain of the polymer, may also be seen by small decreases in the modulus (Figures 4-22, 4-23) and may account for some improvement of impact strength. But these small secondary transitions are not generally of great practical significance (ref 96–99).

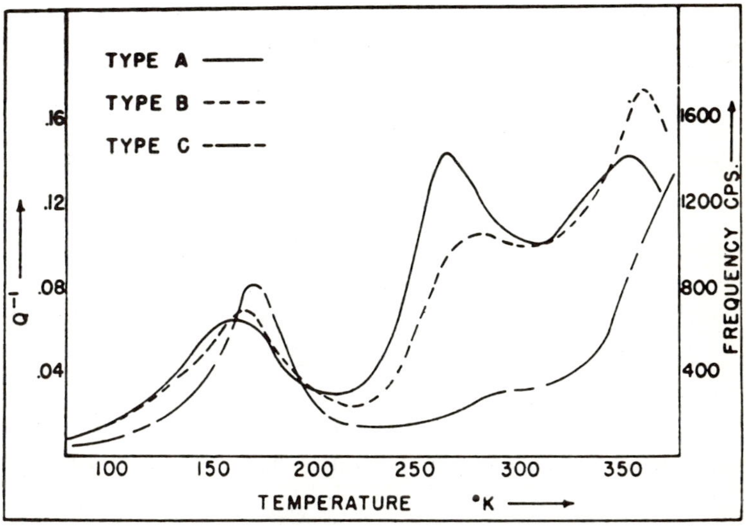

Figure 4-23. Effect of secondary transitions on mechanical damping in polyethylenes (ref 95). Damping of polyethylene as a function of branching. Type A: 3.2 CH_3's/100 CH_2; Type B: 1.6 CH_3's/100 CH_2; Type C: <0.1 CH_3's/100 CH_2.

Glass Transition

With further increase in temperature, thermal vibrations of the atoms and groups in the polymer molecule continue to increase and push the polymer molecules apart, making possible the occasional motion of larger groups of atoms and segments in the polymer molecule; the more flexible the molecular structure, the lower the temperature at which this will occur. At about 2.5 percent of free volume the motion of chain segments of five or more atoms reaches a fairly high level of probability, and the motion of segments this large is sufficient to change mechanical properties from rigid to stiffly flexible or leathery. This temperature is called the glass transition temperature, T_g, and is the most general basic experimental measurement available for estimation and correlation with molecular flexibility (ref 14,77,98,100–104) (Table 4-6).

In nearly all polymers, different main-chain atoms are in more or less hindered (intramolecular) or crowded (intermolecular) positions,

Table 4-6. Glass-Transition Temperature of Polymers (ref 105)

Polymer	Repeat unit	T_g(°C)
Polyethylene	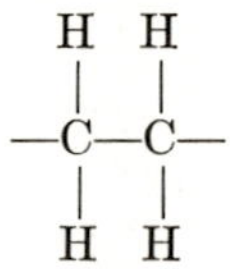	−120
Polypropylene	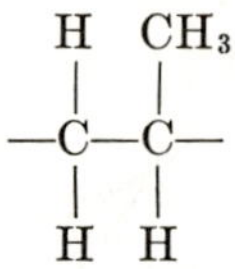	−10, −18
Polyisobutylene	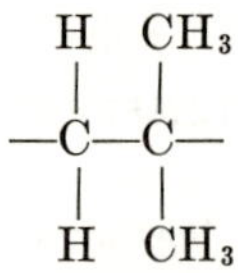	−70, −60
Polyisoprene	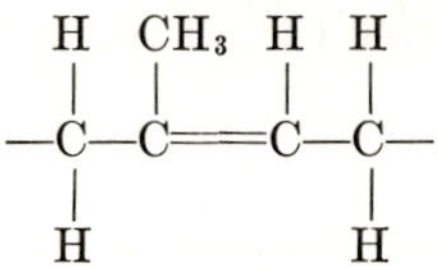	−73
Polybutadiene	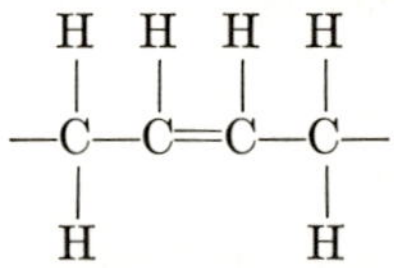	−90
Poly-2-isopropyl-1,3-butadiene	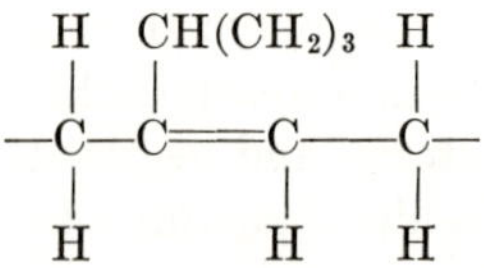	−47

Table 4-6. Glass-Transition Temperature of Polymers (ref 105) (*Continued*)

Polymer	Repeat unit	T_g(°C)
Poly-2-*tert*. butyl-1,3-butadiene	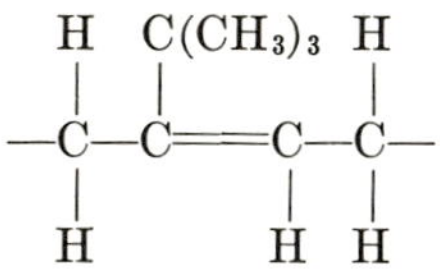	+20
Poly-2-*n*-heptyl-1,3-butadiene		−83
Poly-2-*n*-decyl-1,3-butadiene		−53
Poly-1-butene	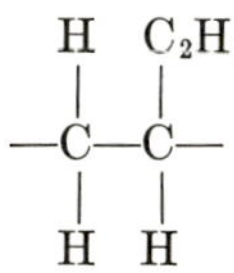	−25
Poly-1-pentene	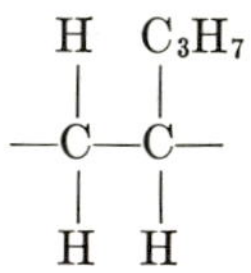	−40, −24
Poly-1-hexene	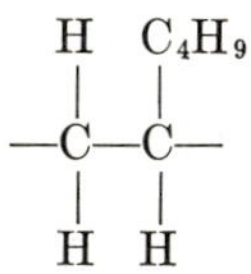	−50
Poly-1-octene	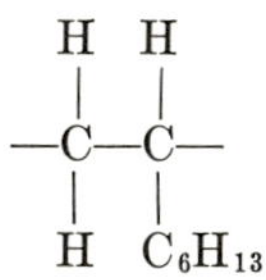	−65
Poly-1-dodecene	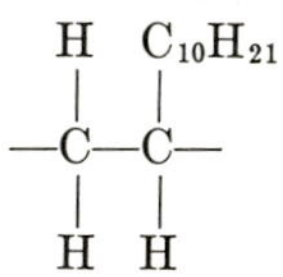	−25 (?)
Poly-4-methyl pentene-1	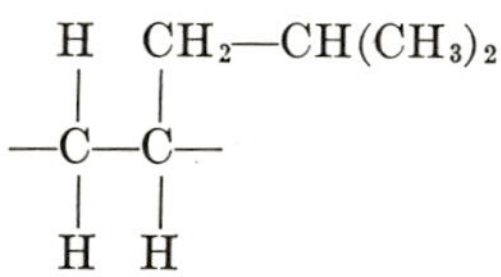	+18 (cryst.) 29 (amorph.)

Table 4-6. Glass-Transition Temperature of Polymers (ref 105) (*Continued*)

Polymer	Repeat unit	T_g(°C)
Polyoxymethylene	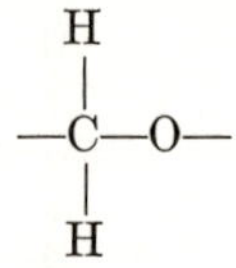	−50 (−85)
Polyvinyl methyl ether	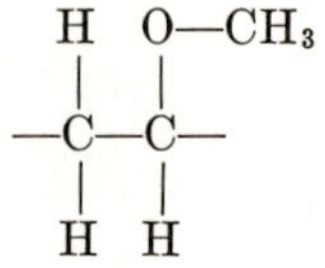	−20, −10
Polyvinyl ethyl ether	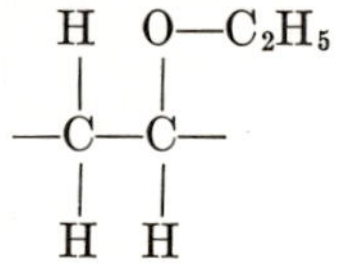	−25
Polyvinyl-*n*-butyl ether	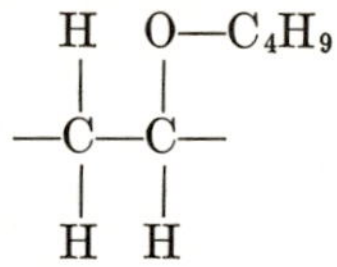	−52
Polyvinyl isobutyl ether	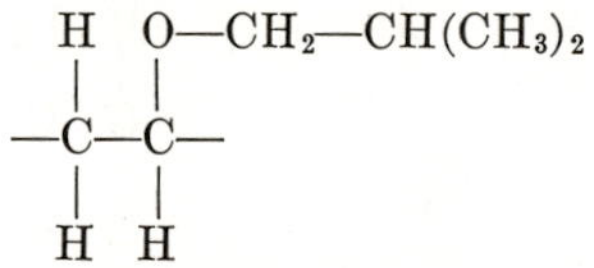	−5, −18
Polyvinyl tert. butyl ether	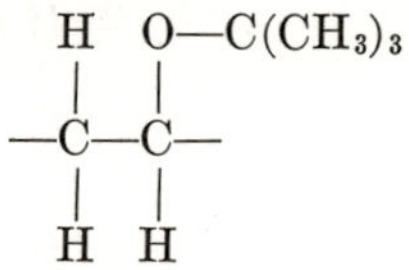	+88
Polydimethyl siloxane	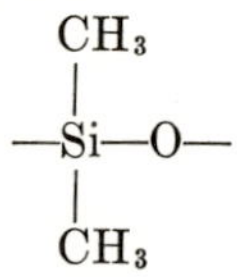	−123
Polyvinyl fluoride	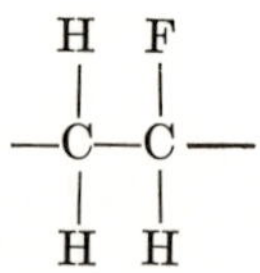	−20

Table 4-6. Glass-Transition Temperature of Polymers (ref 105) (*Continued*)

Polymer	Repeat unit	T_g(°C)
Polyvinyl chloride	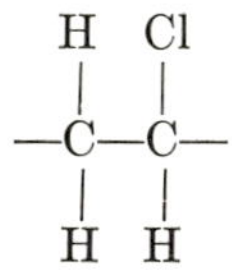	87
Polyvinylidene fluoride	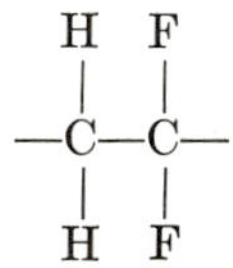	−35
Polyvinylidene chloride	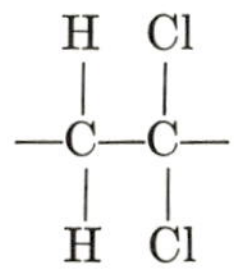	−17
Polyperfluoropropylene	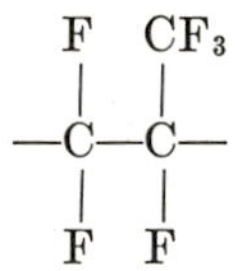	11
Polychloroprene	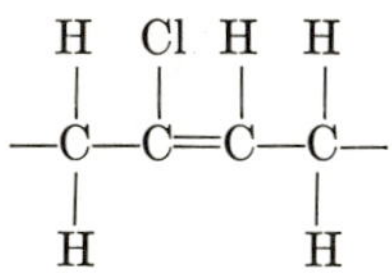	−50
Polychlorotrifluoroethylene	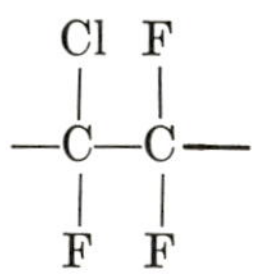	45
Polytetrafluoroethylene	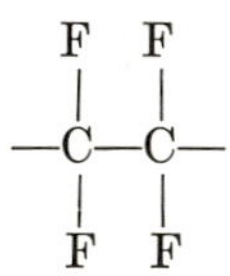	126
Polyacrylonitrile	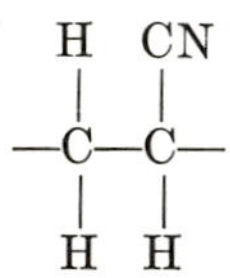	104, 130

Table 4-6. Glass-Transition Temperature of Polymers (ref 105) (*Continued*)

Polymer	Repeat unit	T_g(°C)
Polymethacrylonitrile	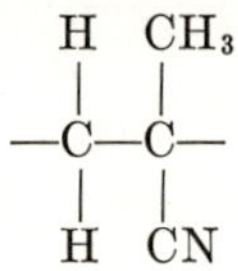	120
Polyvinyl acetate	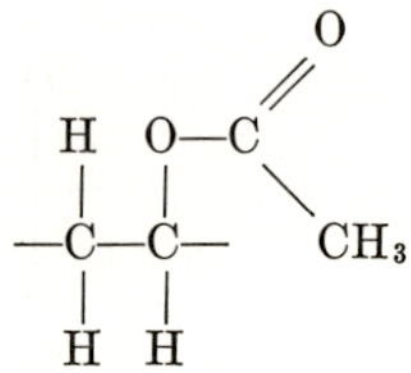	29
Polyvinyl carbazole	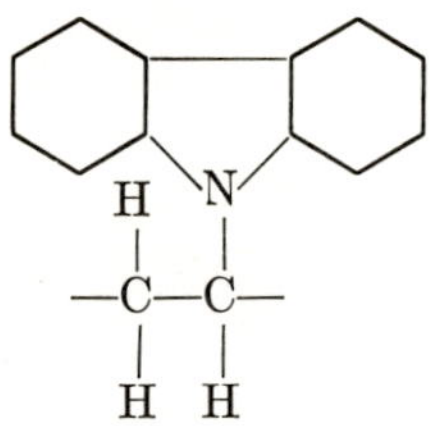	150, 208
Polyvinyl formal	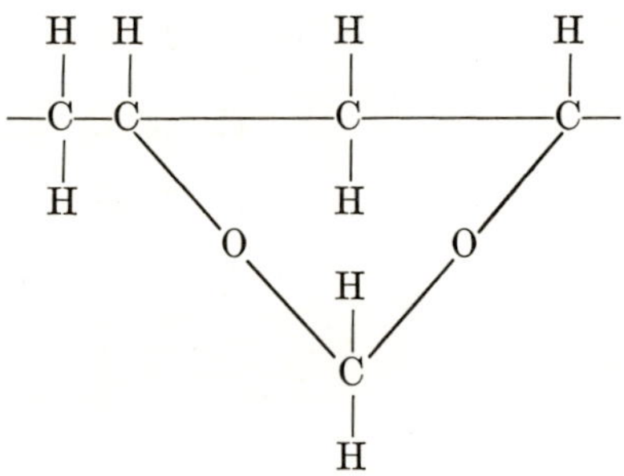	105
Polyvinyl acetal		82
Polyvinyl propional		72
Polyvinyl butyral	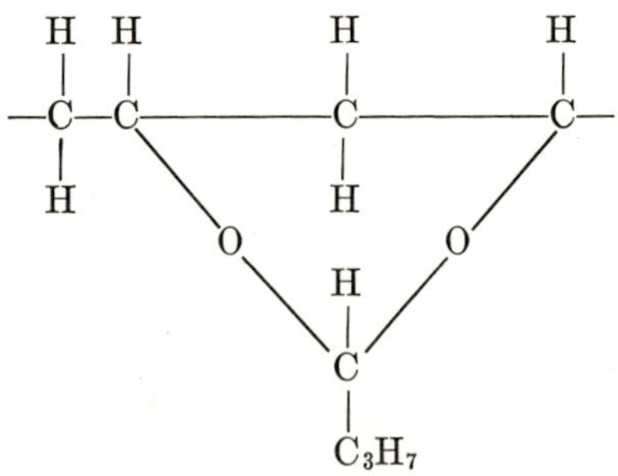	49

Table 4-6. Glass-Transition Temperature of Polymers (ref 105) (*Continued*)

Polymer	Repeat unit	T_g(°C)
Polyvinyl isobutyral		56
Cellulose nitrate	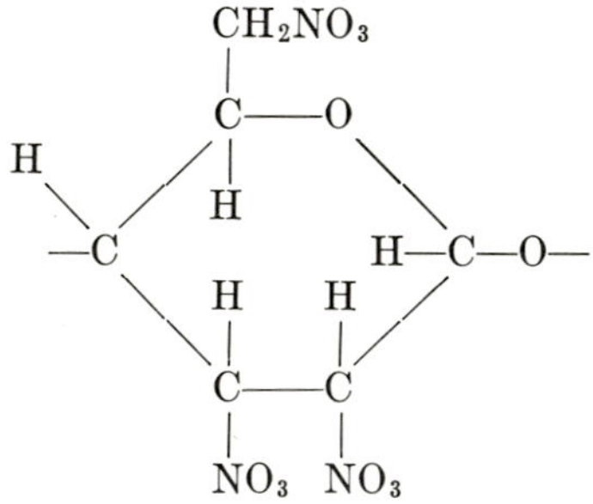	53?
Cellulose triacetate		105, 69?
Cellulose (2.3) acetate		120
Cellulose tributyrate		120
Cellulose (2.6) butyrate		125
Ethyl cellulose		43
Polyvinyl alcohol	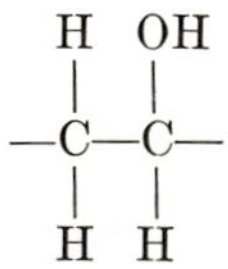	85
Polycarbonate	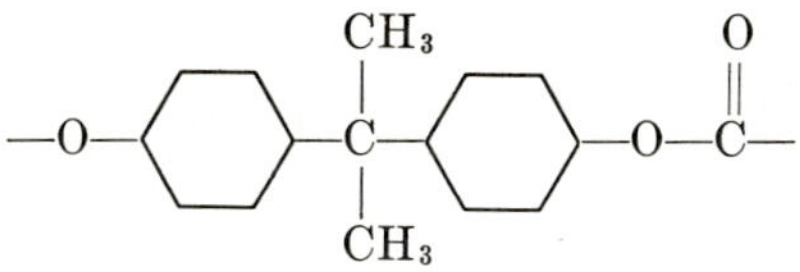	150
Polyethylene tereph-thalate	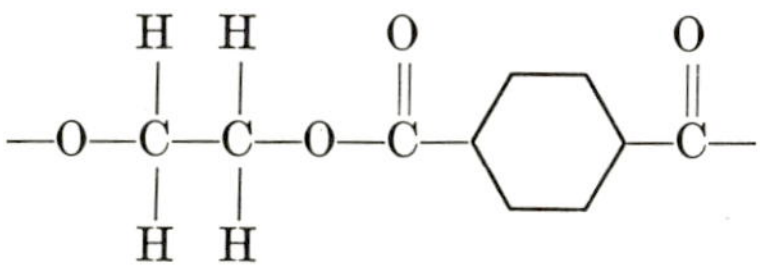	69
Poly(ethylene adipate)	$—(CH_2)_2—O—\overset{\overset{\displaystyle O}{\parallel}}{C}—(CH_2)_4—\overset{\overset{\displaystyle O}{\parallel}}{C}—O—$	−70
Poly(tetramethylene sebacate)	$—(CH_2)_4—O—\overset{\overset{\displaystyle O}{\parallel}}{C}—(CH_2)_8—\overset{\overset{\displaystyle O}{\parallel}}{C}—O—$	−57

Table 4-6. Glass-Transition Temperature of Polymers (ref 105) (*Continued*)

Polymer	Repeat unit	T_g(°C)
Polyhexamethylene adipamide (nylon 6-6)	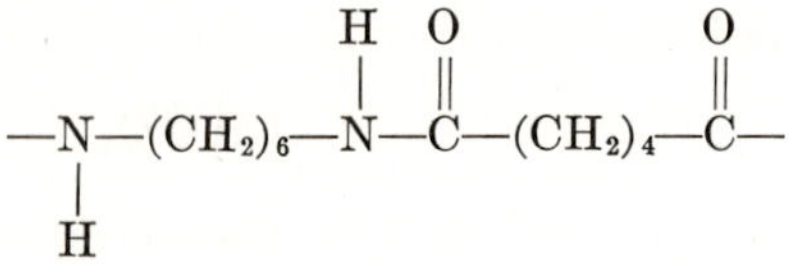	50
Polyhexamethylene sebacamide (nylon 6-10)	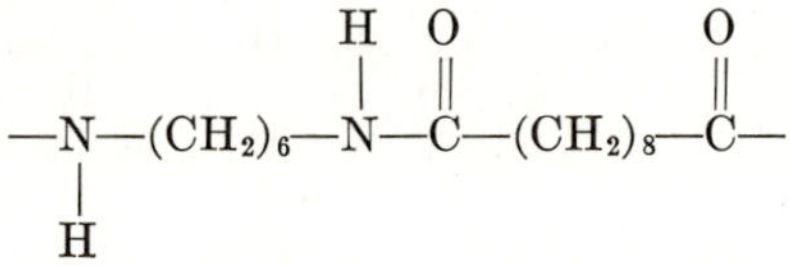	40
Polycaprolactam (nylon 6)	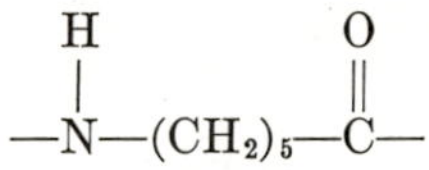	50
Poly-*p*-vinyl benzyl alcohol	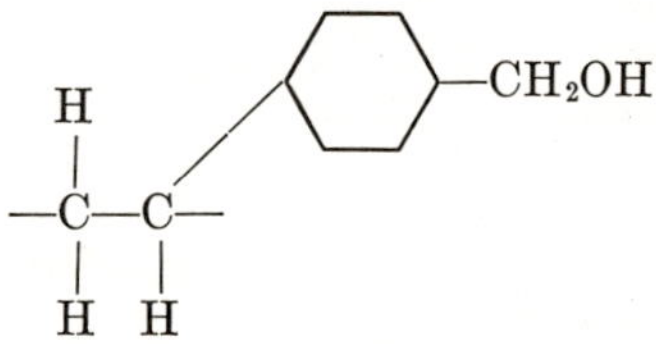	140
Poly-*o*-vinyl benzyl alcohol		160
Poly-*m*-vinyl benzyl alcohol		<125
Polyvinyl pyrrolidone	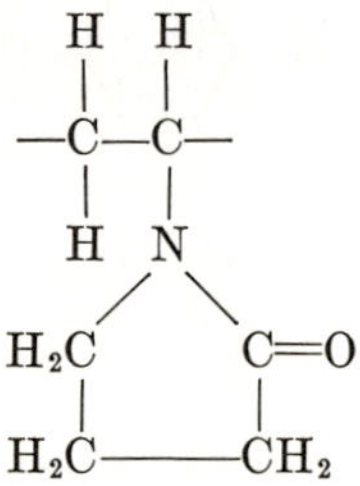	175
Poly-2,2′-dichloromethyl propylene oxide	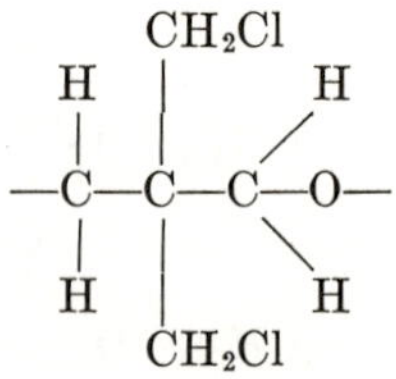	10

and require different amounts of thermal energy to rotate freely. As temperature rises, each reaches the minimal critical energy for rotation at a different temperature, probably spread over a Gaussian distribution, or even a multimodal distribution for the several different species of main-chain atoms in the system. If a single C—C—C angle reaches the energy level required for rotation, it is still tied down by its intra- and inter-molecular neighbors. Only when enough adjacent atoms reach this level can a reasonably large molecular segment move into an adjacent area and produce flow. The glass transition covers a temperature

Table 4-7. Low-Temperature Stiffening and Embrittlement of Elastomers (ref 106,107)

Elastomer	T_g °F	Stiffening temperature °F	Brittle point °F
Silicone	−189	−60 to −160	−90 to −180
cis-Polybutadiene	−130	−30 to −60	−100
Ethylene/propylene		−20 to −50	−90
Polypropylene oxide			−90
Natural rubber		−20 to −50	−80
cis-Polyisoprene	−99	−20 to −50	−80
Butadiene/styrene		0 to −50	−80
Butyl	−76 to −94	−10 to −40	−80
Chlorosulfonated polyethylene		−30 to −50	−60
Polysulfide		−10 to −45	−60
Polyurethane		−10 to −30	−60
Butadiene/acrylonitrile		+30 to −20	−65
Polychloroprene	−58	+10 to −20	−45
Fluoroelastomers		+20 to −30	+10 to −60
Acrylic		+35 to +10	−10

range in which a minimal number of adjacent atoms in a region reach sufficient energy levels to move, and it covers the temperature range for a series of regions which may differ somewhat in constitution or accidental arrangement.

Practically, glass transition temperature is generally assumed to correlate with heat distortion temperature of rigid plastics, and with stiffening and embrittlement on cooling of flexible plastics and elastomers (Table 4-7). In actual practice, these transitions can be complicated, masked, or even eliminated by high crystallinity, high degrees of cross-linking, or heterogeneity. But they are quite generally observed for amorphous linear homogenous polymers, and for polymers with low degrees of crystallinity and little or no cross-linking.

Rubbery Plateau

As the temperature increases still further, larger segments of 10–30 chain atoms become mobile with greater frequency, and mechanical properties soften from leathery to rubbery in nature. The temperature range for such rubbery mobility is inverse to the inherent flexibility of the fundamental molecular structure. Modulus levels out in a rubbery plateau region at about 10^7 dynes/cm^2, in which fairly large segments of the molecule can disentangle and move past each other quite easily and rapidly under rather low mechanical stress. But high-molecular-weight molecules as a whole remain sufficiently entangled so that they do not permit true flow, and so that when the mechanical stress is released the molecules will recoil into a more probable random arrangement of higher entropy, producing fairly true rubbery elasticity (ref 6,10,11, 13,15–18,108). All the common elastomers are in this state at room temperature and use-temperatures. Most rigid plastics pass through a similar state when heated, unless they are too crystalline or cross-linked.

Rubbery Flow and Liquid Flow

At still higher temperatures, the motion of large segments of the polymer molecule becomes so easy that rubbery elasticity begins to change to permanent viscoelastic flow, and this second transitional region is generally described as rubbery flow; it is responsible for creep and permanent set of elastomers, and for shrinkage and melt fracture in thermoplastic processing. Finally, at even higher temperatures, linear molecules can disentangle and flow past each other in fairly true liquid flow (ref 13); this is the normal condition in melt processing. Temperature ranges for rubbery and liquid flow are of course inverse to the inherent flexibility of the fundamental molecular structure, at least in amorphous, linear, homogeneous polymers.

Crystallinity will be discussed fully later, but it should be noted at this point that molecular flexibility also has an inverse effect on crystalline melting point (ref 100,109,110). Here rising temperature signifies the increasing vibration of atoms and segments in the crystalline lattice, until they acquire enough energy to break loose from the rigid lattice structure. More flexible molecules can coil away from the lattice much more readily than rigid molecules, and reach the crystalline melting point at much lower temperatures. Since molecular flexibility lowers both the glass and crystalline transition temperatures, it is not surprising that there is a rough correlation between T_g and T_m (ref 111) (Figure 4-24).

Time-Temperature Superposition

This general relationship between temperature and mechanical properties is quite typically observed for all amorphous, linear, homogeneous

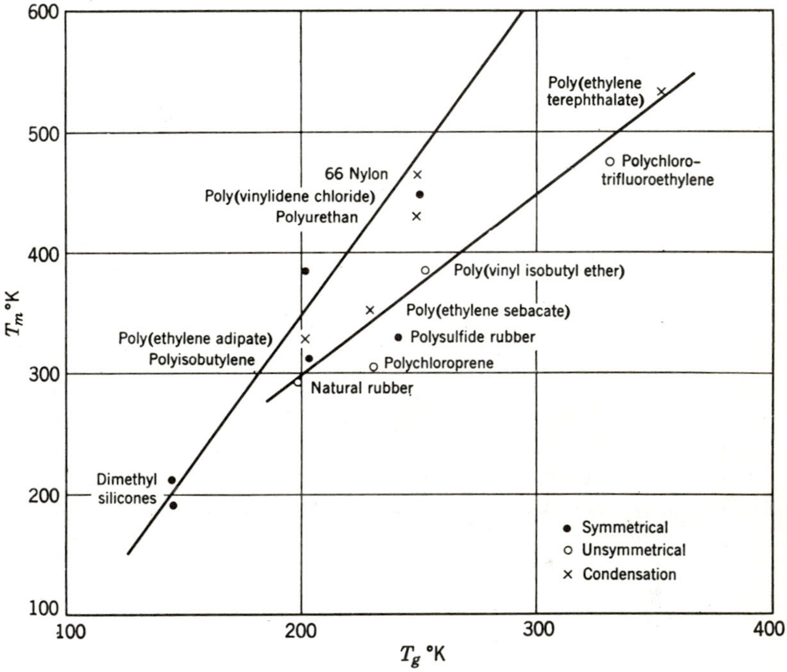

Figure 4-24. Correlation between melting and glass transition temperatures (ref 112).

polymers. In general, polymers whose molecular structures are inherently very flexible will show these phenomena at lower temperatures; while polymers whose molecular structures are inherently stiff, rigid, and inflexible will show these phenomena at higher temperatures. Since the shape of the modulus vs temperature curve is more or less the same for all these materials, it is theoretically possible to superimpose them upon each other simply by sliding the curves horizontally along the temperature axis. Furthermore all of these states are basically measurements of the rate at which segments of polymer molecules find the opportunity to move past each other; thus a slower rate of measurement moves the entire curve to lower temperatures, and a faster rate of measurement moves it to higher temperatures. Thus by changing the rate of testing, it is also possible to move the entire curve along the temperature axis and thus superimpose curves for different materials upon each

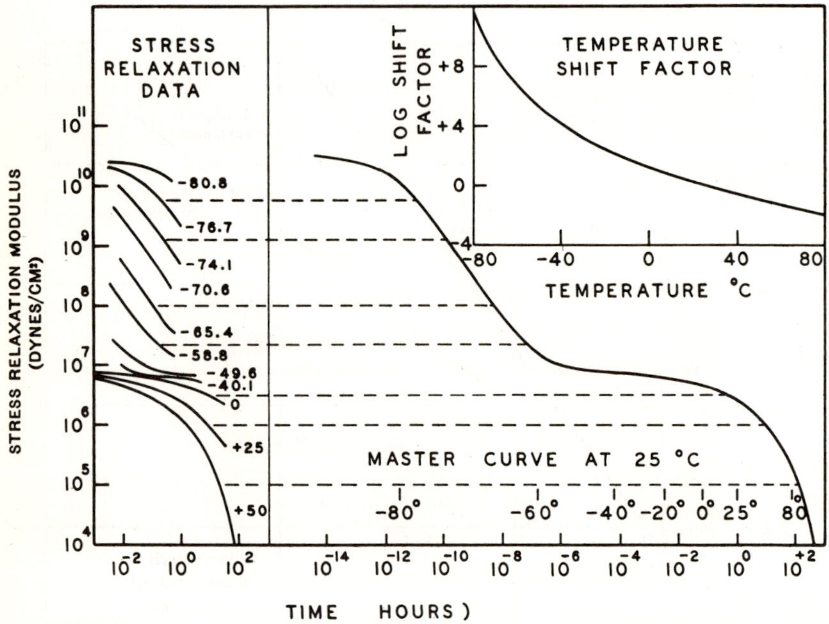

Figure 4-25. Time-temperature superposition principle (ref 117).
Time-temperature superposition principle illustrated with polyisobutylene data. The reference temperature of the master curve is 25°C. The inset graph gives the amount of curve shifting required at the different temperatures.

other. This is known as the time-temperature superposition principle (ref 113–116) (Figure 4-25).

General Applications

The glass transition temperature is thus the most general experimental measurement used in judging: (1) The inherent flexibility of the polymer molecule, (2) Its rate of response to mechanical stress at use temperatures, and (3) The general types of applicaions in which it will be most useful. A typical low-temperature rubber has a glass transition temperature of —85°C, a general-purpose rubber —57°, a latex paint +18°, a medium-impact rigid plastic 68°, and a hard brittle thermoplastic 100° (ref 118); while thermoset and high-temperature plastics go much higher.

Processability

Solid State Processing

Below their glass transition temperatures, polymers may be cut, machined, and bonded by adhesives; but changing their shape by actual

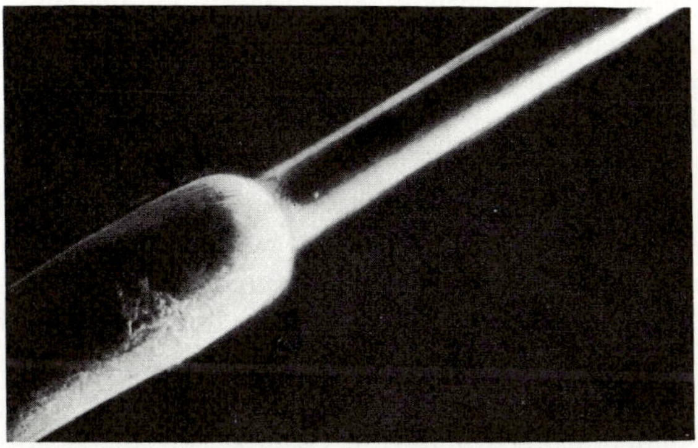

Figure 4-26. Cold drawing → necking of polyethylene fiber (ref 120).

flow is very difficult. Cold forming may be possible by creating sufficient internal friction within the plastic mass to warm it above the glass transition temperature. But it is probably more successfully carried out on semicrystalline polymers or block and graft copolymers, in which at least one micro-phase has enough molecular flexibility to be above its glass transition temperature, thus giving ductility to the entire material (ref 119); this will be discussed later.

Stretching of fibers is possible below the glass transition temperature, but at such low temperatures they do not stretch uniformly; rather, the entire change, from a thick unoriented fiber to a thin oriented one, occurs at a single sharply-moving shoulder or neck, and the process is generally referred to as necking (Figure 4-26). It is believed that when all of the stress and stretching are concentrated at this one narrow shoulder, the internal friction generates enough heat to raise this shoulder locally above the glass transition temperature and thus permit it to flow and stretch. The heat generated and the stress concentration then move continually to the next adjoining microsegment of unstretched fiber, and thus the stretching neck moves along the fiber as it progresses. Such cold-stretching probably works best on polymers whose molecules are linear and flexible.

Leathery State Processing

Stretching is generally carried out above the glass transition temperature of the particular molecular structure, where reasonably large segments of molecules have enough freedom to slip past each other under mechanical stress. On the other hand, the stretching temperature must not be so high that rapid motions of large segments of molecules occur spontaneously; because if they did, then Brownian motion of these large segments would negate any orientation that had been accomplished. In commercial practice, stretching is therefore carried out at a temperature moderately above the glass transition, and the stretched film or fiber is then chilled back below the glass transition temperature to freeze in the orientation which has been produced. Alternatively, in a heat-crystallizable material such as polyethylene terephthalate, the stretching of the amorphous polymer may be carried out slightly above the glass transition temperature, and the stretched material then exposed briefly to a higher temperature to induce crystallization and heat-setting which will also freeze in the stretched oriented structure that has just been created.

Melt Processing

Rubbery Flow Rubbery flow can be a distinct problem in the processing of molten polymers. If the molecular weight is too high or the temperature is not high enough for the molecules to disentangle flexibly and flow past each other easily, the flow of large segments of the molecules may be somewhat restrained by the remaining entanglements of other segments of the same molecules. When the rate of flow imposed on such a polymer is greater than the rate at which its segments can respond, melt fracture may occur, producing irregular rippled extrusions or even complete crumbling and breakage of the extrudate (Figure

4-27). In extrusion of film and extrusion coating of paper, rubbery flow also leaves residual strains which cause postshrinkage.

Liquid Flow In true flow of polymer melts and solutions, the flexibility of the individual polymer molecule determines the fluidity of the entire material. The more flexible the polymer chain, the less viscous the flow will be (ref 100,109,122,123). Highly flexible molecules, such

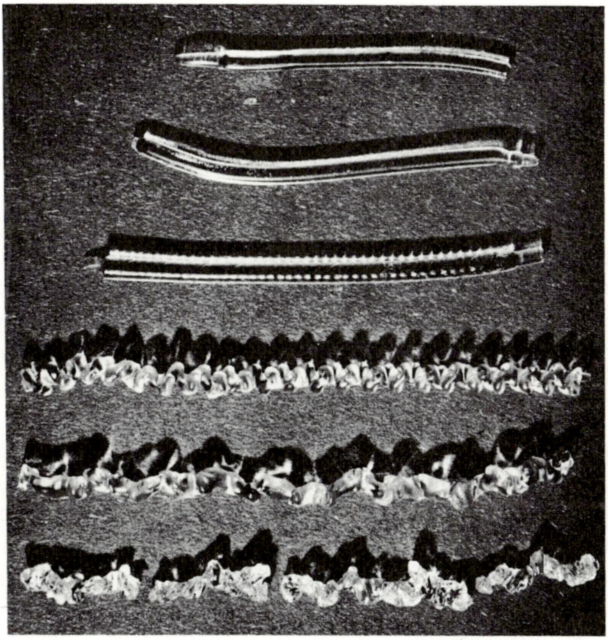

Figure 4-27. Melt fracture in extrusion of polymethyl methacrylate at increasing rates of stress (ref 121).

as polyethylene and nylon, form very fluid low-viscosity melts. Stiff polymer molecules such as polytetrafluoroethylene, polyphenylene oxides, polycarbonate, polyimides, and polybenzimidazole form very viscous melts of very difficult processability. In general the activation energy of melt viscosity may be a good measure of the stiffness of the polymer molecule (Table 4-8).

Latex Paint In the application of polyvinyl acetate latex paints, gradual evaporation of water leaves $\sim 1\mu$ particles of polyvinyl acetate,

Table 4-8. Activation Energy of Melt Viscosity (ref 124)

Polymer	Activation energy, kcal/gm mole
Polyethylene	11.0–12.8
Polyisobutylene	15.7–16.4
GR-S rubber	20.8
Polystyrene	22.0–23.0
Polyvinyl butyral	25.9
Polyvinyl chloride acetate	35.0 and 60.0
Cellulose acetate	70.0

which must flow together under the driving force of their surface tension, in order to form a continuous homogeneous paint film. The molecular flexibility of polyvinyl acetate (T_g = 86°F) is insufficient to permit such flow in normal or cool weather. It can be improved by adding liquid monomeric plasticizers, or preferably by copolymerization with flexibilizing acrylic esters, to increase molecular flexibility and lower T_g below the temperature at which the paint will be applied (Table 4-9).

Pseudoplasticity and Thixotropy The flow of molten polymers is generally non-Newtonian, producing lower viscosities at higher rates of flow and/or over longer periods of time. This may be explained by two mechanisms: (1) With increasing rates of flow, individual polymer molecules become more and more disentangled, and better able to flow past each other readily; and (2) The individual randomly-coiled poly-

Table 4-9. Copolymerization of Vinyl Acetate with Acrylic Esters to Decrease Glass Transition Temperature T_g and Minimum Temperature for Film Formation T_f (ref 125)

Acrylate Comonomer	Ethyl		Butyl		Ethyl hexyl	
%	T_f	T_g	T_f	T_g	T_f	T_g
10	65	60	58	55	60	52
15	62	53	54	45	53	40
20	58	50	50	36	47	29
25	56	47	47	28	41	20
30	54	44	43	22		

mer molecule may be considered as a rolling sphere, which undergoes doughy deformation at low speed, but rolls like a hard ball at higher speed (ref 126). Both these mechanisms obviously depend upon the structural flexibility of the polymer molecules.

Crystallizability and rate of crystallization are directly related to the flexibility of the polymer molecule and its ability to move into the precise positions required in the crystal lattice (ref 127). This subject will be discussed later.

Mechanical Properties

In general, the flexibility of the individual polymer molecule is directly related to the flexibility of the entire polymeric material (ref 100,103). This is a primary relationship in (1) reversible mechanical properties, and a major contributing factor in (2) ultimate mechanical properties and failure.

Reversible Mechanical Properties

The temperature and flexibility of the polymer molecule together determine the size of the molecular segments which are in motion, and the frequency with which they occur, and thus together determine the rate at which the polymer molecule can respond to mechanical stress. Thus modulus is a function not only of temperature and molecular flexibility, but also of the speed or frequency with which the mechanical load is applied (ref 128). Probably the most extreme example of this phenomenon is the fact that, in aircraft flying at supersonic speeds, the impact of raindrops occurs at such high rate of speed that even the monomeric water molecules cannot flow past each other rapidly enough for the raindrop to act as a normal liquid, and instead it produces sharp high impacts which cause serious rain erosion of the aircraft surface.

The stiff molecules of rigid glassy polymers, which are normally below their glass transition temperatures, can respond to mechanical stress only by bending and stretching of the individual interatomic bonds (ref 129). These are very short, fast, elastic responses up to the point of brittle failure (ref 130). Polystyrene and polymethyl methacrylate are typical of such structures and behavior.

In fairly stiff molecules which are normally in the transitional region around T_g, molecular segments can slip past each other to some extent. This can produce undesirable creep under sustained long-term mechanical load (ref 131,132); it also converts brittle polymers into tough ductile ones. Polyvinyl acetate and polymethyl acrylate are typical of this range.

With somewhat higher segment mobility the polymer develops a leathery flexibility. In this transitional region much of the energy from the applied mechanical stress is converted into internal friction and heat. Such restricted mobility in polyurethanes and plasticized polyvinyl chloride produces considerable energy absorption, which is useful in crash padding.

In flexible molecules which are normally above the glass transition temperature, reasonably large segments can disentangle and slip past each other quite rapidly in response to mechanical stress, and the polymer is softly flexible and rubbery in behavior; all elastomers and highly plasticized plyvinyl chloride belong to this class. When mechanical stress is applied, molecular segments slip past each other, and random coils disentangle and stretch in response to the stress. When the stress is released, the potential energy stored in the uncoiled molecule is released by recoiling back to a more probable arrangement, releasing the energy and producing rubbery elasticity (ref 29,133,134). Actually, even here, some of the mechanical energy is absorbed by viscous flow and dissipated by conversion to heat. In pneumatic tires, for example, the elastic energy produces springiness. While the viscous dissipation is responsible for a smoother ride (ref 137), it also causes overheating in large bus and truck tires.

Mechanical Failure

Molecular flexibility in general permits the polymer to respond more readily to lower mechanical stresses. Thus flexible polymer molecules, such as amorphous polyolefins, diene polymers, and polyacrylic esters, generally have low tensile yield strengths and ultimate tensile strengths, and high ultimate elongations. Stiff polymer molecules, such as polystyrene and polymethyl methacrylate, have higher ultimate strengths and low ultimate elongations. Correspondingly, slow rates of testing will permit the molecule to respond flexibly and demonstrate softer properties, while fast rates of testing will simulate stiffer molecules and harder properties (ref 138). Ultimate failure, however, is a complex phenomenon and depends upon many factors in addition to simple molecular flexibility, so these generalizations are not very precise and cannot be carried too far.

Brittle response to the rapid application of mechanical stress is a still more complex phenomenon. In stiff polymers such as polystyrene and polymethyl methacrylate, which are normally below the glass transition temperature, molecular segments are completely unable to disentangle and respond to the rapid application of mechanical stress, and impact produces brittle failure. In flexible polymers such as elastomers and plasticized vinyls, which are normally above the glass transition temperature, large segments of the molecule can disentangle and respond rapidly to mechanical stress, and the polymer has high impact strength. In the intermediate leathery transition region of stiffly flexible molecules, impact strength depends upon the relative rate of application of mechanical stress versus rate of response of molecular segments. Thus impact strength and brittleness are directly related to the combined effects of molecular flexibility and rate of test. At low rates of impact, relatively

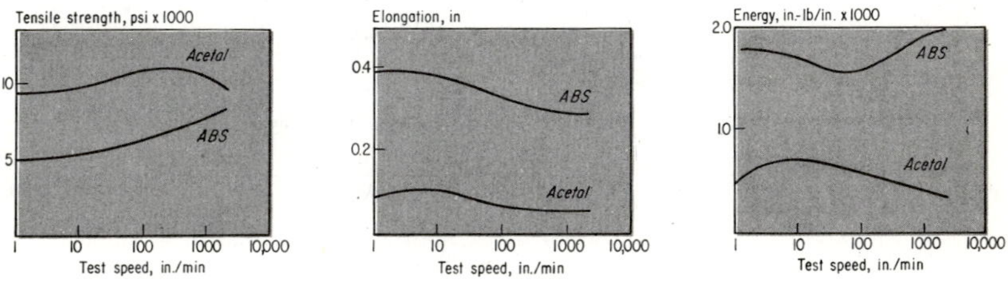

Figure 4-28. High-speed tensile tests (ref 139).
High speed tensile tests on high-impact ABS and acetal show that energy, or product of tensile and elongation, is strongly affected by strain rate.

stiff leathery materials can still have good impact strength; while at very high rates of test, even rubbery materials may exhibit brittle failure.

These simple fundamental principles become very complicated and often masked in practice. The basic theory applies primarily to amorphous, linear, homogeneous polymer systems. In practice, most commercial systems are modified by intermolecular crystallinity, hydrogen-bonding, and cross-linking, and by supermolecular heterogeneity. These effects include but overshadow the simple effects of molecular flexibility alone. The mechanics of testing and the interpretaton of test measurements may also complicate the understanding of both fundamental and practical properties. For example, the effects of rate of test on tensile and puncture properties may be very complex and difficult to interpret (ref 139) (Figures 4-28, 4-29). Thus many more fundamental studies on simplified model systems will be required to define clearly what molecular flexibility does to overall polymer properties.

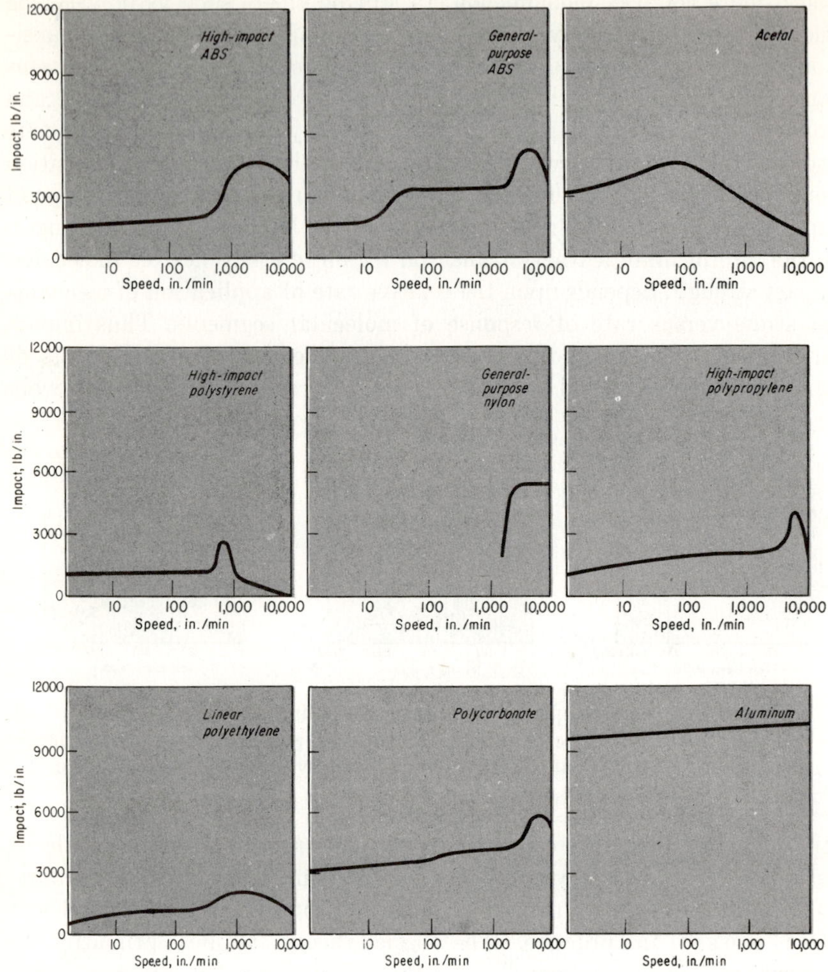

Figure 4-29. High-speed puncture impact tests (ref 139). Puncture impact resistance for the nine materials at strain rates up to 10,000 in./min. Except for aluminum, energy absorption reaches a maximum at some critical speed.

Acoustic Properties

Impact between two rigid materials produces rapid short-range bending and stretching of interatomic bonds and rapid elastic response, thus producing high-frequency sound waves in the adjacent air. In many cases such noise can be disturbing or even painful. Whether it is the clatter of leather heels on hard floors, the moving of furniture, dropping of dishes, or operation of motors or high-speed machinery, the replacing or padding of one of these hard surfaces with a softer type of polymeric material, either leathery or rubbery in nature, converts some of the rapid mechanical energy into a slower viscoelastic uncoiling of the polymer molecules, thereby converting some of the mechanical energy into heat and thus deadening the sound or at least lowering its frequencies below the pain or even the hearing level (ref 137). This is particularly useful in flooring, machinery mountings, and door gasketing.

When noise already exists, it can be damped, absorbed, and minimized by a dam of viscoelastic material whose molecules respond sluggishly at the frequencies of the objectionable sound waves, thus converting much of the sound energy into heat. Since energy absorption also depends upon mass, the best noise control is often achieved by suspending a high-mass heavy material, such as lead, in a viscoelastic polymer such as polyurethane foam or plasticized vinyl sheet (ref 140–142). Important applications are in factories, public buildings, apartment houses, and aircraft.

On the other hand, in applications, such as loud speakers, where it is desirable to reflect sound completely and precisely, polymers of low molecular flexibility and high rigidity, such as polystyrene and phenolic resins, are the most desirable.

Electrical Properties

Molecular flexibility has a number of distinct effects upon the (1) Resistance and (2) Dielectric Properties of polymeric materials.

Resistance

Conduction of electricity through most polymers is due to the electrolytic motion of ionic impurities. Thus it requires both the presence of ions and sufficient mobility for them to migrate through the material. Low mobility is essential in electrical insulation (ref 143). In fundamental measurements resistivity in analogous to viscosity (ref 144). Addition of liquid plasticizers to polyvinyl chloride produces increased mobility, as evidenced by the increase in mechanical flexibility and the correspondng decrease in electrical resistance (Table 3-3). Similar phenomena can probably be observed in comparison of rubbery versus rigid materials (ref 167), use of plasticizers in other polymers, and effects of known concentrations ionic impurities.

Dielectric Properties

When an electric field is applied to a polymer molecule, polar groups in the molecule tend to orient in the direction of the field (Figure 4-30). If the polar groups are sufficiently flexible to orient in this way, they produce a material high in dielectric constant. If they are not, the material has low dielectric constant. If they have intermediate flexibility which permits only a slow partial response, some of the electrical energy is converted into heat and measured as dielectric loss. In an alternating

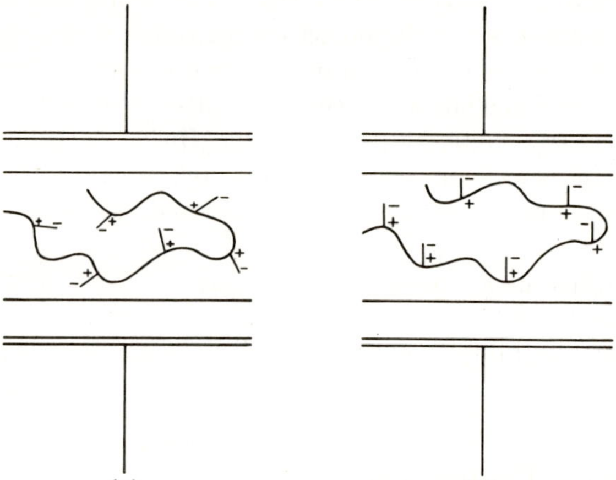

Figure 4-30. Orientation of polar groups on a polymer in an electric field (ref 145). (a) No field. (b) Field applied, polymer becomes polarized.
Turner Alfrey and Edward F. Gurnee, *Organic Polymers*, © 1967. Reprinted by permission of Prentice-Hall, Inc., Englewood Cliffs, New Jersey.

electric field (Figure 4-31), a particular polar group will be flexible enough to respond at low frequency but not flexible enough to respond at high frequency, and will exhibit most loss in the transitional region where it is changing from high-response-at-low-frequency to low-response-at-high-frequency (ref 164,165) (Figure 3-50). The relationship between molecular flexibility and polarizability is most readily observed by increasing the temperature to increase the molecular flexibility, and then noting the corresponding increase in dielectric constant, ability to respond at higher frequencies, and the accompanying shift in dielectric loss peaks (ref 147,148) (Figure 4-32).

When the polar groups are in or on the main-chain of the polymer molecule, the electrical transition occurs at about the glass transition temperature, dielectric constant is analogous to elastic compliance, and electrical loss is analogous to mechanical loss (ref 144,150,151). When the polar group is on a side chain of the polymer molecule, it may remain flexible to much lower secondary transition temperatures (ref 144,150–152).

Thus the molecular flexibility of the polymer must be carefully chosen and controlled to obtain the specific dielectric properties required for any particular electrical application. In actual practice, however,

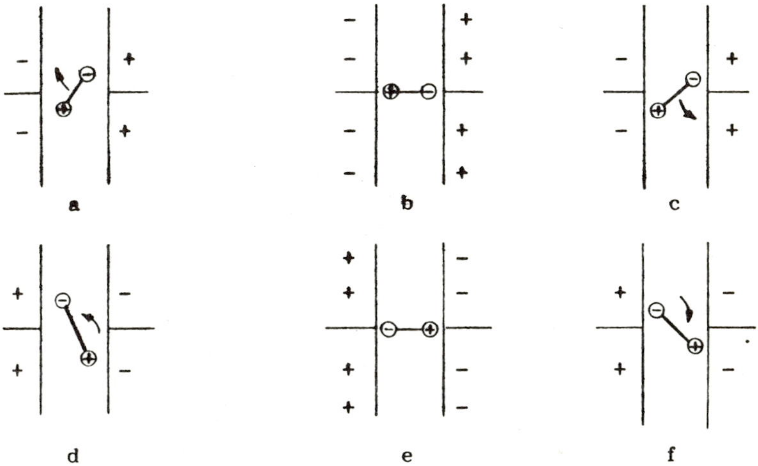

Figure 4-31. Response of polar groups on a polymer to an alternating electric field (ref 146).

dielectric properties may be much more complex (Figures 4-33, 4-34), indicating that much further study, both theoretical and applied, is needed for fuller understanding and control.

Optical Properties

The bending, stretching, and rotation of individual interatomic bonds occur at frequencies comparable with the infrared region of the optical spectrum. Different individual interatomic bonds have different characteristic frequencies and absorb different characteristic infrared

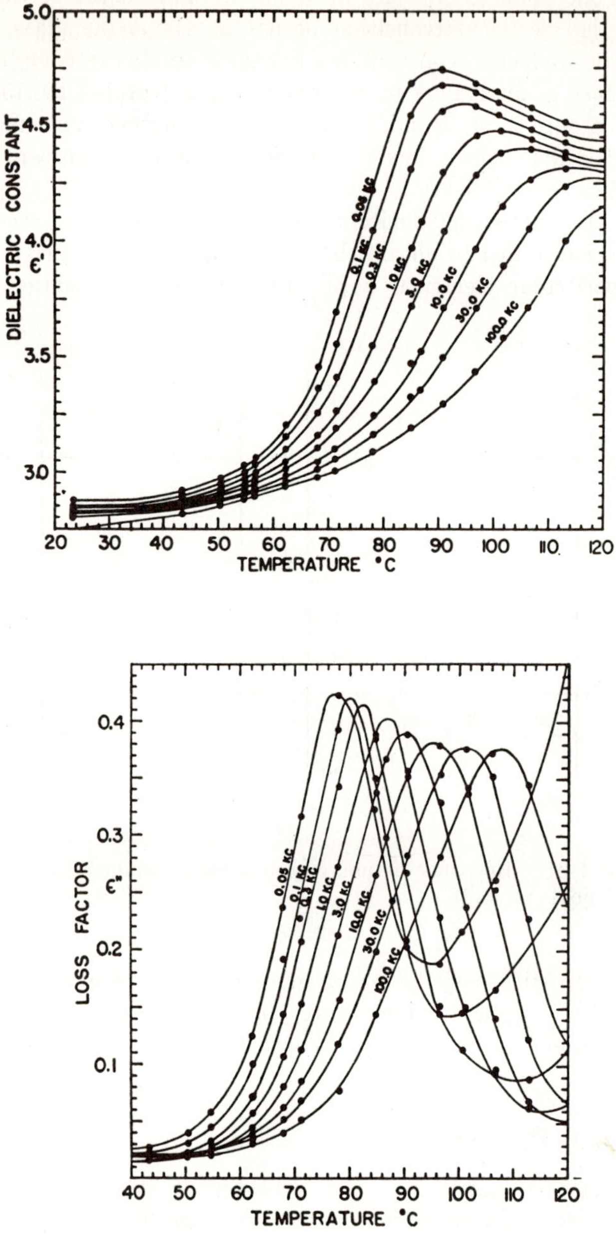

Figure 4-32. Temperature → molecular flexibility → dielectric response for polyvinyl butyral (ref 149).

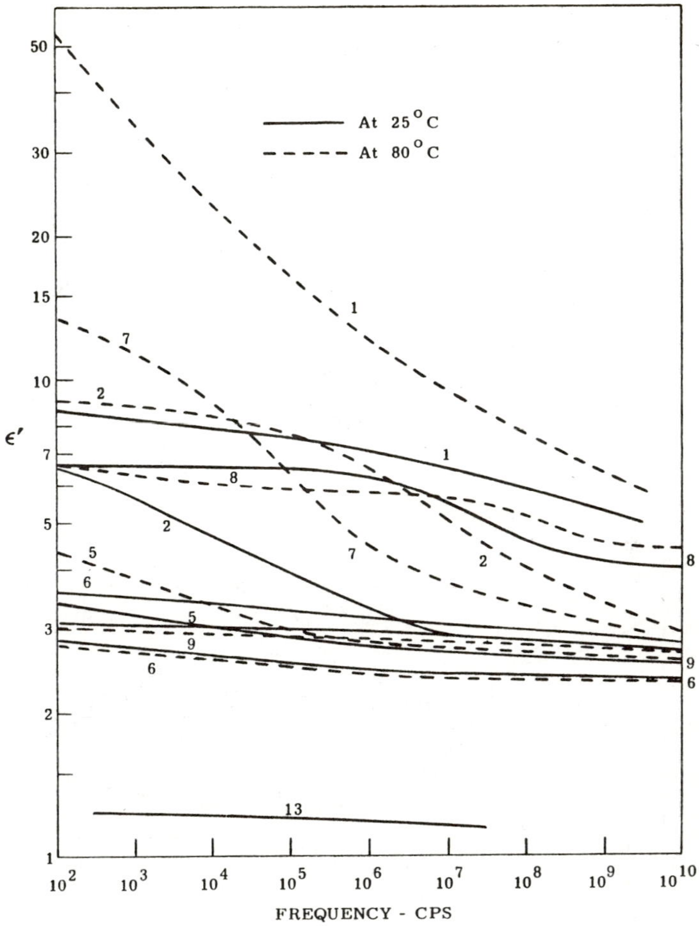

Figure 4-33. Dielectric constants of commercial polymers (ref 153). See figure 4-34 for legend.

wavelengths, (Figure 4-35). This is particularly useful in the applications of infrared spectroscopy to the analysis and characterization of polymer structure (ref 156).

Chemical Properties

Absorption of low-molecular-weight monomeric species into polymers depends upon some mobility of the polymer chains to accept

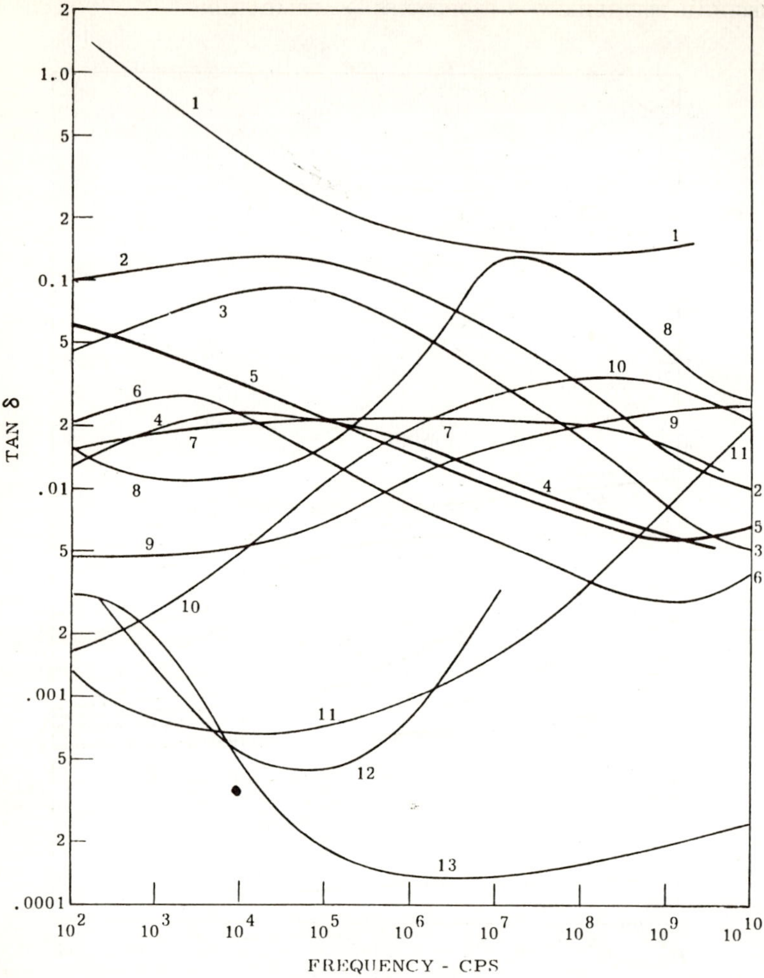

Figure 4-34. Dissipation factors of commercial polymers (ref 154).

1—Clear cast phenolic resin
2—Plasticized polyvinyl chloride
3—Copolymer-polyvinyl and vinylidene chloride (Dow's Saran)
4—Unplasticized polyvinyl chloride
5—Polymethylmethacrylate
6—Polychlorotrifluoroethylene
7—Hexamethylene-adipamide (nylon)
8—Compounded poly-2-chlorobutadiene-1,3 (neoprene)
9—Plasticized ethyl cellulose
10—Epoxy casting resin (Ciba's "Araldite B")
11—Silicone rubber (GE's SE–550)
12—Polyurethane foam (density = .033)
13—50% polystyrene-50% chlorinated diphenyl

these foreign molecules. Thus chain stiffness may retard absorption, particularly of larger molecules (ref 157).

More generally, however, it is diffusion and permeability which are primarily dependant upon molecular flexibility (ref 166). Small molecules diffuse through a polymeric matrix by jumping into the transient holes which cause the free volume in the polymer structure. To the extent that the polymer molecules are flexible and their segments are

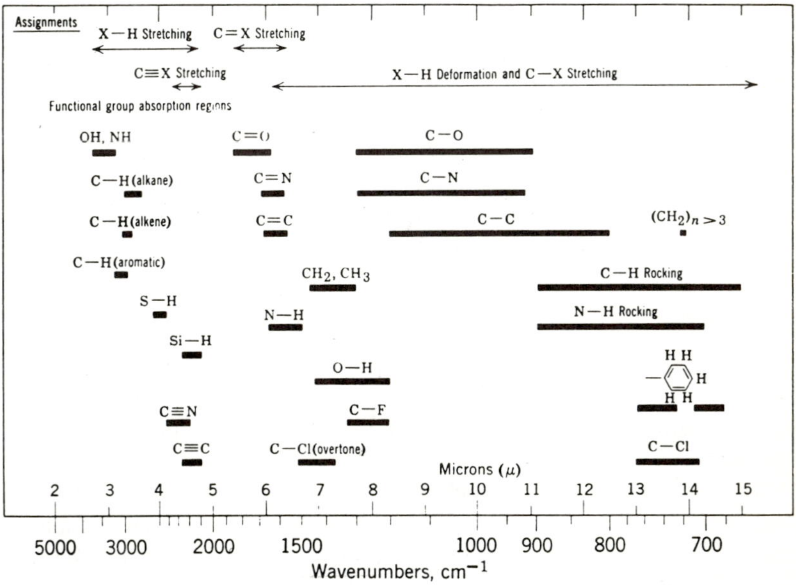

Figure 4-35. Infrared absorption bands used in characterizing polymers (ref 155).

in thermal motion, the location of these holes moves randomly like Brownian motion. The small diffusing molecules can jump from one hole to the other, moving with the gradient from high to low concentration, and thus producing diffusion and permeation through the polymeric material (ref 158–161) (Table 4-10). Thus permeability is greater in polymers above their glass transition temperatures, and increases with temperature. Practical diffusion data confirm this theory roughly, but show many complicating effects due to polarity, intermolecular, and supermolecular structure.

Table 4-10. Permeability Data for Various Polymers: Values for $P \times 10^{10}$ cm.^{3} sec.$^{-1}$ mm. cm.$^{-2}$ cm. Hg^{-1} (ref 162)

Polymer	Permeability				Nature of Polymer
	N_2(30°C)	O_2(30°C)	CO_2(30°C)	90 % R.H. H_2O(25°C)	
Poly(vinylidene chloride)	0.0094	0.053	0.29	14	Crystalline
PCTFE	0.03	0.10	0.72	2.9	Crystalline
Poly(ethylene terephthalate)	0.05	0.22	1.53	1,300	Crystalline
Rubber hydrochloride (Pliofilm ND)	0.08	0.30	1.7	240	Crystalline
Nylon 6	0.10	0.38	1.6	7,000	Crystalline
PVC (unplasticized)	0.40	1.20	10	1,560	Slight crystalline
Cellulose acetate	2.8	7.8	68	75,000	Glassy
Polyethylene (d = 0.954–0.960)	2.7	10.6	35	130	Crystalline
Polyethylene (d = 0.922)	19	55	352	800	Some crystalline
Polystyrene	2.9	11	88	1,200	Glassy
Polypropylene	—	23	92	680	Crystalline
Butyl rubber	3.12	13.0	51.8	—	Rubbery
Methyl rubber	4.8	21.1	75	—	Rubbery
Polybutadiene	64.5	191	1,380	—	Rubbery
Natural rubber	80.8	233	1,310	—	Rubbery

Price

There is no primary fundamental relationship between molecular flexibility and price. There have, however, been some major historical factors at different times in the commercial development of the polymer industry.

In the earlier years of polymer technology, the ready availability of natural rubber at low cost fulfilled most of the need for flexible materials. In Germany during World War I, and in most of the world during World War II, natural rubber was not available; and the initial research, development, and production of synthetic rubber and flexible plastics were quite expensive. Public support, and continued development and improvement in production, eventually made flexible synthetics very readily available at very modest cost.

The early development of common thermosetting plastics made rigid plastic materials readily and inexpensively available in the early part of the twentieth century. Easy-processing thermoplastics were somewhat less rigid, particularly at higher use temperatures. The development of new engineering thermoplastics, useful to higher temperatures, has progressed far enough to show that this is perfectly feasible, but not yet far enough to make them low in price. Even more advanced development of ultra-high-temperature materials, such as polytetrafluoroethylene and especially the new polyheterocycles for aerospace applications, is still in such an elementary state that they are not readily available, difficult to process, and very high in cost. By extrapolation from past historical experience, however, there is no reason why they cannot ultimately become just as readily available and low in cost as the common commodity materials.

Balance of Properties

Each particular application requires certain specific properties, many of which depend at least partly upon molecular flexibility. When these demands conflict, the choice of optimum molecular flexibility must compromise its effects upon different properties.

In most polymer systems, molecular flexibility is sufficient for normal processing. In high-temperature polymers, however, such as polytetrafluoroethylene, polyphenylene oxides, and especially the ultra-high-temperature heterocycles, the molecular rigidity which provides hot strength also causes very high melt viscosities and thus very difficult melt processing.

More often the conflict comes between the need for molecular rigidity to provide high modulus, strength, and heat distortion temperature, vs the need for molecular flexibility to provide resistance to brittle failure under high-speed impact. This conflict is most commonly resolved by use of either: partly crystalline flexible molecules; or heterogeneous systems containing either rubbery particles or fibers to absorb the shock of impact on the rigid plastic matrix. These types of structures will be discussed later.

As has been pointed out several times, the effects of molecular flexibility are frequently complicated by interaction with other structural features. This makes them more difficult to understand in theory, but it offers a greater variety of ways to achieve the desired balance of practical properties for any particular application. These will be observed repeatedly throughout this study.

REFERENCES

1. Pauling, Linus, *The Nature of the Chemical Bond,* Cornell University Press, 1942, pp 81–85.
2. Schmidt, A. X., & Marlies, C. A., *Principles of High Polymer Theory and Practice,* McGraw-Hill, 1948, p 27.
3. Billmeyer, F. W., Jr., *Textbook of Polymer Science,* Interscience, 1962, p 9.

4. *Ibid.*, pp 27–29.
5. Flory, P. J., *Principles of Polymer Chemistry,* Cornell University Press, 1953, Chapter 10.
6. Meares, Patrick, *Polymers: Structure and Bulk Properties,* Van Nostrand, 1965, Chapter 7.
7. *Ibid.*, p 186.
8. Miller, M. L., *The Structure of Polymers,* Reinhold, 1966, p 254.
9. Meares, *op. cit.*, p 185.
10. Billmeyer, *op. cit.*, p 10.
11. Meares, *op. cit.*, pp 243–49.
12. *Ibid.*, pp 257–58.
13. Tobolsky, A. V., *Properties and Structure of Polymers,* Wiley, 1960, pp 74–77.
14. Brydson, J. A., *Plastics Materials,* Iliffe, 1966, pp 34–37, 45–51.
15. Golding, Brage, *Polymers and Resins,* Van Nostrand, 1959, pp 551–52.
16. Billmeyer, *op. cit.*, p 181.
17. Flory, *op. cit.*, Chapter 11.
18. Tobolsky, *op. cit.*, pp 23–26.
19. Miller, *op. cit.*, p 255.
20. Brydson, *op. cit.*, p 47.
21. Miller, *op. cit.*, p 259.
22. Brydson, *op. cit.*, p 48.
23. Miller, *op. cit.*, pp 255–59.
24. Billmeyer, *op. cit.*, pp 212–20.
25. Miller, *op. cit.*, pp 253–54.
26. Golding, *op. cit.*, p 547.
27. *Ibid.*, p 553.
28. Miller, *op. cit.*, p 299.
29. Meares, *op. cit.*, p 163.
30. Nielsen, L. E., *Mechanical Properties of Polymers,* Reinhold, 1962, p 38.
31. Alfrey, Turner, & Gurnee, E. F., *Organic Polymers,* Prentice-Hall, 1967, pp 37–40.
32. Nielsen, *op. cit.*, pp 36–38.
33. Alfrey, *op. cit.*, pp 37–39.
34. Billmeyer, *op. cit.*, pp 217–18.
35. Brydson, *op. cit.*, pp 48–50.
36. MacFarlane, S. B., *Technology of Synthetic Fibers,* Fairchild Publications, 1953, pp 87–88.
37. Nielsen, *op. cit.*, pp 37–38.
38. Billmeyer, *op. cit.*, p 218.
39. Frazer, A. H., *High Temperature Resistant Polymers,* Interscience, 1968, pp 275–82.
40. Alfrey, *op. cit.*, p 37.
41. Billmeyer, *op. cit.*, p 215.
42. Nielsen, *op. cit.*, pp 38–39.

43. Pauling, *op. cit.,* p 85.
44. Miller, *op. cit.,* pp 299–300.
45. Nielsen, *op. cit.,* p 22.
46. Meares, *op. cit.,* p 165.
47. Miller, *op. cit.,* p 131.
48. Billmeyer, *op. cit.,* pp 218–20.
49. Golding, *op. cit.,* p 555.
50. Meares, *op. cit.,* pp 259–61.
51. Miller, *op. cit.,* p 300.
52. *Ibid.,* pp 288, 531.
53. Nielsen, *op. cit.,* p 25.
54. Billmeyer, *op. cit.,* pp 219–20.
55. Brydson, *op. cit.,* pp 48–49.
56. Miller, *op. cit.,* p 534.
57. White, R. A., & Deanin, R. D., U.S. Patent 3,253,932 (1966).
58. White, R. A., & Deanin, R. D., U.S. Patent 3,305,507 (1967).
59. Billmeyer, *op. cit.,* p 219.
60. Golding, *op. cit.,* pp 555–56.
61. Meares, *op. cit.,* p 258.
62. Miller, *op. cit.,* p 294.
63. Lenz, R. W., *Organic Chemistry of Synthetic High Polymers,* Interscience, 1967, pp 743–44.
64. Miller, *op. cit.,* p 66.
65. *Ibid.,* pp 530–31.
66. Nielson, *op. cit.,* pp 15–25.
67. *Ibid.,* p 24.
68. Miller, *op. cit.,* p 299.
69. Nielsen, *op. cit.,* p 33.
70. *Ibid.,* pp 22–24.
71. Billmeyer, *op. cit.,* p 125.
72. Flory, *op. cit.,* pp 629–37.
73. Miller, *op. cit.,* Chapter 12.
74. *Ibid.,* p 587.
75. Davidson, R. L., & Sittig, Marshall, *Water-Soluble Resins,* Reinhold, 1968, pp 10–11.
76. Golding, *op. cit.,* pp 543–45.
77. Brydson, *op. cit.,* pp 45–51.
78. Modern Plastics Encyclopedia 1968, McGraw-Hill, 1967, pp 14–15.
79. Tobolsky, *op. cit.,* pp 78–79.
80. *Ibid.,* p 79.
81. Modern Plastics Encyclopedia, *op. cit.,* p 15.
82. Nielsen, *op. cit.,* pp 171–72.
83. *Ibid.,* p 171.
84. Meares, *op. cit.,* pp 261–63.
85. *Ibid.,* p 263.

86. Miller, *op. cit.*, p 287.
87. Wakeman, R. L., *The Chemistry of Commercial Plastics,* Reinhold, 1947, pp 322–25.
88. *Ibid.*, pp 340–44.
89. Monsanto Technical Bulletin, *Plasticizers.*
90. Wakeman, R. L., *op. cit.*, p 341.
91. *Ibid.*, p 343.
92. Brydson, *op. cit.*, p 51.
93. Baer, Eric, *Engineering Design for Plastics,* Reinhold, 1964, p 547.
94. Billmeyer, *op. cit.*, p 203.
95. Nielsen, *op. cit.*, p 181.
96. Billmeyer, *op. cit.*, pp 203–04.
97. Miller, *op. cit.*, pp 300–01.
98. Nielsen, *op. cit.*, pp 44–45.
99. *Ibid.*, pp 177–80.
100. Billmeyer, *op. cit.*, pp 208–09.
101. Miller, *op. cit.*, p 281.
102. *Ibid.*, p 284.
103. *Ibid.*, pp 298–300.
104. Modern Plastics Encyclopedia, *op. cit.*, p 14.
105. Nielsen, *op. cit.*, pp 16–21.
106. Larsen, P. J., *Product Engineering,* pp 54–71 (1/8/62).
107. *Mat. Des. Eng.*, pp 119–34 (8/65).
108. Golding, *op. cit.*, p 545.
109. Miller, *op. cit.*, p 278.
110. Nielsen, *op. cit.*, pp 36–39.
111. Billmeyer, *op. cit.*, p 212.
112. *Ibid.*, p 213.
113. Alfrey, *op. cit.*, pp 73–74.
114. Baer, *op. cit.*, pp 164–72.
115. Nielsen, *op. cit.*, pp 89–92.
116. Tobolsky, *op. cit.*, pp 141–59.
117. Nielsen, *op. cit.*, p 90.
118. Tobolsky, *op. cit.*, p 80.
119. Nielsen, *op. cit.*, pp 172–73.
120. Billmeyer, *op. cit.*, p 165.
121. *Ibid.*, p 177.
122. Brydson, *op. cit.*, p 57.
123. Miller, *op. cit.*, p 280.
124. Berhardt, E. C., *Processing of Thermoplastic Materials,* Reinhold, 1959, p 40.
125. Burrell, Harry, *The Glass Transition in Coating Formulations, Official Digest of the Federation of Societies for Paint Technology,* February, 1962.
126. Meares, *op. cit.*, pp 359–60.
127. Alfrey, *op. cit.*, pp 36–37.

128. Miller, *op. cit.*, pp 280–81.
129. Billmeyer, *op. cit.*, p 172.
130. Alfrey, *op. cit.*, pp 91–92.
131. Nielsen, *op. cit.*, p 47.
132. *Ibid.*, pp 53–54.
133. Billmeyer, *op. cit.*, p 230.
134. Brydson, *op. cit.*, pp 34–35.
135. *Ibid.*, p 45.
136. Meares, *op. cit.*, pp 161–63.
137. Miller, *op. cit.*, p 237.
138. *Ibid.*, p 305.
139. Furno, F. J., Webb, R. S., & Cook, N. P., "High-Speed Tests Evaluate Toughness of Plastics," in *Product Engineering*, August 17, 1964.
140. International Lead Zinc Research Organization, Technical Bulletins.
141. Ferro Corporation, Cordo Division, Technical Data Bulletin on Coustifab.
142. Soundcoat Company, Technical Bulletins.
143. Volk, M. C., Lefforge, J. W., & Stetson, Russell, *Electrical Encapsulation*, Reinhold, 1962, p 14.
144. Nielsen, *op. cit.*, p 210.
145. Alfrey, *op. cit.*, p 121.
146. Baer, *op. cit.*, p 531.
147. *Ibid.*, pp 539–44.
148. Nielsen, *op. cit.*, pp 211–13.
149. *Ibid.*, p 212.
150. Brydson, *op. cit.*, p 90.
151. Nielsen, *op. cit.*, p 214.
152. Tobolsky, *op. cit.*, pp 172–73.
153. Baer, *op. cit.*, p 551.
154. *Ibid.*, p 548.
155. Billmeyer, *op. cit.*, p 91.
156. *Ibid.*, pp 90–96.
157. Baer, *op. cit.*, pp 649–51.
158. *Ibid.*, pp 623–25.
159. *Ibid.*, pp 651–52.
160. Brydson, *op. cit.*, pp 80–81.
161. Meares, *op. cit.*, Chapter 12.
162. Brydson, *op. cit.*, p 82.
163. Stille, J. K., *Introduction to Polymer Chemistry*, Wiley, 1962, p 171.
164. Baer, *op. cit.*, pp 533–34.
165. Nielsen, *op. cit.*, pp 210–14.
166. Baer, *op. cit.*, p 647.
167. Kinney, G. F., *Engineering Properties and Applications of Plastics*, Wiley, 1961, p 225.

Part Four

INTERMOLECULAR STRUCTURE

The size and flexibility of the polymer molecule tell us how an individual molecule would behave if it were completely isolated from its neighbors. Such isolated molecules are encountered only in theoretical studies on dilute solutions. In actual practice, polymer molecules always occur in a mass, and the behavior of each individual molecule is very greatly affected by its relationship to adjacent polymer molecules in the mass. These intermolecular relationships between adjacent molecules may be divided into two types: intermolecular order, which describes the geometrical arrangement of adjacent molecules in space; and intermolecular bonding, which describes the attractive forces between adjacent molecules in the polymeric mass. Together these two types of intermolecular relationships modify the simple effects of molecular size and flexibility, and determine the overall behavior of homogeneous polymeric materials.

Chapter Five

Intermolecular Order

Intermolecular order is the geometrical relationship between adjacent polymer molecules and their arrangement in the total mass of polymeric material. At present, polymer science recognizes three distinctly different states of intermolecular order (Figure 5-1).

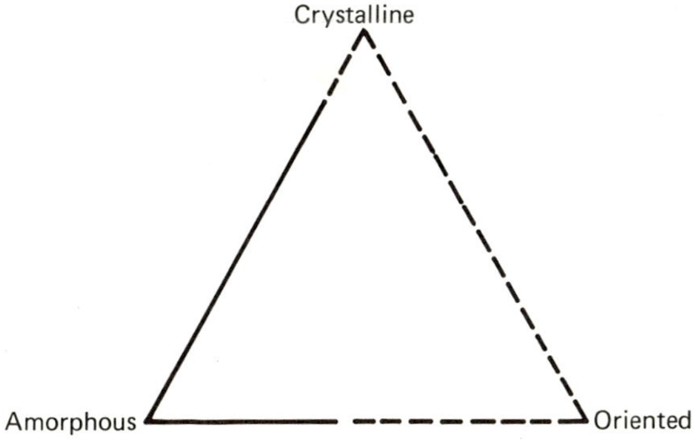

Figure 5-1. States of intermolecular order.

When polymer molecules are arranged in completely random, intertwined coils, this completely unordered structure is known as the amorphous state. When polymer molecules are so neatly arranged that each of their atoms falls into precise position in a tight-packed repeating regular structure, this highly ordered structure is described as the crystalline state. When polymer molecules are stretched into a rather linear conformation and lie fairly parallel to each other in the mass, this partially ordered structure is described as the oriented state.

In practice, some polymers are completely amorphous; others are partly amorphous and partly crystalline; others are partly amorphous and partly oriented; and still others are partly amorphous, partly crystalline, and partly oriented. At present it seems unlikely that polymers can be obtained in completely crystalline or completely oriented state. Each of these three states has very distinctive effects upon polymer properties, and polymers which contain combinations of these states show their combined effects.

5-1 AMORPHOUS STATE

The preceding discussion of molecular flexibility (Chapter 4) described the random coiling and behavior of individual polymer molecules. Implicit in the entire discussion, however, was the fact that these randomly-coiled molecules are all intertwined and intertangled with each other, and that their behavior in response to any type of stress is largely a measure of the ease or reluctance with which they can disentangle to relieve that stress. Typical commercial polymers include polystyrene, polyvinyl acetate, polymethyl methacrylate, and most random copolymers and elastomers.

Thus the discussion of the flexibility of the individual polymer molecule has already implicitly described the amorphous state in considerable detail, and no further discussion is necessary at this point. Rather, what is now important is to consider the ways in which the crystalline and oriented states differ from the random amorphous state in their structure and their effects upon polymer properties.

5-2 CRYSTALLINITY

The Nature of Crystallinity in Polymers

Conventional Crystallinity in Monomeric Materials

In a single pure monomeric species at low temperature, the individual molecules tend to pack into a neat orderly crystalline arrangement of high density, sharp melting point, and low solubility. The stability of such a highly ordered structure at low temperatures and energy levels is demonstrated by the heat of fusion or enthalpy ΔH which is liberated during the formation of the crystal, and which must be supplied in order to melt and disperse the molecules out of the crystal. At higher temperatures and energy levels, on the other hand, vibrations and Brownian motions of the molecules tend to drive them apart, destroying the crystals and dispersing them into a highly random and disordered state containing many degrees of freedom as represented by the entropy ΔS. The balance between these two factors determines the melting point of each particular material. Thus $Tm = \Delta H/\Delta S$ (ref 1,2).

Complexity → Restricted Crystallinity in Polymeric Materials

These same forces favoring crystallization at low temperatures and low energy levels, and favoring melting at high temperatures and high energy levels, fundamentally apply to polymers in the same way as they do to monomers. In practice, however, the polymer molecule is much larger than simple monomeric species, often by 3 or 4 orders of magnitude, perhaps some 10^4 Å in size (ref 3). When a polymer crystallizes, it is not the whole molecule which fits into the crystal lattice, but rather the individual atoms of the molecule, each of which finds its own particular place at specific points in the crystal lattice. As compared with conventional monomeric systems, polymers are very complex, containing a variety of structural features and molecular species. Their long-chain nature restricts the mobility of their individual atoms, and the rate at which they can move into the required locations in the crystal lattice. Thus it is natural that polymers crystallize more slowly and less completely, and form crystals which are smaller and less perfect, than typical monomeric materials.

Molecular Conformation in the Crystal Lattice

A randomly-coiled polymer molecule cannot fit neatly into a regular geometrical crystalline lattice, and must change to a much more linear and regular shape. The normal tetrahedral 109° angle of the carbon atom prevents the polymer molecule from becoming completely linear in conformation. In most cases polymers crystallize by assuming either a zigzag or a helical conformation of the polymer (Figures 5-2, 5-3), both of which are more linear and regular than the random coil (ref 6,7) (Table 5-1).

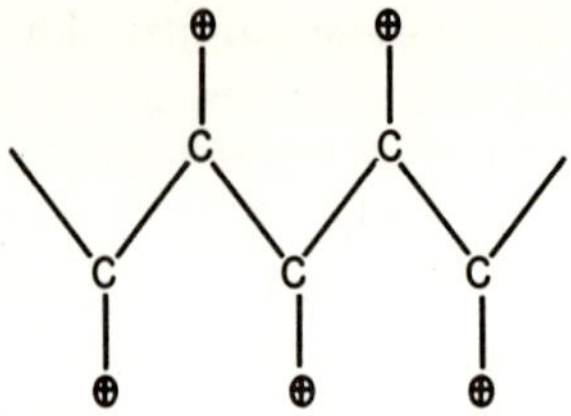

Figure 5-2. Zigzag conformation of polyethylene chain in the crystal (ref 4). ⊕ indicates the position of the hydrogen atoms, one above and one below the plane of the paper.

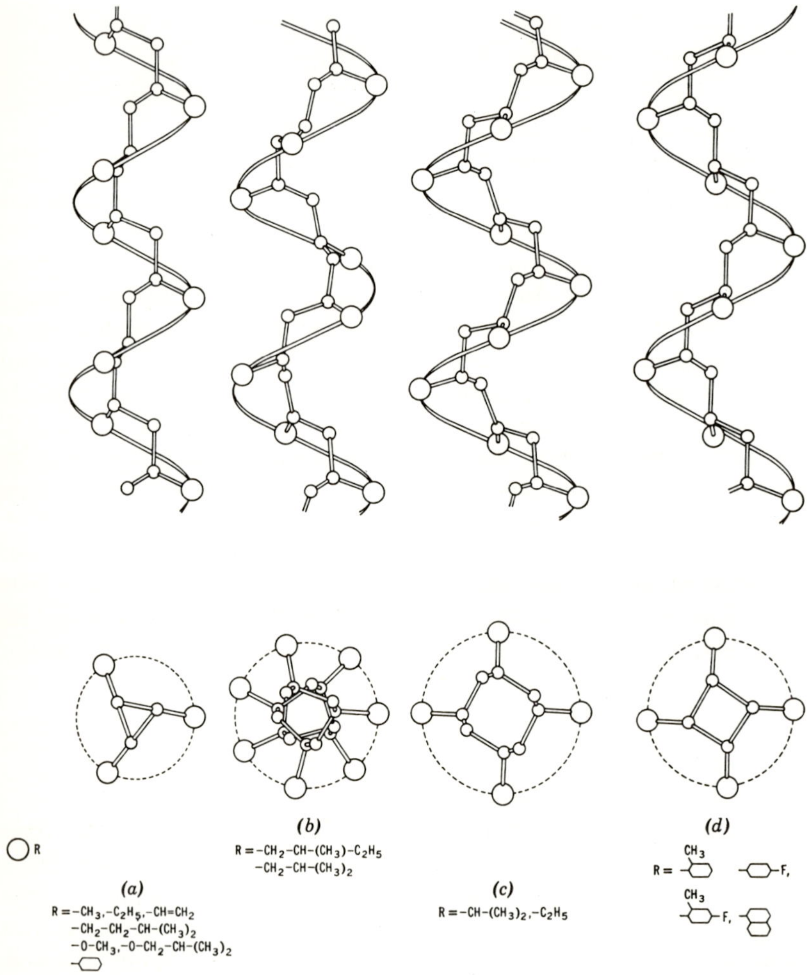

Figure 5-3. Helical conformations of isotactic vinyl polymers (ref 5).

Table 5-1. Conformations of Polymer Molecules in Crystalline Form (ref 8)

Polymer	Conformation*
Polyethylene	Z
Polypropylene, isotactic	$H,3_1$
Polypropylene, syndiotactic	$H,2_1$
Poly-1-butene	$H,3_1$
1,2-polybutadiene, isotactic	$H,3_1$
1,2-polybutadiene, syndiotactic	Z
1,4-trans-polybutadiene	Z
1,4-cis-polybutadiene	Z
Poly-3-methyl-1-butene	$H,4_1$
Poly-4-methyl-1-pentene	$H,7_2$
Poly-5-methyl-1-hexene	$H,3_1$
Polyvinylcyclohexane	$H,4_1$
Polystyrene	$H,3_1$
Polytetrafluoroethylene	H
Poly-α-vinylnaphthalene	$H,4_1$
Poly(vinyl chloride)	Z
Poly(vinyl alcohol)	Z
Poly(methyl methacrylate)(isotactic)	$H,5_2$
Poly(isobutyl acrylate)	$H,3_1$
Poly(sec-butyl acrylate)	$H,3_1$
Poly(tert-butyl acrylate)	$H,3_1$
Poly(vinyl isobutyl ether)	$H,3_1$
Poly(oxymethylene)	$H,5_1$
Polyacetaldehyde	$H,4_1$ or 4_3
Polypropionaldehyde	$H,4_1$ or 4_3
Poly-*n*-butyraldehyde	$H,4_1$ or 4_3
Poly(ethylene oxide)	$H,7_2$
d,1-poly(propylene oxide)	Z
Poly(methyl vinyl ketone)	$H,7_2$
Polyacetone	$H,7_2$
Poly(ethylene terephthalate)	Z
Polycarbonate(from bisphenol A)	Z
Poly(*N*,*N*-di-*n*-butylacrylamide)	$H,3_1$

* *Z* indicates zigzag. *H* indicates helix. Numbers following *H* indicate number of monomer units and number of turns required to make a complete repeat unit in the helical structure.

Structure of Polymer Crystals

The exact way in which polymer molecules are built into a crystalline structure is still in doubt. Since polymer micro-crystals are much smaller than the dimensions of total polymer molecules, early workers believed that each polymer molecule passes through successive amor-

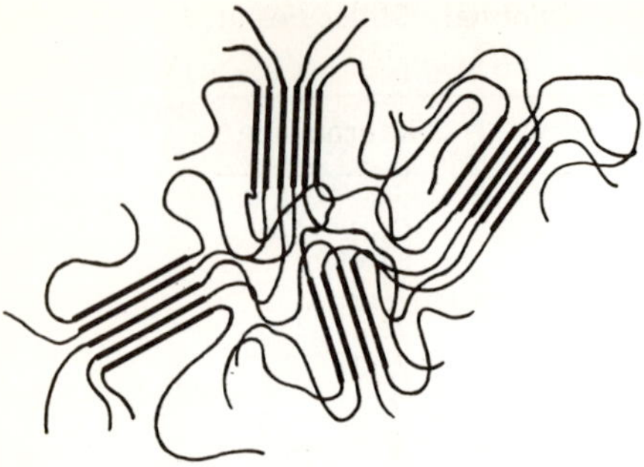

Figure 5-4. Fringed micelle concept of partly crystalline polymer (ref 14).

phous and crystalline areas in a fringed micelle type of structure (ref 9–13) (Figure 5-4). This explained very neatly the fact that polymers could exist in partly amorphous, partly crystalline states, and also explained many of the effects of such partial crystallinity upon polymer properties. More recent analytical studies have suggested, however, that the polymer micro-crystal is formed by folding a single polymer molecule back and forth upon itself in a folded lamella type of structure (ref 9,13,15–18) (Figures 5-5, 5-6); this newer concept of polymer crystalline structure is more in accordance with recent analytical data, but makes it more difficult to explain partial crystallinity and the effects of crystallinity upon properties. Most recently there have been attempts to combine the two types of concepts into a more complex and comprehensive

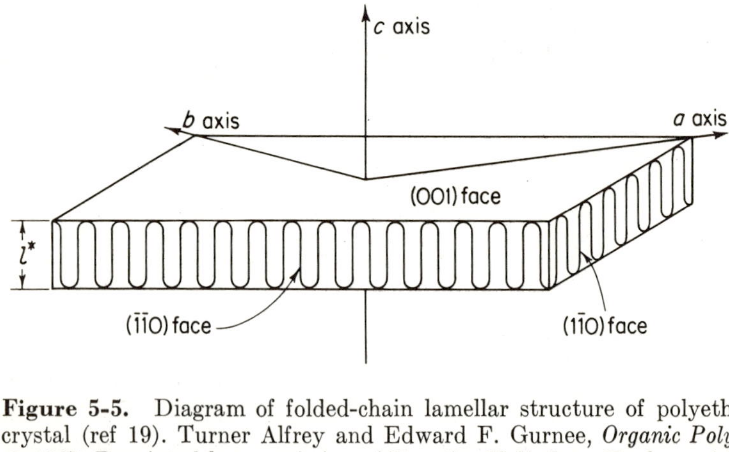

Figure 5-5. Diagram of folded-chain lamellar structure of polyethylene crystal (ref 19). Turner Alfrey and Edward F. Gurnee, *Organic Polymers*, © 1967. Reprinted by permission of Prentice-Hall, Inc., Englewood Cliffs, New Jersey.

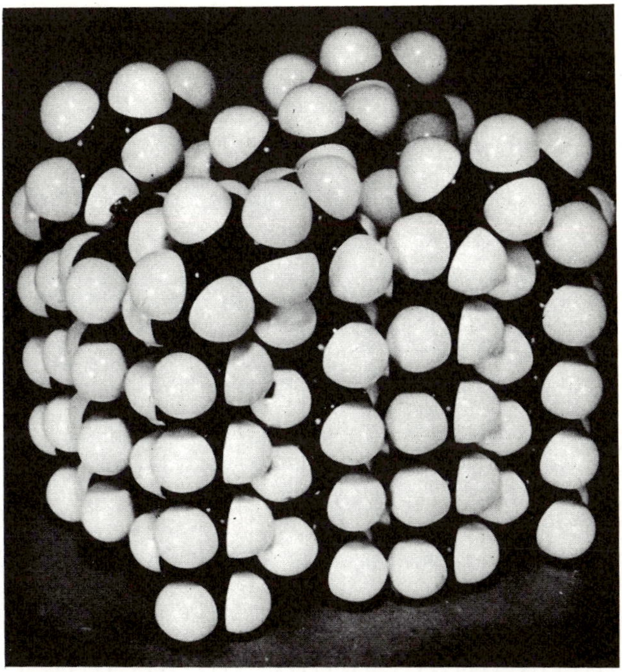

Figure 5-6. Molecular model of folded-chain lamellar structure of polyethylene crystal (ref 20).

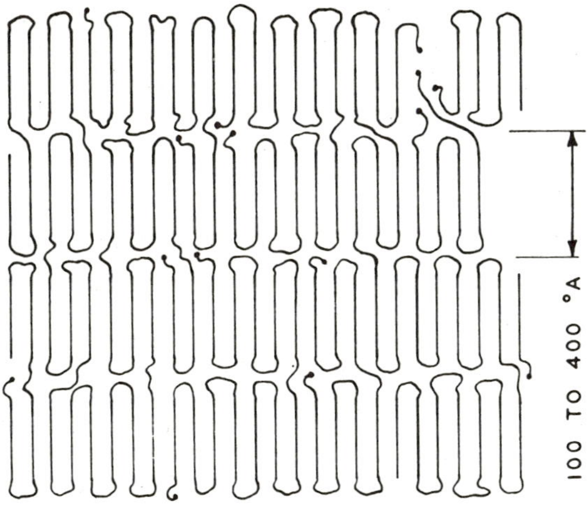

Figure 5-7. Interweaving of polymer molecules through folded-chain lamellae (ref 25).

understanding of partial crystallinity and its effects upon polymer properties (ref 9,21–24) (Figures 5-7, 5-8).

The relationships between polymer structure, crystallinity, and properties are best understood by examining first the factors affecting crystallinity, and then the ways that crystallinity affects properties and applications.

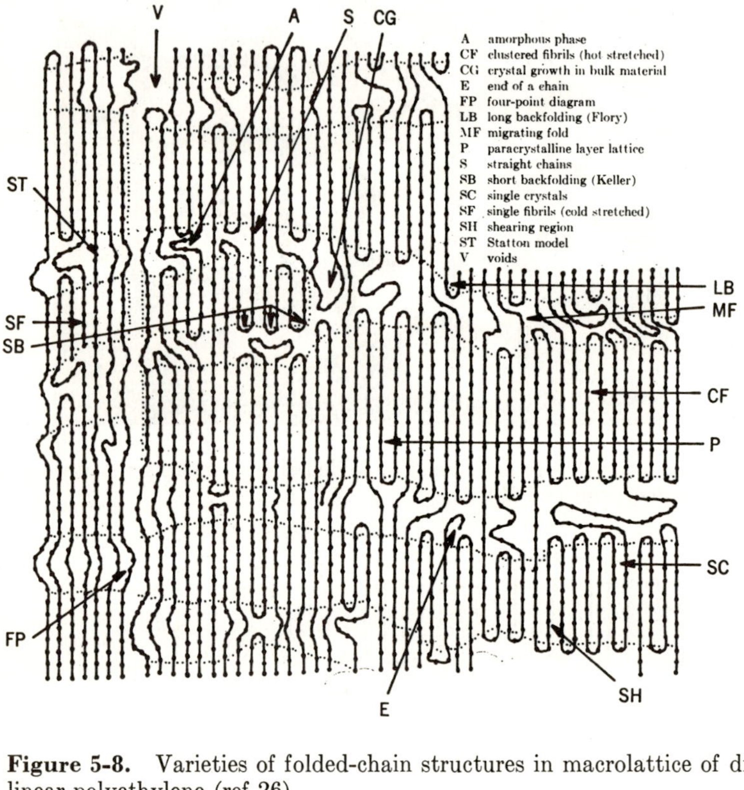

Figure 5-8. Varieties of folded-chain structures in macrolattice of drawn linear polyethylene (ref 26).

Factors Determining Crystallinity

The extent to which a polymer will crystallize is determined first by the thermodynamic forces favoring maximum potential crystallinity at equilibrium, and second by the kinetic forces determining the rate and extent to which the polymer may actually approach such a theoretical maximum degree of crystallinity.

Thermodynamic Forces Favoring Maximum Potential Crystallinity

There are two distinct types of structural factors favoring maximum potential crystallinity at equilibrium: (1) Regularity of molecular structure makes it possible for the molecule to fit into a regular crystalline lattice. (2) Polarity increases the attraction between adjacent chains, and therefore the attraction tending to draw them into the crystalline lattice and hold them firmly there.

Regularity In the crystallization of small simple monomeric molecules, purity is required to insure that all of the successive units entering the crystalline lattice will be similar and occupy similar points in a

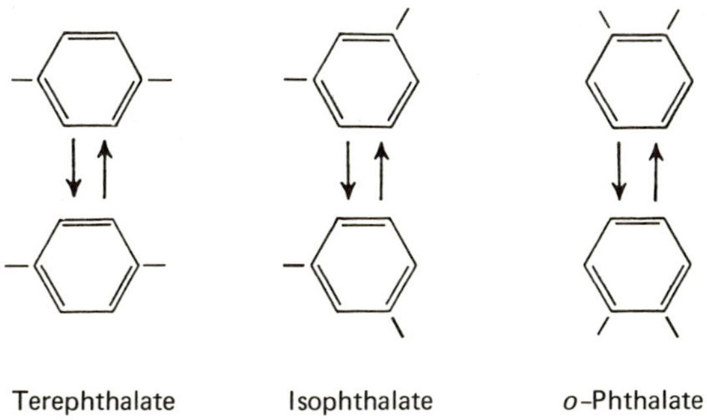

Figure 5-9. Symmetry in the Phthalates

regular repeating structure. In polymeric crystallization, regularity of molecular structure is required to insure that successive units entering the crystal lattice will all fill it in the same way and thus permit the regular repeating structure to form. Some typical aspects of such structural regularity may be considered, ranging from individual atoms and monomer units up to entire polymer molecules, monomeric additives, and impurities.

Symmetry The monomer units in the polymer molecule do not have to be completely symmetrical in structure, but such symmetry does favor higher crystallinity (ref 27,28). For example, the *p*-phenylene ring in terephthalates is more symmetrical than the *m*- or *o*-rings in isophthalates and *o*-phthalates (Figure 5-9); on rotation, the less symmetrical *m*- and *o*-phthalates form new equivalent conformations, gain more entropy, and therefore melt at lower temperature. The high symmetry and crystallinity of the terephthalates produce high strength and high melting points which are most useful in fibers and films; while the lower symmetry, crystallinity, and melting points of the isophthalates and orthophthalates are more useful in plastics, coatings, and adhesives.

Even vs Odd Number of Atoms in the Monomer Unit Repeating monomer units which contain an even number of atoms tend to produce

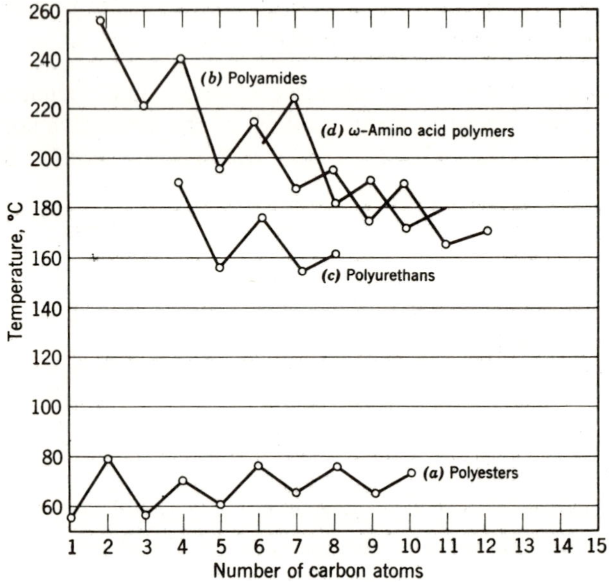

Figure 5-10. Effect of number of main-chain carbon atoms in the monomer unit upon crystalline melting point of the polymer (ref 31).

Dependence of crystalline melting point on spacing of polar groups. Number of carbon atoms refers to (*a*) acid for polyesters made with decamethylene glycol; (*b*) diamine for polyamides made with sebacic acid; (*c*) diamine for polyurethanes made with tetramethylene glycol; and (*d*) ω-amino acid polymers.

higher crystallinity and higher crystalline melting points than repeating monomer units which contain an odd number of atoms in their chain (ref 29,30) (Figure 5-10). This is because they form two different types of crystal structure, and the even-numbered type is the more compact (Figure 5-11).

Condensation Polymers Most linear condensation polymers are inherently regular in structure and crystallize quite readily (ref 33). Typical commercial polymers are polyethylene terephthalate and the nylons.

Vinyl Polymers Most vinyl polymers contain an asymmetric carbon atom

in each monomer unit, and most commercial polymerization reactions scatter these in a random atactic sequence which is noncrystallizable

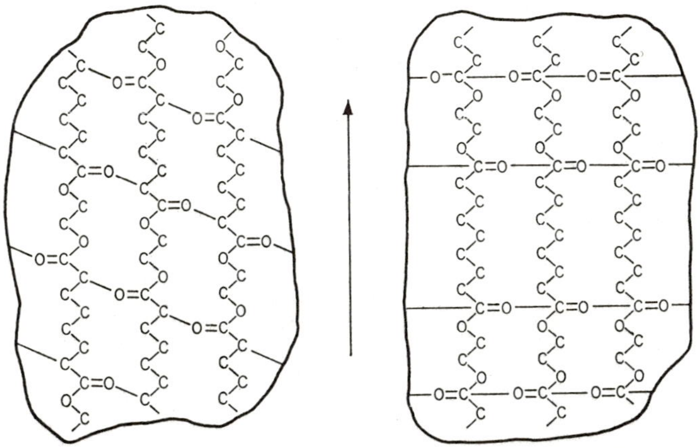

(a) ETHYLENE SUBERATE (b) ETHYLENE AZELATE

Figure 5-11. Effect of even and odd numbers of carbon atoms in the monomer unit upon packing of the polymer molecule in crystalline form (ref 32).

Molecular arrangement in crystallites of polyesters formed from ethylene glycol and a dicarboxylic acid with (*a*) an even number of carbon atoms and (*b*) an odd number of carbon atoms.

and amorphous (Figure 5-12). Polystyrene, polyvinyl acetate, and polymethyl methacrylate are typical commercial polymers. Stereospecific catalysts or large bulky side groups may be used to orient these asymmetric carbon atoms in a uniform fashion, producing isotactic or syndiotactic regularity, and frequently helical coil structures, which permit crystallization. Polypropylene, poly-1-butene, and poly (4-methyl pentene-1) are typical. Carbon atoms with small groups, as in CH_2, CHOH, CF_2, and C=O, are so similar in size that the specific group does not matter, and they can all fit into the same crystal lattice and form very similar crystalline structures (ref 35,36). Polyacrylonitrile contains

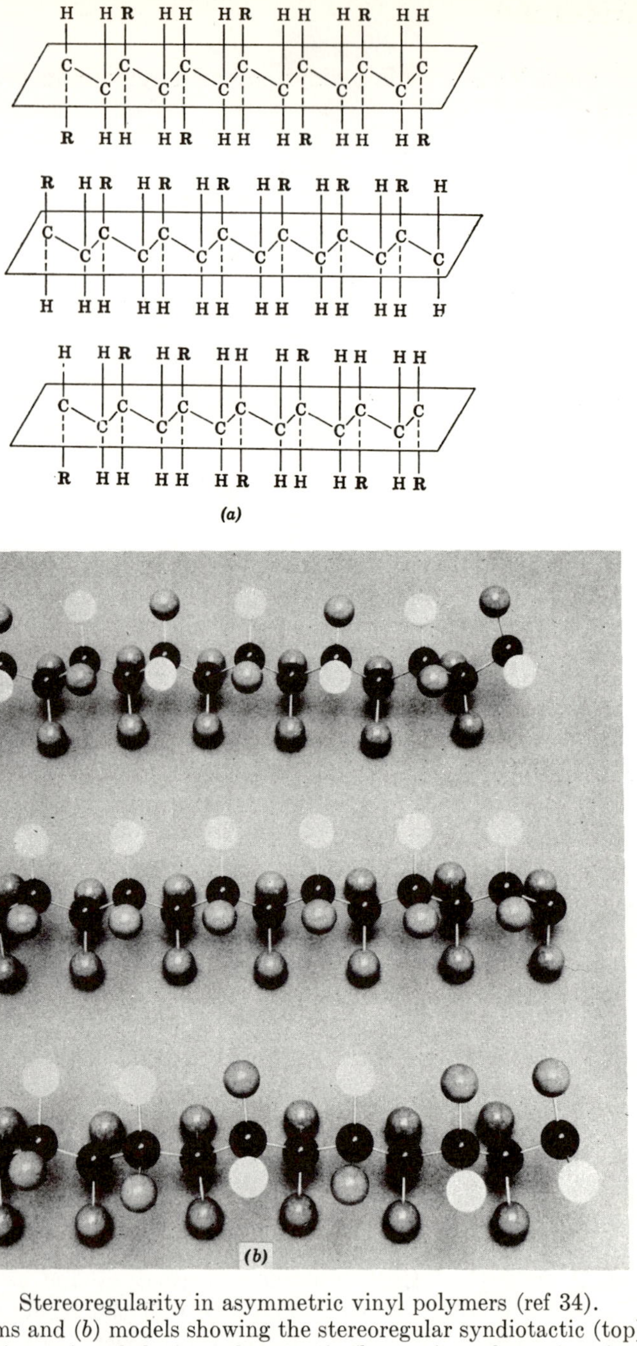

Figure 5-12. Stereoregularity in asymmetric vinyl polymers (ref 34).

(*a*) Diagrams and (*b*) models showing the stereoregular syndiotactic (top) and isotactic (center) and the irregular atactic (bottom) configurations in a vinyl polymer. The main carbon-carbon chain is depicted in the fully extended (all *trans*) planar zigzag conformation.

bulky polar —C≡N side groups, whose steric hindrance and dipole repulsion make the molecule rod-like and therefore able to fit into a fairly regular crystalline structure (ref 35). Even without crystallization, stereo-regular polar polymers may pack slightly better, giving higher density and glass transition temperature and lower solubility, than the corresponding atactic isomers (ref 37) (Table 5-2).

Table 5-2. Density of Amorphous Tactic Polymers (ref 38)

Polymer of	Tacticity	Density ($g\ ml^{-1}$)
Styrene	Atactic	1.045
	Isotactic	1.054
Methyl methacrylate	Atactic	1.188
	Isotactic	1.22
	Syndiotactic	1.19
Butadiene	1,2-isotactic	0.96
	1,2-syndiotactic	0.96
p-methylstyrene	Atactic	1.011
	Isotactic	1.015

Diene Polymers Polymerization of dienes can produce 1,4- addition in cis and trans structures, as well as 1,2- and 3,4- addition in various tacticities (ref 39–41) (Figure 5-13). The regular repeating cis-1,4 structure in natural rubber permits crystallization on stretching. The irregular sequence of isomers in most free-radical-polymerized diene elastomers prevents crystallization. Use of newer stereospecific catalysts produces more regular synthetic elastomers, which again can be crystallized on stretching. Trans configuration around the double bond, in balata and gutta percha, crystallizes and packs more uniformly and densely than the less regular cis configuration, giving higher melting and transition temperatures; the resulting toughness has proved particularly useful in gold balls (ref 41) (Table 5-3). Similarly, 1,2 addition creates vinyl side groups which also produce denser packing, lower free volume, and higher glass transition temperature (Figure 5-14).

Branching Short irregular branches tend to decrease regularity and therefore lower crystallinity (ref 48); polyethylene is the most important example (Figure 5-15). Regular repetitive short branching or side groups produce coiling in a very regular helical coil which favors crystallization,

cis-1,4

trans-1,4

1,2

3,4

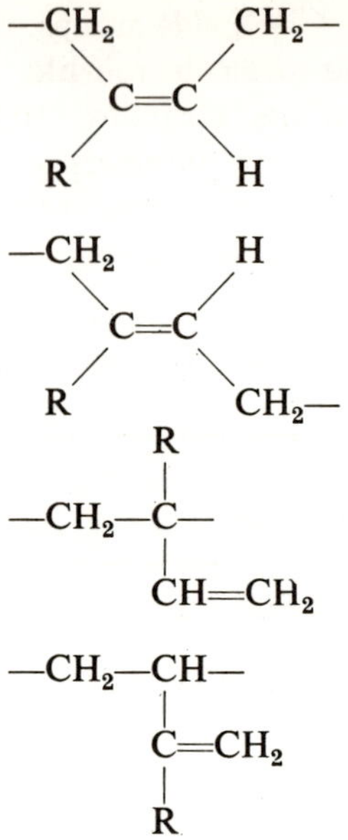

Figure 5-13. Structural configurations of diene polymers (ref 42).

for example in polypropylene, poly-1-butene, and poly-(4-methylpentene-1). Longer side chains, particularly over 5–14 atoms in length, favor side-chain crystallization (ref 45–47); this effect has been observed in polyolefins, vinyl and acrylic ester polymers.

Very long branching has relatively little effect upon crystallinity, because the long branches resemble whole polymer molecules; only the branch points and the end groups of these branches create minor imperfections which decrease local crystallinity, as in low-density polyethylene (ref 48). Similarly cross-linking creates occasional local imperfections

Table 5-3. Melting Points and Glass-Transition Temperatures of Cis- and Trans-Polymers (ref 43)

Polymer	Melting Point (°C)		Glass-transition Temperature (°C)	
	cis-1,4	trans-1,4	cis-1,4	trans-1,4
Polyisoprene	−73	+67	−73	−38
Polybutadiene	0 to 6	135 to 145	−103 to −95	−38

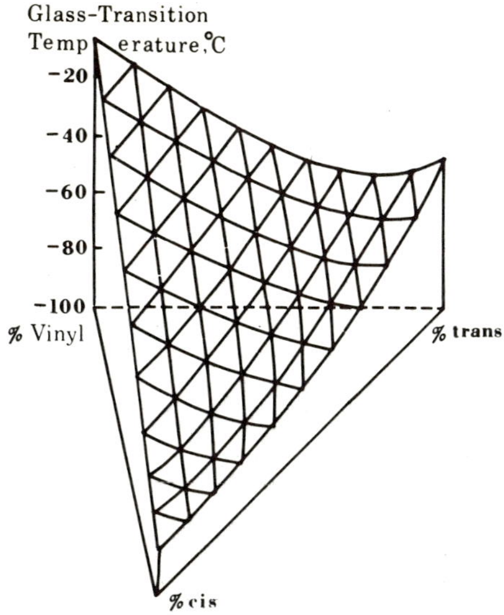

Figure 5-14. Glass transition temperatures of polybutadienes containing various percentages of cis-, trans-, and vinyl (atactic) isomers (ref 44).

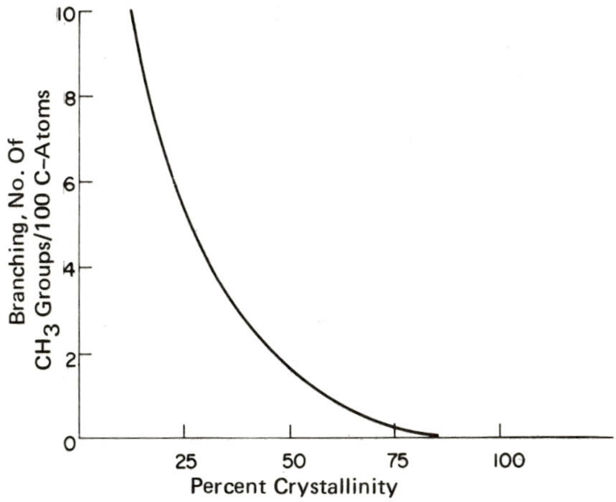

Figure 5-15. Effect of chain branching upon crystallinity of polyethylene (ref 55).

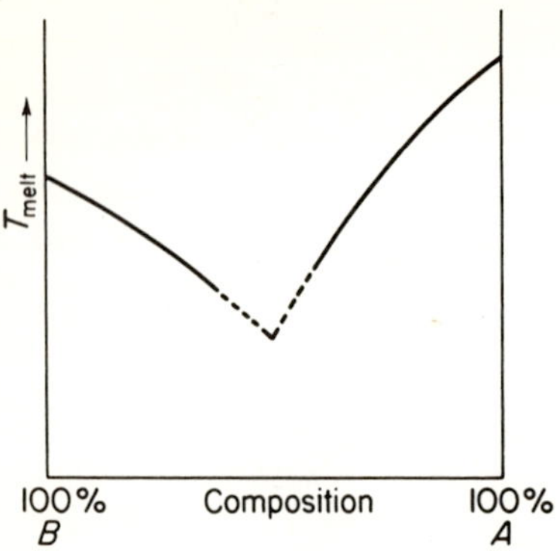

Figure 5-16. Effect of copolymerization upon crystallinity (ref 56). Turner Alfrey and Edward F. Gurnee, *Organic Polymers*, © 1967. Reprinted by permission of Prentice-Hall, Inc., Englewood Cliffs, New Jersey.

which likewise lower crystallinity somewhat (ref 39,49,50), particularly in polyethylene and stereoregular elastomers.

Random Copolymerization Conventional random copolymerization makes a polymer molecule less regular in structure, and therefore less

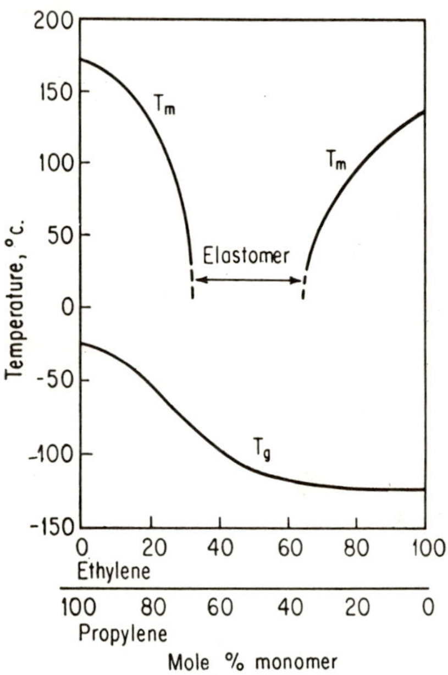

Figure 5-17. Melting and glass transition temperatures of ethylene/propylene copolymers (ref 52).

Reprinted by permission of Modern Plastics Encyclopedia, McGraw-Hill, Inc.

able to crystallize. Percent crystallinity decreases and the size of the crystallites also decreases (ref 51–54,74) (Figure 5-16). Thus $1/T_m - 1/T_m^\circ = -R\ ln\ X_A/\Delta H_u$, where the melting point T_m of the copolymer is lower than T_m° of the homopolymer A, ΔH_u is the heat of fusion of polymer A per mol of monomer unit, R is the gas constant, and X_A is the mol fraction of monomer A (ref 47,57,58).

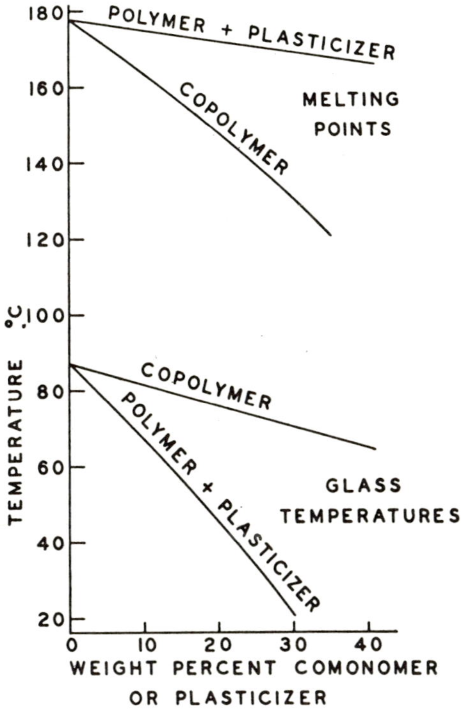

Figure 5-18. Relative effects of comonomer vs plasticizer upon melting and glass transition temperatures (ref 59).

Random copolymerization to lower the melting point is used commercially in making EPR ethylene/propylene rubber (Figure 5-17), FEP fluorinated ethylene/propylene, and polyester and polyamide hot-melt adhesives.

As compared with plasticizer molecules, comonomer units occupy less volume and cause more disruption of regularity, so they have a greater effect on melting point (Figure 5-18). On the other hand, plasticizer molecules have more mobility than copolymerized comonomer units, and therefore have a greater effect on glass transition temperature and flexibility (ref 60).

Isomorphous Copolymers There are some specialized cases in which random copolymerization does not lower crystallinity.

In rare cases two comonomer units may be isomorphous, fit interchangeably in the crystal structure without disrupting it, and give

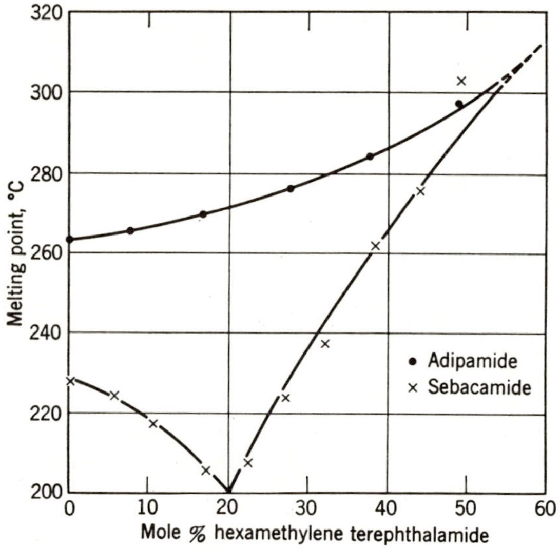

Figure 5-19. Crystalline melting points of hexamethylene adipamide/terephthalamide and hexamethylene sebacamide/terephthalamide copolymers (ref 62).

intermediate melting points as well (ref 1,47,57,61). Hexamethylene adipamide/terephthalamide is one such isomorphous copolymer (Figure 5-19), whereas hexamethylene sebacamide/terephthalamide is typical of the more common mutual repression of crystallinity in random copolymers.

If the comonomer side groups are very small (as in ethylene copolymers with tetrafluoroethylene, vinyl alcohol, and carbon monoxide) these CF_2, CHOH, and C=O groups may not change the size and shape of the polymer molecule sufficiently to interfere with crystallization (ref 35).

Regular Copolymers In certain systems, copolymerization may produce structures of high enough regularity to permit good crystallization. These include (1) Alternating Copolymers, and (2) Block and Graft Copolymers.

Alternating Copolymers If copolymerization reactivities favor a regularly alternating copolymer, this alternating structure may be sufficiently regular to permit crystallization (ref 63). Typical systems include

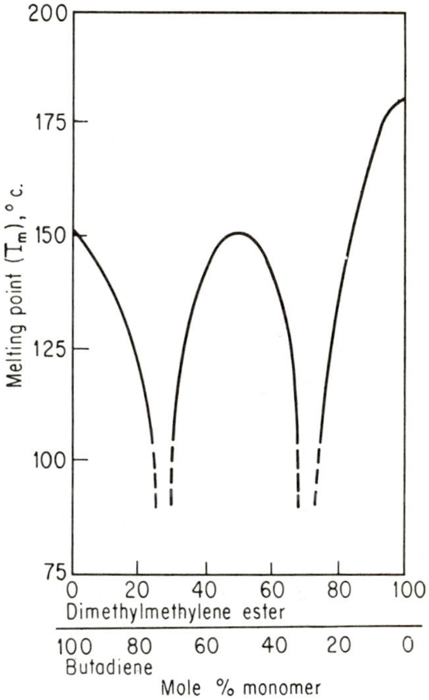

Figure 5-20. Alternating copolymer of butadiene with dimethyl methylene malonate (ref 64).

Reprinted by permission of Modern Plastics Encyclopedia, McGraw-Hill, Inc.

butadiene/dimethyl methylene malonate (Figure 5-20), copolymers of vinylidene cyanide, and ethylene copolymers with carbon monoxide and with sulfur dioxide. None of these has achieved commercial importance.

Block and Graft Copolymers Stereo-block polymerization of a single monomer, and block and graft copolymerization of two or more monomers, may produce sufficiently long sequences of a single regular repeating monomer unit in the polymer structure, so that these regular portions of the polymer molecule may still enter into crystallization (ref

58,63,65,66) (Figures 5-21, 5-22). While the practical potential of such crystalline copolymers is obviously tremendous (ref 67) (Figure 5-23), present commercial applications are limited mainly to propylene/ethylene block copolymerization for improvement of impact strength (ref 68), and certain polyamide block copolymers for specialty applications (ref 69).

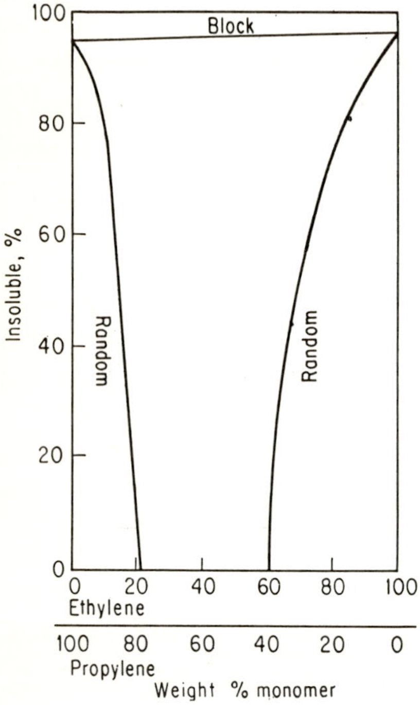

Figure 5-21. Crystalline insolubility of block vs random copolymers (ref 64). Reprinted by permission of Modern Plastics Encyclopedia, McGraw-Hill, Inc.

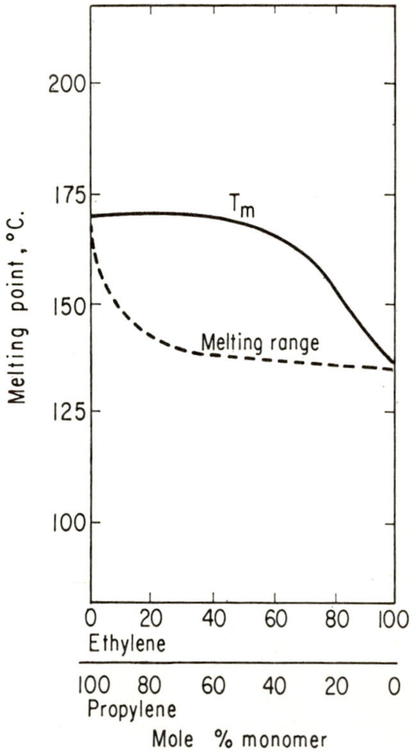

Figure 5-22. Crystalline melting points of block copolymers (ref 63). Reprinted by permission of Modern Plastics Encyclopedia, McGraw-Hill, Inc.

Molecular Weight The end groups of the polymer molecule are different from the rest of the molecular chain and thus constitute impurities. They are also more mobile. These are factors which interfere with regular packing and crystallization, lowering the melting point of the crystalline phase. Thus

$$\frac{1}{T_m} - \frac{1}{T_m^\circ} = \frac{2RM_o}{\Delta H_u \bar{M}_n} \qquad \text{or} \qquad T_m = \frac{T_m^\circ(\bar{M}_n/M_o)}{B + \bar{M}_n/M_o}$$

where T_m and T_m° are the melting points of a polymer in °K at $\bar{M}_n$ and infinite molecular weights, respectively, M_o the molecular weight of the monomer unit, ΔH_u the heat of fusion per mol of repeating unit in the crystalline polymer, R the gas constant, and B an empirical constant (ref 58,70) (Figure 3-46). This effect is most noticeable at low molecular

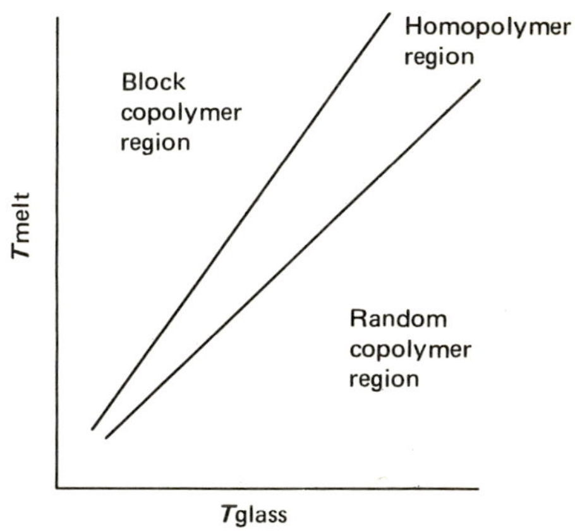

Figure 5-23. Melting vs glass transition temperatures for block copolymers, homopolymers, and random copolymers (ref 67).

Turner Alfrey and Edward F. Gurnee, *Organic Polymers*, © 1967. Reprinted by permission of Prentice-Hall, Inc., Englewood Cliffs, New Jersey.

weights, and diminishes rapidly at higher molecular weights in common commercial use.

Plasticization Addition of compatible monomeric plasticizer or solvent molecules to a polymer attracts these monomeric units to portions of the polymer molecule, producing a less regular structure which is much less able to fit into a regular crystalline lattice, and thus reduces crystallinity and crystalline melting point. Thus $1/T_m - 1/T_m^\circ = (R/\Delta H_u)(V_u/V_1)(v_1 - X_1 v_1^2)$, where T_m and T_m° are the melting points in °K of the plasticized and pure polymers, respectively, V_u the molar volume of the repeating unit in the polymer, V_1 the molar volume of the plasticizer, v_1 the volume fraction of plasticizer, ΔH_u the heat of fusion per mol of repeating unit in the polymer, R the gas constant, and X_1 a first-neighbor interaction parameter (ref 71). X_1 is 0 for good solvents or highly compatible plasticizers, positive for poor ones; thus good ones are more efficient in lowering the melting point. Since plasticizers and solvents occupy a larger unit volume than copolymerized comonomer units, they are therefore less efficient in lowering the melting point (ref 60) (Figure 5-18). Thus in commercial practice, plasticization of polyvinyl chloride, cellophane film, and some polyamide hot melt adhesives all show the effects of plasticizer as a moderate depressor of crystallinity; but larger effects are more easily obtained by copolymerization as in ethylene/propylene rubber, vinyl chloride/vinyl acetate copolymers, vinylidene chloride copolymers, copolyester films, and copolyester and copolyamide hot melt adhesives.

Impurities In crystallization of conventional monomeric materials, impurities are rejected by the growing lattice, and the pure phase forms perfect crystals. In crystallization of polymers, noncrystallizable impurities include both irregularities in the polymer structure and monomeric additives and accidental impurites. As the polymer crystals form, they also try to reject these impurities, leaving them in the remaining amorphous areas between the growing crystals. At high crystallinity, the rejected impurities remain as boundary lines and pockets between the crystalline areas, and constitute areas of weakness in the final structure, incipient points of attack for mechanical or chemical failure in latter use (ref 72,73).

These are typical of the structural factors which influence regularity in the structure of the polymer molecule, and therefore its ability to enter and pack into a regular repeating crystalline lattice.

POLARITY At low temperatures and energy levels, where there is little atomic motion, molecules tend to settle and pack into regular repeating crystalline lattices. As has just been observed, the polymer molecule must be regular enough so that successive atoms and groups in the molecule can fit into successive points in the regular repeating lattice structure. In addition, if the polar attractive forces between atoms and groups in adjacent chains are sufficiently high, the forces favoring crystallization will be greater, crystallinity will be higher, and the crystals will retain their identity up to higher temperatures indicating that they require the input of greater thermal motion and energy in order to overcome the bonding forces in the crystal structure (ref 75–79) (to be covered later). Hydrogen-bonding is a particularly common and powerful example of such interchain forces operating in crystallization of polymers (ref 29,66). Conversely the addition of plasticizers, which hydrogen-bond

Table 5-4. Crystalline Melting Points of Polymers (ref 83)

Polymer	Melting Point (°C)	Polymer	Melting Point (°C)
Polyethylene	137	Polypropylene oxide	75
Polypropylene	176	Poly-1-methoxybutadiene	118
Polybutene-1	126	Polyvinyl methyl ether	144
Polypentene-1	75	Polyvinyl ethyl ether	86
Polyhexene-1	−55	Polyvinyl-*n*-propyl ether	76
Polyheptene-1	−40	Polyvinyl isopropyl ether	190
Polyoctene-1	−38	Polyvinyl-*n*-butyl ether	64
Polydodecene-1	45	Polyvinyl isobutyl ether	115 (165)
Polyoctadecene-1	76	Polyvinyl *tert.* butyl ether	260
Poly-3-methyl butene-1	310	Polyvinyl neopentyl ether	216
Poly-4-methyl pentene-1	250	Polyvinyl benzyl ether	162
Poly-4-methyl hexene-1	188	Polyvinyl-2-chloroethyl ether	150
Poly-5-methyl hexene-1	130	Polyvinyl-2-methoxyethyl ether	73
Polyisoprene, cis (natural rubber)	28 (36)	Polyisopropyl acrylate (isotactic)	162
Polyisoprene, trans (gutta percha)	74	Polytertiary butyl acrylate	193
Polyvinyl *tert.* butane	>350	Polymethyl methacrylate (isotactic)	160
1,2-polybutadiene (syndiotactic)	154	Polymethyl methacrylate (syndiotactic)	>200
1,2-polybutadiene (isotactic)	120	Poly(lauryl acrylate)†	2
1,4-trans-polybutadiene	148 (92)	Poly(tetradecyl acrylate)†	22
1,4-trans-poly-2,3-dimethyl butadiene	260	Poly(hexadecyl acrylate)†	33
Polyisobutylene	128 (1.5)	Poly(octadecyl acrylate)†	42
Polyvinyl cyclohexane	305	Poly(lauryl methacrylate)†	−34
Polystyrene (isotactic)	240	Poly(tetradecyl methacrylate)†	−2
Poly-*o*-methyl styrene	>360	Poly(hexadecyl methacrylate)†	20
Poly-*m*-methyl styrene	215	Poly(octadecyl methacrylate)†	34
Poly-2,4-dimethyl styrene	310 (350)	Polyethylene terephthalate	267
Poly-2,5-dimethyl styrene	340	Polytrimethylene terephthalate	233
Poly-3,5-dimethyl styrene	290	Polytetramethylene terephthalate	232
Poly-3,4-dimethyl styrene	240	Polypentamethylene terephthalate	134
Poly-*o*-fluorostyrene	270		
Poly-*p*-fluorostyrene	265		
Poly-2-methyl-4-fluorostyrene	360		
Poly-alpha-vinyl naphthalene	360		
Poly-*p*-xylene	375		
Polyoxymethylene	181		
Polyethylene oxide	66		

† Alkyl side-chain only crystallizes.

Table 5-4. Crystalline Melting Points of Polymers (ref 83) (*Continued*)

Polymer	Melting Point (°C)	Polymer	Melting Point (°C)
Polyhexamethylene terephthalate	160	Nylon 11	194
Polyoctamethylene terephthalate	132	Polyhexamethylene adipamide (nylon 6-6)	265
Polynonamethylene terephthalate	85	Polyhexamethylene sebacamide (nylon 6-10)	227
Polydecamethylene terephthalate	138	Nylon 9-9	175
Polyethylene isophthalate	240	Nylon 10-9	214
Polytrimethylene isophthalate	132	Polydecamethylene sebacamide (nylon 10-10)	210 (216)
Polytetramethylene isophthalate	152	Cellulose triacetate	306
Polyhexamethylene isophthalate	140	Cellulose tripropionate	234
Polyethylene sebacate	76	Cellulose tributyrate	183 (207)
Polytetramethylene sebacate	64	Cellulose trivalerate	122
Polydecamethylene sebacate	80	Cellulose tricaproate	94
Polyethylene adipate	50	Cellulose triheptylate	88
Polytrimethylene adipate	38	Cellulose tricaprylate	86 (116)
Polydecamethylene adipate	80	Cellulose tricaprate	88
Polytrimethylene succinate	47	Cellulose trilaurate	91
Polytrimethylene glutarate	39	Cellulose trimyristate	106
Polytrimethylene pimelate	37	Cellulose tripalmitate	105
Polytrimethylene suberate	41	Cellulose trinitrate	>725
Polytrimethylene azelate	50	Polyvinyl chloride	212
Polydecamethylene azelate	69	Polyvinylidene chloride	198
Polycaproamide (nylon 6)	225 (215)	Polychloroprene	80
		Polyvinyl fluoride	200
		Polytetrafluoroallene	126
		Polychlorotrifluoroethylene	220
		Polytetrafluoroethylene	327
		Polyacrylonitrile	317
		Polycarbonate (Bis Phenol-A)	220 (267)
		Poly-N-isopropyl acrylamide	200
		Poly-3,3′-bischloromethyl oxacyclobutane	180

to the polymer, interferes with hydrogen-bonding between adjacent polymer chains, and is capable of destroying low degrees of crystallinity in this way (ref 80).

Regularity of the repeating unit structure in the polymer molecule is necessary to permit crystallization, and polar attractive forces between molecules add to the strength of the crystal that is formed. These are the factors which determine the maximum potential crystallinity of a polymer, and the stability of the crystalline phase, at thermodynamic equilibrium. Their interaction in particular polymer structures is complex and has not yet been fully analyzed (ref 81,82). Pending such complete analysis, comparative study of crystalline melting points in a series of commercial and experimental polymers can be very instructive and thought-provoking (Table 5-4).

Kinetic Factors Affecting Rate and Extent of Crystallization

When a small monomeric molecule crystallizes, it is relatively easy for all of its individual atoms to find their appointed places in the crystalline lattice structure. When a polymer molecule crystallizes, again each atom seeks its own preferred location in the growing crystal lattice. But since each atom in a polymer molecule is attached to adjacent atoms in a long chain structure, these restrict its mobility and its ability to move into this preferred position in the lattice structure. This restriction of mobility slows the formation of polymer crystals, and usually restricts them to micro-crystalline dimensions perhaps 50–500 Å in size (ref 84–87).

Thus the actual rate at which polymers crystallize, and the extent to which this process actually occurs in practice, depend upon kinetic factors affecting the ability of successive atoms in the polymer molecule to move into successive positions in the crystal lattice despite their attachment and entanglement in the polymer chain. These kinetic factors involve both (1) Molecular Flexibility of the long chain structure, and (2) Crystallization Conditions which exist both during the processing history by which the polymer is converted into an end-product, and the environmental conditions in which it is used.

Molecular Flexibility For a polymer to crystallize, the molecule must have sufficient mobility so that successive individual atoms and groups in the polymer molecule can move into precise positions in the crystalline lattice which is being formed. Thus the polymer must be above the glass transition temperature, to permit reasonable segmental mobility (ref 88). This accounts for the easy crystallizability of flexible molecules like polyethylene and polyoxymethylene, and the difficult crystallizability of stiff molecules like polyvinyl chloride and many regular but bulky aromatic condensation polymers.

Low molecular weight favors mobility and thus produces faster and more thorough crystallization (ref 79). Conversely, it has just been observed that low molecular weight produces a large number of end groups which are imperfections in the final crystalline structure and thus hinder the ultimate extent of crystallization at equilibrium (see Molecular Weight this Chapter). Thus these two factors tend to oppose each other.

Addition of a trace of monomeric plasticizer tends to improve mobility, and thus speeds crystallization to higher levels at equilibrium (ref 89). When this occurs, it is sometimes referred to as an antiplasticizer. It is particularly noticeable in stiff polymer molecules such as polyvinyl chloride and bisphenol polycarbonate.

In the wet-spinning of infusible fibers such as polyacrylonitrile and viscose rayon, crystallization occurs while the wet fiber is still swollen with solvent, producing sufficient mobility for the molecules to organize before the solvent has left completely (ref 90).

In all of these cases, molecular flexibility and mobility are required to permit each atom and group in the polymer molecule to move readily and rapidly into the precise position that will permit it to become part of the crystalline lattice that is being formed.

CRYSTALLIZATION CONDITIONS DURING PROCESSING AND USE Energy must be supplied to move the segments of the polymer molecule into the precise position so that they can enter the growing lattice and crystallize. This energy may be supplied either thermally or mechanically.

Thermal Energy While low temperature and low thermal energy favor the tendency to pack and remain neatly and stably in crystalline structures, high temperature and high thermal energy produce the mobility which permits the polymer molecule to find its way initially into position to form such structures (Figure 5-24). The balance of these two counter-opposed factors generally produces maximum crystallization rate at some intermediate temperature (ref 91–93) (Figures 5-25 to 5-27).

Above the melting point, the molecules have too much energy and their packing is too loose to permit crystallization. They remain in constant spontaneous random motion.

Just below the melting point, the high mobility of the polymer molecules produces faster growth to larger and more perfect crystals of higher melting point, and a higher overall degree of crystallinity. Such conditions are obtained by cooling the polymer slowly from the molten state, or by rewarming the solid polymer close to the melting point (ref 97–99).

Maximum rate of crystallization often occurs at about $\frac{9}{10}$ of the melting point in °K (ref 100). When a polymer is hard to crystallize, this is approximately the optimum temperature to use.

Lowering the temperature toward the glass transition temperature produces faster nucleation, larger numbers of nuclei, and therefore smaller crystals (ref 101–103) (Figure 5-28). This technique is particularly useful in extrusion of fast-crystallizing polymers like polyethylene. It is also important in speeding the initiation of crystallization in more sluggish polymers during injection molding, by use of fairly cool molds.

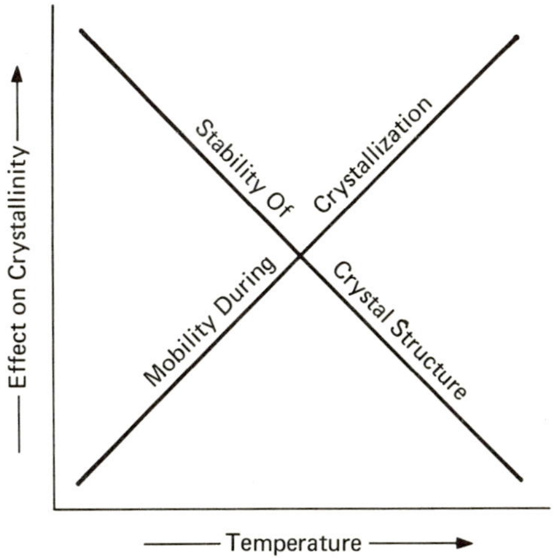

Figure 5-24. Conflicting effects of temperature upon rate of crystallization vs stability of crystal structure.

Quenching the molten polymer suddenly below the glass transition temperature reduces mobility to the point at which crystallization can no longer occur (ref 93,104). In stiff polymer molecules which crystallize slowly, such as isotactic polystyrene and polyethylene terephthalate, quenching below the glass transition temperature produces an amorphous product. In flexible polymer molecules which crystallize readily, such as linear polyethylene, crystallization may occur so rapidly during quenching that a large number of fine crystals can still form before the entire mass of polymer has cooled to low temperature.

In practical processing, quenching of easily-crystallized polymers generally produces many small fine crystals which may be less perfect and exhibit a broader, lower melting range (ref 97). When a quenched

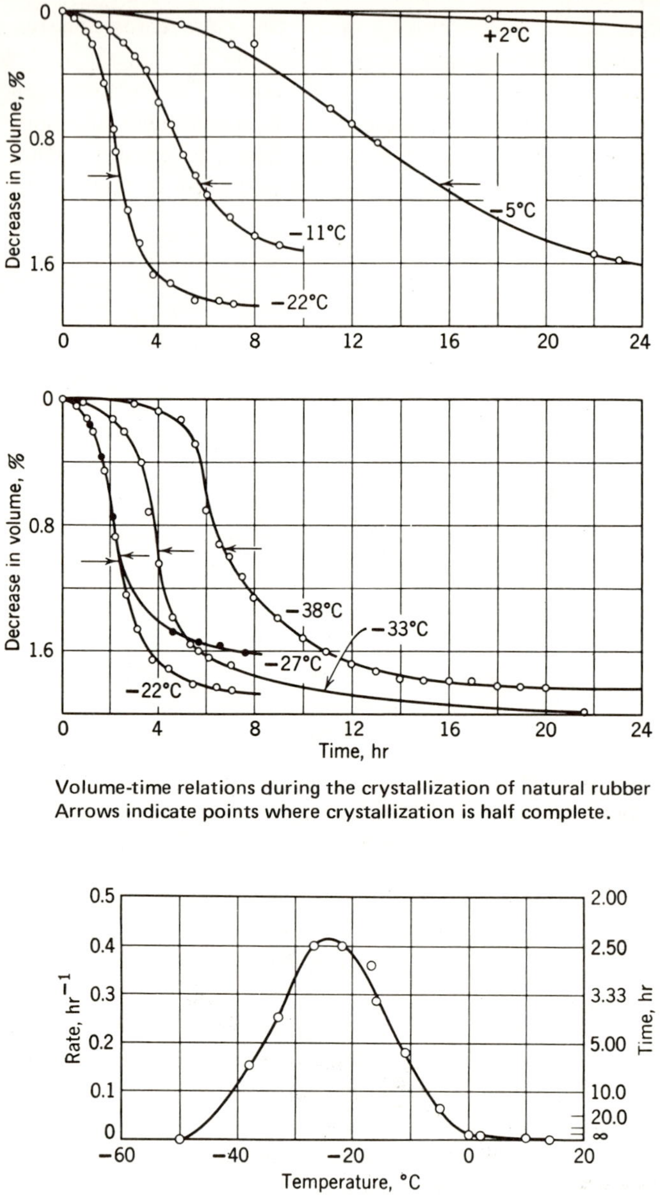

Figure 5-25. Effect of temperature upon rate of crystallization of natural rubber (ref 94).

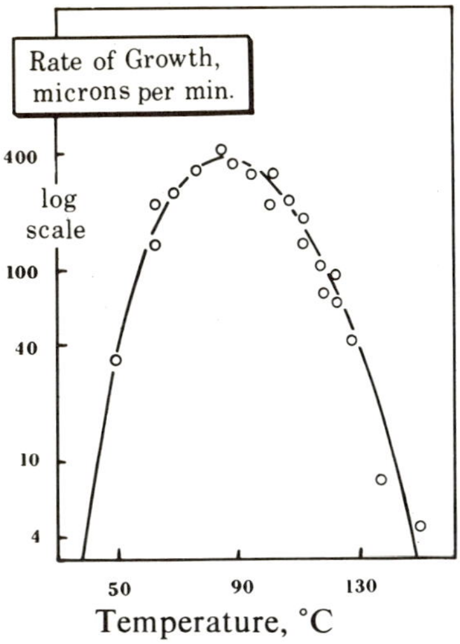

Figure 5-26. Rate of growth of crystalline spherulites in poly(oxymethylene) (ref 95).

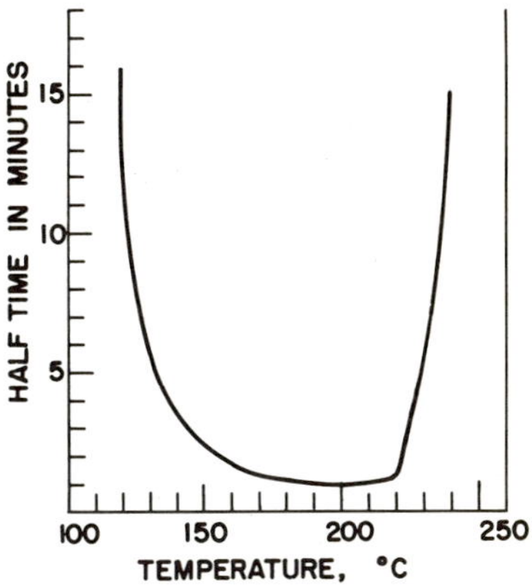

Figure 5-27. Half time for crystallization of polyethylene terephthalate (ref 96).

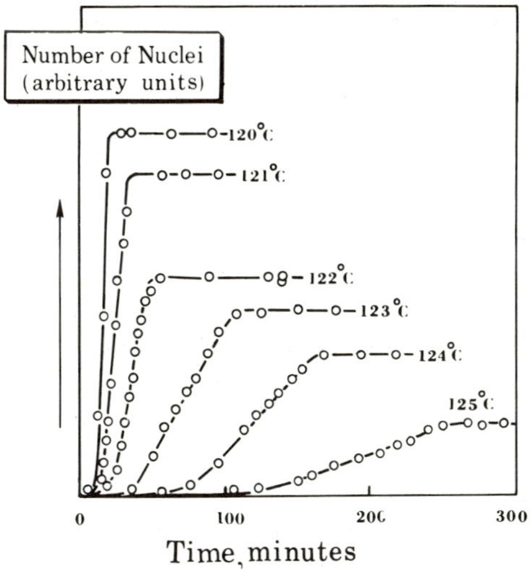

Figure 5-28. Rate of nucleation in poly(decamethylene terephthalate) (ref 95).

polymer has not crystallized completely and has a glass transition near room temperature, storing it at room temperature may produce gradual crystallization and post-shrinkage. In poly-1-butene there is even a slow polymorphic transition, over a 10-day period, from a meta-stable initial crystalline Form II, to a stable final crystalline Form I, causing an additional shrinkage and change in properties (ref 105).

To overcome the dichotomy between fast nucleation at low temperature as compared with fast crystal growth at high temperature, there has been a recent trend to the addition of small amounts of heterogeneous nucleating agents in crystalline polymers. These can produce fast nucleation even at temperatures close to the melting point of the polymer,

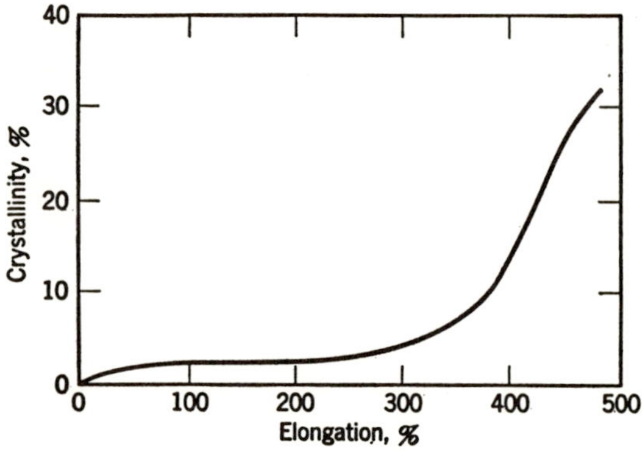

Figure 5-29. Crystallization of rubber by stretching (ref 110).

whereupon it can then crystallize rapidly, thus making the processing cycle shorter and more economical. Such nucleating agents are frequently fine inorganic crystalline powders. They may also be high-melting organic polymers of high cohesive energy density, which will crystallize rapidly even at high processing temperatures, and thus provide nuclei as sites for crystallization of the major polymer (ref 88). Their use is particularly important in injection molding of polypropylene and nylons.

Mechanical Stress Directional mechanical stress can also be applied to the polymer molecules to produce motion and align them in parallel position favoring easier crystallization.

All synthetic fibers are stretched monoaxially during production to align the polymer molecules parallel to each other and produce maximum crystallinity (ref 105–107). Most crystalline polymers can be drawn 4 to 5 fold in this way (ref 108).

Even molecules with high mobility, such as elastomers, if they are sufficiently regular in structure, will crystallize on stretching to high

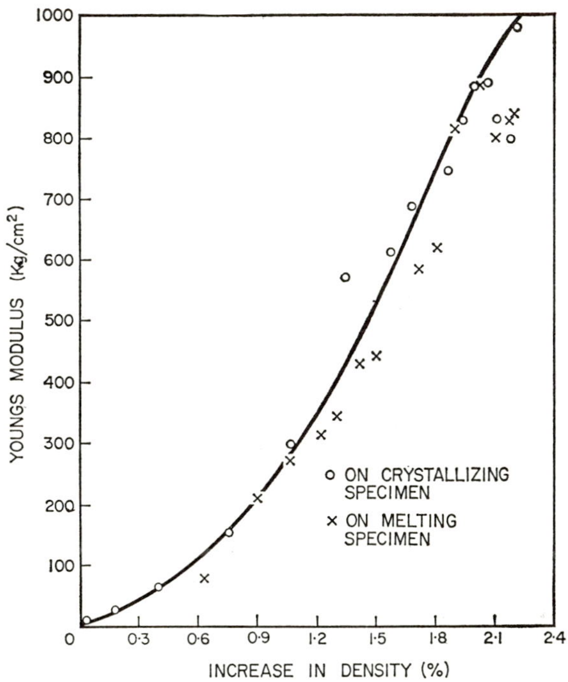

Figure 5-30. Effect of crystallinity on modulus of natural rubber (ref 111). Increase in Young's modulus of natural rubber as a function of crystallinity. (1% increase in density is approximately equivalent to 10% crystallinity.)

degrees of elongation, at which their molecules become reasonably parallel (ref 109) (Figure 5-29). This crystallization at high elongation produces the excellent strength of the more regular elastomers (ref 114) (Figures 5-30 to 5-32).

Biaxial stretching during production of films, such as polyethylene terephthalate, also uses mechanical energy to align the polymer molecules, and produces an increase in crystallinity (ref 115).

Hydraulic stresses during molten flow in injection molding and calendering also align the polymer molecules parallel to each other and thus favor crystallization (ref 116).

In all of these examples, directional mechanical stress produces motion to align the polymer molecules parallel to each other and thus helps to promote crystallization.

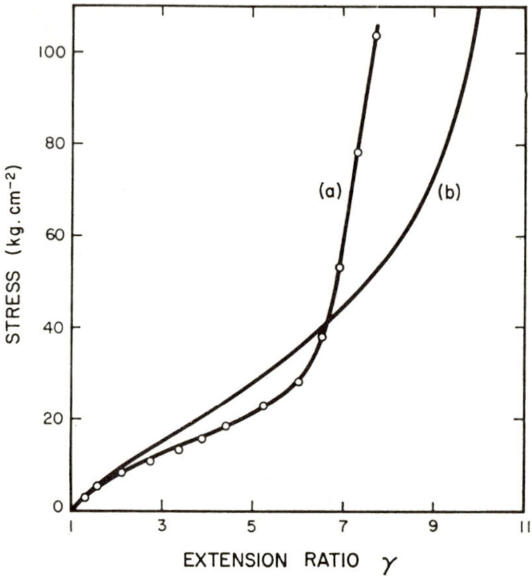

Figure 5-31. Effect of stretch-crystallization upon stress-strain behavior of rubber (ref 112).

Stress versus strain for elongation of latex rubber at 50°C showing effect of crystallization at large extensions. Curve (*a*): experimental latex rubber lightly vulcanized. Curve (*b*): theoretical.

Compressive stress on some plastics may apply sufficiently high pressure to produce thermodynamic factors favoring the denser, more crystalline form, and thus increase crystallinity (ref 105,116). In this case, the effect of stress is not so much the directional effect on alignment, but rather the thermodynamic effect on equilibrium.

Crystallinity → Properties

When polymer molecules crystallize, this high degree of organization becomes a major factor in overall structure. In particular, the forces holding the polymer molecule into the crystalline lattice greatly restrict its mobility and thus affect most of its properties. Thus most properties depend upon the percent of crystallinity and the size of the crystals present in the polymer.

Processability

It has already been observed that processing conditions affect both the percent of crystallinity and the size and structure of the crystals

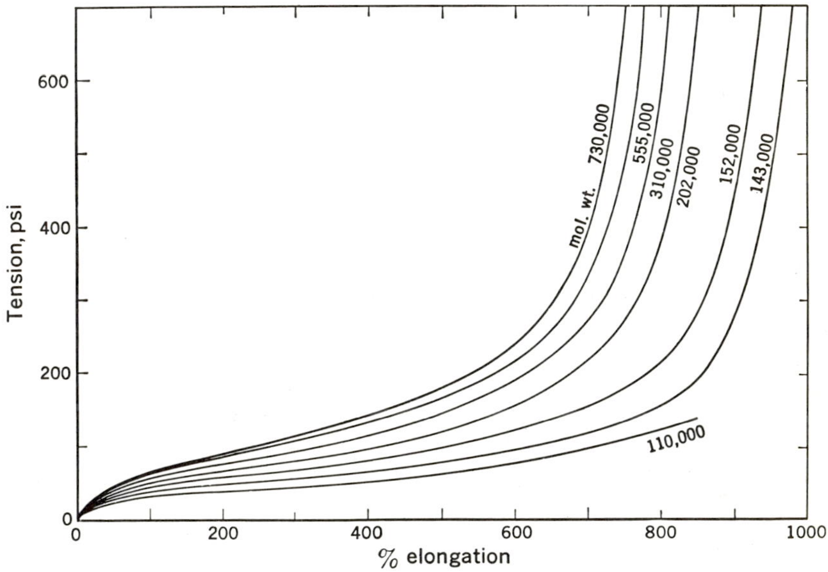

Figure 5-32. Effect of stretch-crystallization upon stress-strain behavior of butyl rubber (ref 113).

which are formed. It is also important to note that the crystallinity of a polymer may affect its processing.

MOLDING Thus a sharp melting point from rigid to fluid state is useful in injection molding of acetals and nylons; but in low-density polyethylene short branching is preferred to lower crystallinity and produces easier flow in molding (ref 52). Fast and complete crystallization to equilibrium must occur in the mold, so that the molded piece can be ejected rapidly without later distortion or shrinkage; thus addition of nucleating agents to polypropylenes and nylons has proved very beneficial in shortening their molding cycles, whereas the slow polymorphic transition in poly-1-butene has proved to be a major problem.

THERMOFORMING Thermoforming requires a tough pliable state at processing temperature, rather than a sharp crystalline melting point from rigid to molten polymer. In crystalline materials, such a tough pliable state can be achieved by partial crystallinity, for example by partial atacticity or copolymerization; or by using high-molecular-weight material of high melt viscosity.

HEAT-SEALING Heat-sealing of polyethylene film is sensitive to crystallinity. High crystallinity produces higher and sharper melting point and thus narrows the range of useful heat-sealing temperatures (ref 116). Consequently lower crystallinity favors easier heat-sealability.

ORIENTATION Polymers which crystallize too readily may become so highly crystalline that they are difficult to orient. If the molten polymer can be quenched to a low degree of crystallinity, this may still permit orientation before crystallization is complete. This technique is particularly useful in making polyethylene terephthalate film.

EXTRUSION OF FIBER AND FILM In general, fiber-forming polymers should desirably melt below a temperature of 300°C to permit effective melt spinning (ref 117). Strong crystalline forces may make the melting temperature of a crystalline polymer higher than its thermal decomposition temperature, and thus prevent thermal processing. Solution processing into fibers and films is still possible, but very often the same high forces of crystallinity which produce high melting temperatures also produce low solubility in volatile common solvents. Polymers such as polyacrylonitrile and cellulose are too high-melting for thermal processing and too insoluble for processing in volatile solvents, and they can only be processed by solution in highly polar solvents of high boiling point.

SOLID-PHASE POLYMERIZATION Some of the new ultra-high-temperature plastics are so high-melting that they cannot even be polymerized to completion in molten form, and must be finished by solid-phase polymerization. This makes fabrication into end products extremely difficult.

Mechanical Properties

Crystallization reduces molecular mobility and this clearly stiffens the reversible mechanical properties. The effect on ultimate and more complex mechanical properties is not as simple or as well understood.

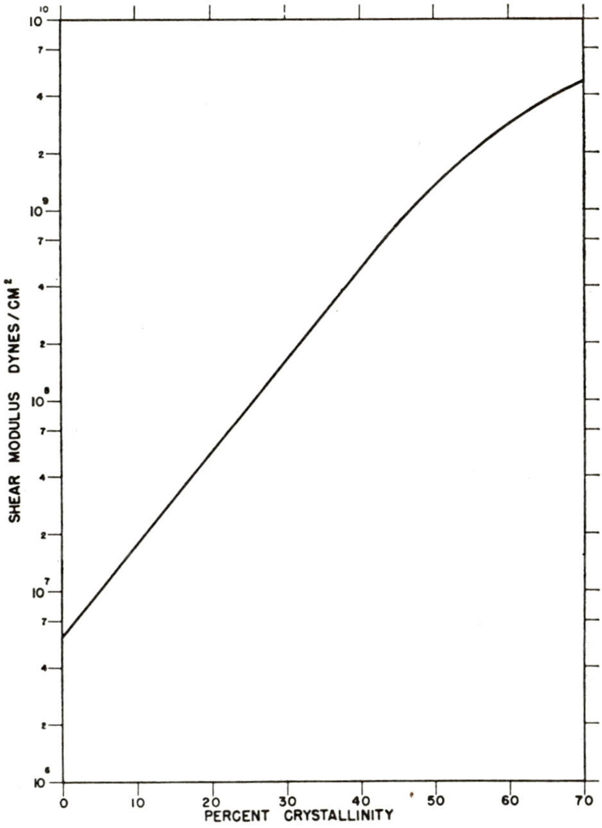

Figure 5-33. Effect of crystallinity on modulus (ref 118). Dynamic shear modulus at one cycle per second of semicrystalline polymers as a function of the degree of crystallinity.

Reversible Mechanical Properties Organization of polymer molecules into a crystalline lattice restricts their mobility greatly, increasing their modulus (Figures 5-33 and 5-34), stiffness (Figures 5-35 to 5-37), hardness (Figure 5-38), and rigidity.

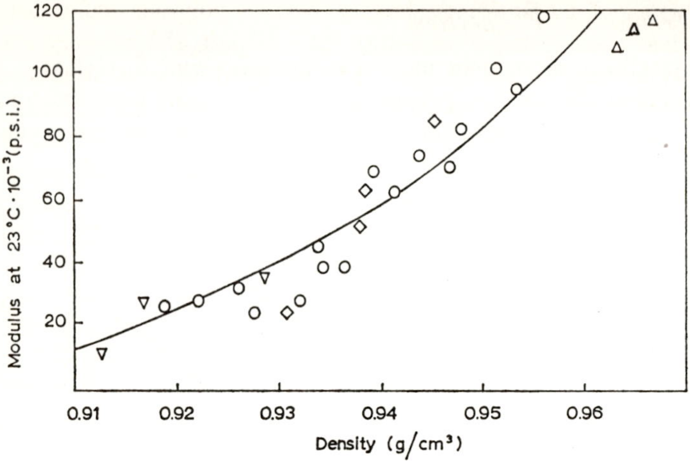

Figure 5-34. Effect of crystallinity upon modulus of polyethylene (ref 119). Modulus versus density for linear and branched polyethylenes, polyethylene blends, and copolymers. ○, Copolymers; △, linear polyethylenes; △, branched polyethylenes; ◇, polyethylene blends. Note that density increases with crystallinity.

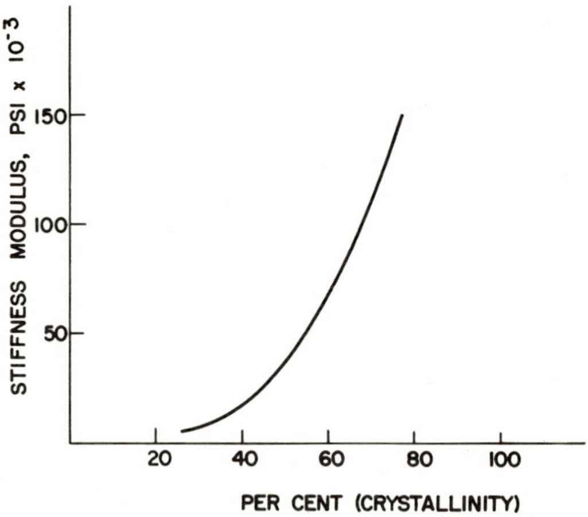

Figure 5-35. Effect of crystallinity upon stiffness (ref 120). Stiffness modulus of polyethylene as a function of crystallinity at 25°C.

Degree of Crystallinity Highly flexible polymer molecules (above their glass transition temperature) may be soft and rubbery even after the introduction of low degrees (5% to 15%) of crystallinity; in fact, these occasional points of immobility, in an otherwise highly flexible matrix, resemble the occasional cross-links in vulcanized rubbers, and actually provide the basis for strength and elasticity in systems such as plasticized polyvinyl chloride. Flexible polymer molecules become tough and leathery at intermediate degrees (20% to 60%) of crystallinity, for example in low-density polyethylene; and stiff to hard at

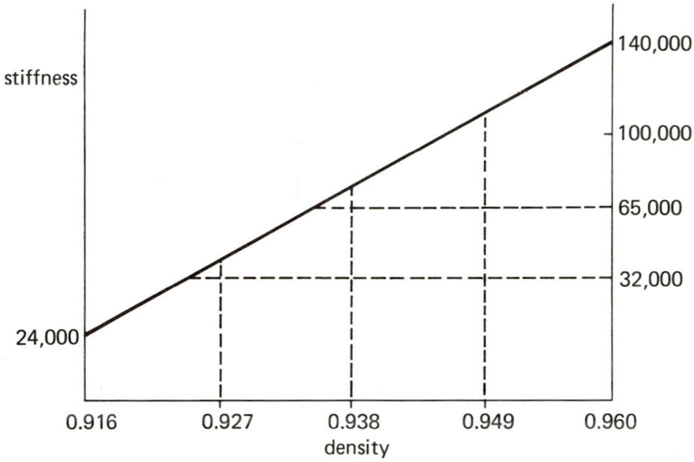

Figure 5-36. Effect of crystallinity upon stiffness of polyethylene (ref 121). Note that crystallinity increases with density.

high degrees (70% to 90%) of crystallinity, as in high-density polyethylene, polyoxymethylene, and nylons (ref 123–125). Probably the most detailed understanding of the relationships between molecular weight, molecular flexibility and crystallinity have been achieved in the study of polyethylenes (Figure 5-39).

Interaction between Crystallinity and Other Structural Features Stretching of regular amorphous rubbery molecules makes them parallel enough to permit some crystallization at high degrees of elongation, and this increases their modulus as they are stretched (ref 127) (Figures 5-29 to 5-32). Even low degrees of crystallinity also retard molecular motion sufficiently to give longer slower relaxation times and broader flatter modulus vs frequency curves, characteristic of plasticized polyvinyl chloride; and these effects become much more pronounced in more highly crystalline polymers (ref 128,129) (Figures 5-40 and 5-41).

In glassy amorphous polymers (below their glass transition temperatures), the molecules already have very little flexibility and mobility,

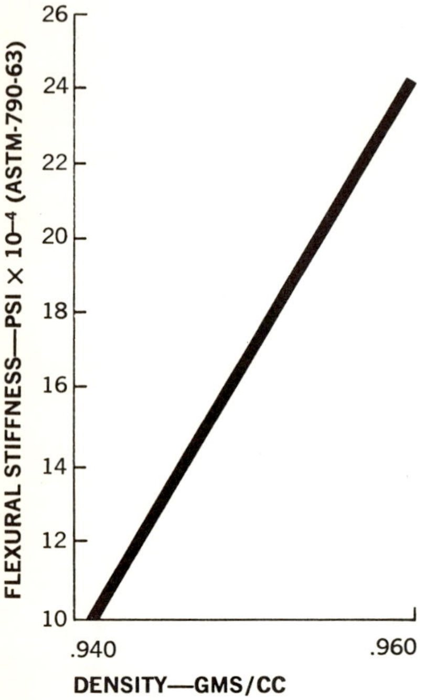

Figure 5-37. Effect of crystallinity upon flexural stiffness of polyethylene (ref 122). Note that density increases with crystallinity.

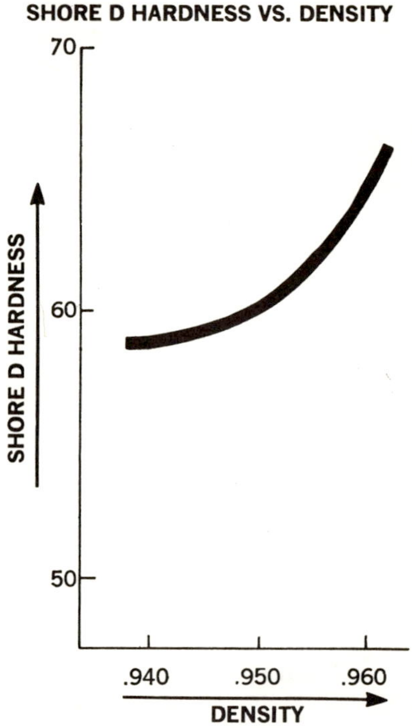

Figure 5-38. Effect of crystallinity upon hardness of polyethylene (ref 122). Note that density increases with crystallinity.

and the polymer already has a high modulus; with increasing degrees of crystallinity it may become tougher, more horn-like, and still harder and higher in modulus (ref 123). Molecules with a helical conformation generally give crystals with a modulus of about 10^{11} dynes/cm^2, while molecules with a zigzag conformation pack more neatly and tightly and give crystals with a modulus of about 10^{12} dynes/cm^2 (Table 5-5); both types are strong enough to be very useful in forming synthetic fibers (ref 132).

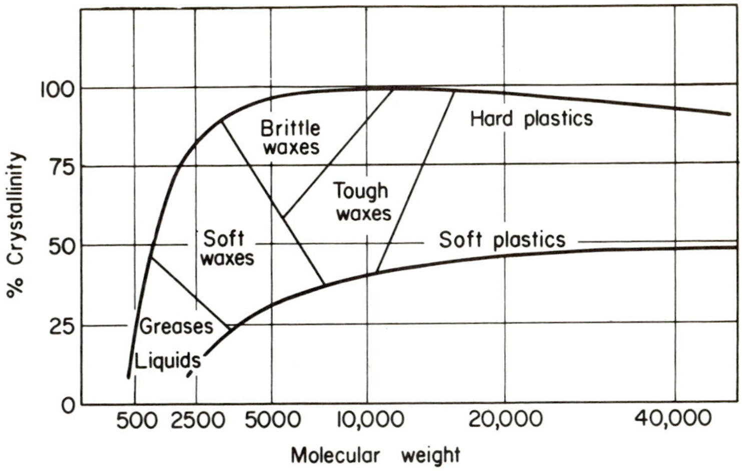

Figure 5-39. Effect of molecular weight and crystallinity upon mechanical properties of polyethylene (ref 126).
Turner Alfrey and Edward F. Gurnee, *Organic Polymers*, © 1967. Reprinted by permission of Prentice-Hall, Inc., Englewood Cliffs, New Jersey.

Thus the packing of polymer molecules into crystalline lattices generally restricts their mobility and stiffens their reversible mechanical properties.

ULTIMATE MECHANICAL PROPERTIES The effects of crystallinity upon ultimate mechanical properties are more diverse and less clearly understood.

Table 5-5. Conformation and Modulus of Crystalline Polymers (ref 129)

Polymer	Chain Conformation	Lattice Modulus $E \times 10^{11}$ dynes/cm^2
Polyethylene	Zigzag	23.5
Poly(vinyl alcohol)	Zigzag	25.0
Polypropylene (isotactic)	Helix	4.1
Polystyrene	Helix	1.2
Poly(oxymethylene)	Helix	5.3
Poly(ethylene terephthalate)	Helix	7.5
Poly(vinylidene chloride)	Helix	4.1
Polytetrafluoroethylene	Helix	7.6

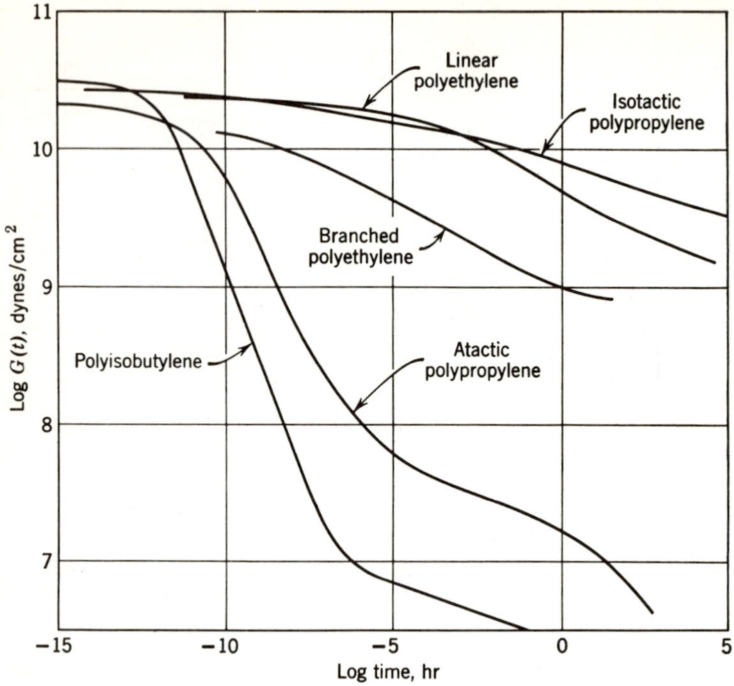

Figure 5-40. Stress relaxation of crystalline and amorphous polyolefins (ref 130).

Yield Strength The stress at the yield point, generally measured as tensile yield strength, increases markedly with crystallinity (ref 133–137) (Figures 5-42 and 5-43). This is probably the point at which the amorphous regions have flowed as far as possible, and beyond which the crystalline areas themselves must flow and re-orient to permit further deformation (Figure 5-44).

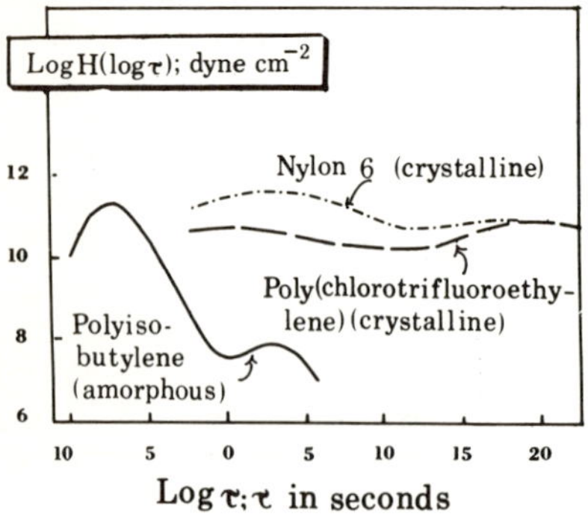

Figure 5-41. Relaxation time spectra of crystalline polymers (ref 131).

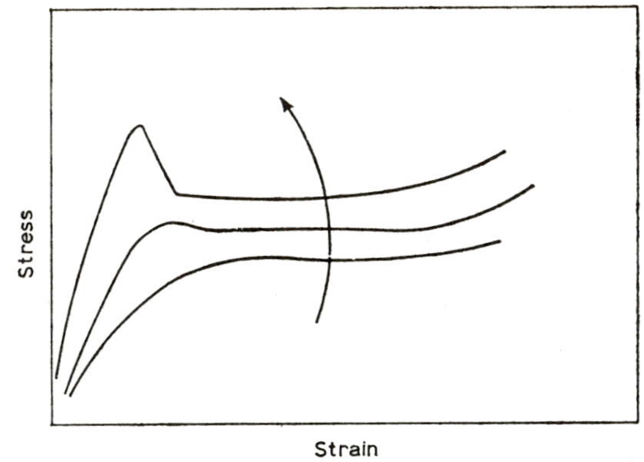

Figure 5-42. Effect of crystallinity upon yield point of polyolefins (ref 138). The arrow indicates increasing crystallinity. The growing peak indicates the appearance of a prominent yield point.

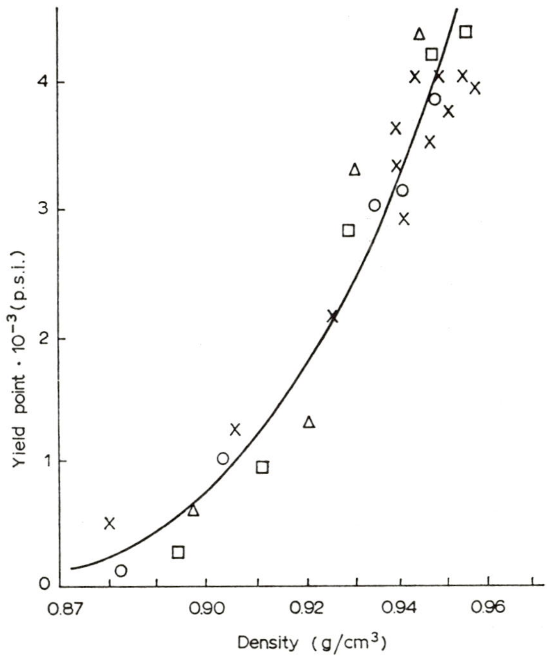

Figure 5-43. Effect of crystallinity upon yield point of polyethylenes (ref 139).

Yield point-density relationship for ethylene–1-olefin systems. ○, Ethylene–butene-1 copolymers; ×, ethylene–octene-1 copolymers; △, polyethylene–amorphous polypropylene blends; □, polyethylene–polyhexene blends. Note that crystallinity increases with density.

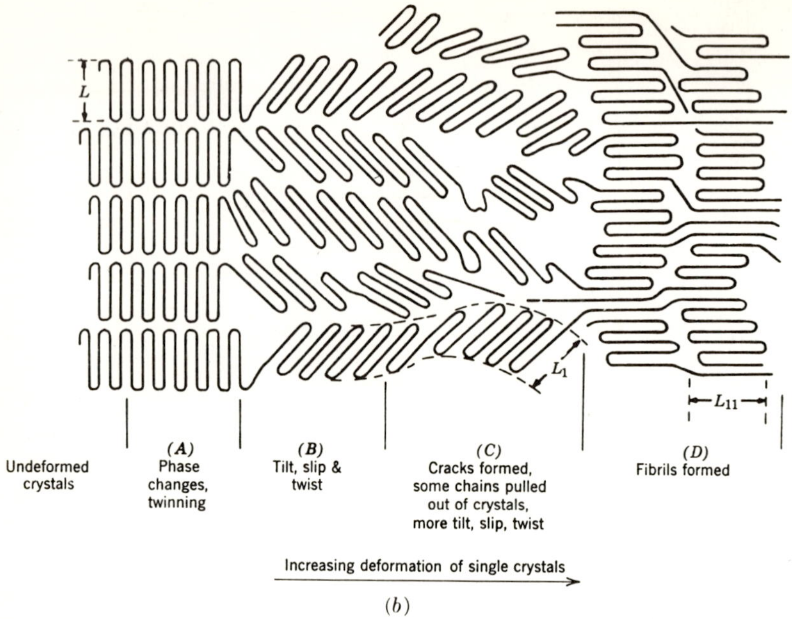

Figure 5-44. Reorientation during deformation of crystalline polymer (ref 140).

Ultimate Strength Ultimate tensile strength may be related to crystallinity, but the relationships are complex and not entirely understood. In some systems ultimate tensile strength may increase with crystallinity (ref 84,141–143) (Table 5-6) (Figure 5-45), particularly if the polymer is near or above its glass transition temperature (ref 145). In other studies there appears to be no relationship (ref 146); or crystallinity

Table 5-6. Crystallinity and Ultimate Tensile Strength of Polyethylenes (ref 144)

Density		Crystallinity	Ultimate Tensile Strength
Low (Branched Polymer)	0.920	65%	2000 psi
Medium (Copolymer)	0.935	75	2500
High (Copolymer)	0.950	85	4000
High (Homopolymer)	0.960	87	4500
Ultra-High (Homopolymer)	0.965	95	5500

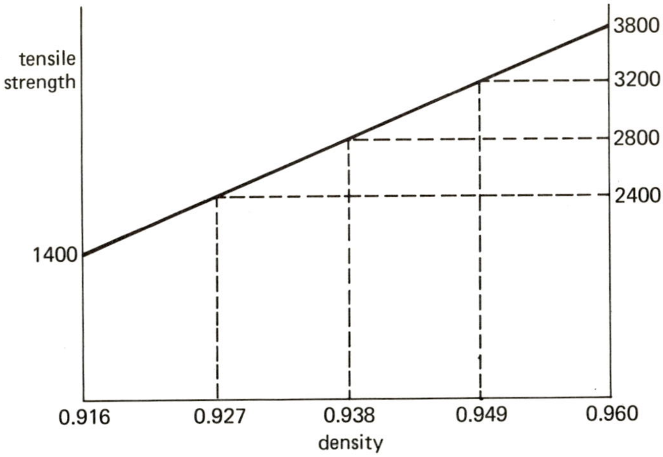

Figure 5-45. Effect of crystallinity upon ultimate tensile strength of polyethylene (ref 121). Note that crystallinity increases with density.

may actually decrease strength by causing stress-concentration and brittle failure (ref 145,147). In one theory, failure in a crystalline polymer is explained thermodynamically by localized melting at the point of fracture (ref 148). Finally, in the stretching of stereoregular elastomers, such as natural and butyl rubbers, temporary parallel orientation of the molecules produces temporary crystallization which greatly increases the ultimate tensile strength (ref 149,150) (Figure 5-46); when tearing

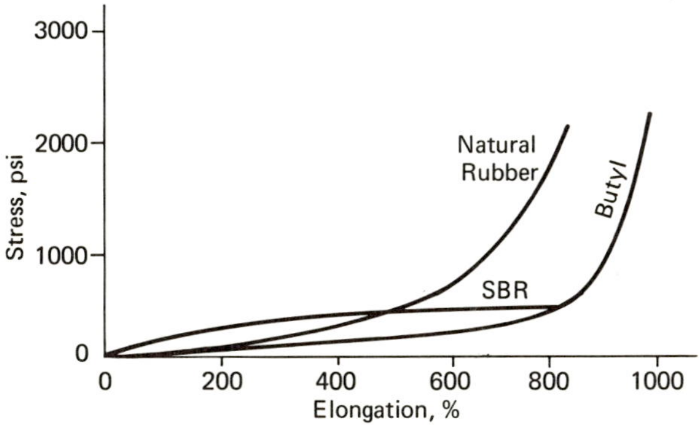

Figure 5-46. Stress-strain curves of vulcanized pure gum elastomers (ref 151).

failure finally does occur, it proceeds transversely through the amorphous phase, but stops when it encounters the much stronger oriented crystalline phase (ref 152). Since the mechanism of tensile failure itself is not yet entirely understood, it is not surprising that its relationship to crystallinity should also await further study and elucidation.

Ultimate Elongation Ultimate elongation, in polymers that are near or above the glass transition temperature, generally decreases as crystallinity increases (ref 145); this is due to the decreasing mobility of the

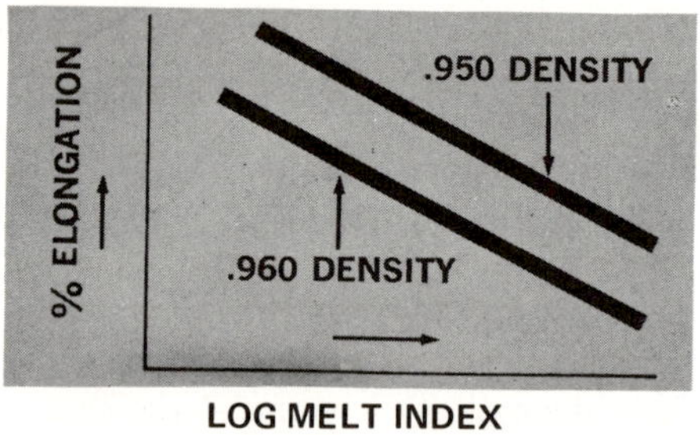

Figure 5-47. Effect of crystallinity upon ultimate elongation of polyethylenes (ref 122). Note that increasing density indicates increasing crystallinity.

system. Whereas polymers of medium crystallinity (20% to 60%) will cold draw beyond the yield point, polymers of high crystallinity (70% to 90%) become brittle and break (ref 128) (Figure 5-47); stress-concentration at the weak crystalline interfaces probably accounts for such premature failure.

Creep and Compliance Creep and compliance of polymers, above the glass transition temperature, are greatly reduced by crystallinity (ref 145,153,154). Small amounts of crystallinity ($\leq$15%) act like cross-links in vulcanized rubber, and prevent sagging and deformation in plasticized polyvinyl chloride film and sheet products. Beyond 40 percent crystallinity the crystallites are in continuous contact throughout the polymer mass, and reduce compliance greatly. These effects extend the rubbery plateau of stress relaxation over a much broader time span, so that the Boltzmann super-position principle no longer applies. Even so, creep is greater than in glassy or highly cross-linked systems, particularly at higher temperatures; this is because mechanical strain on imperfect crystalline areas causes local temporary melting and

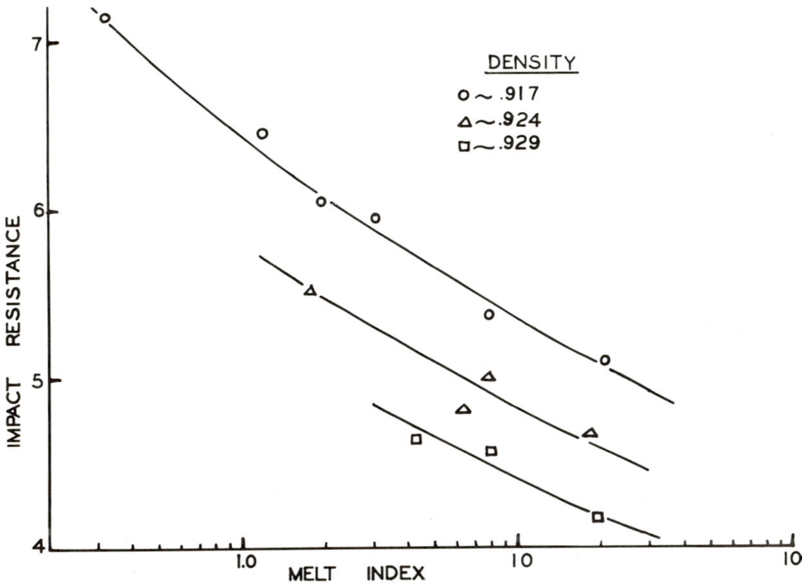

Figure 5-48. Effect of crystallinity upon impact resistance of polyethylenes (ref 160). Note that increasing density indicates increasing crystallinity.

recrystallization in less strained orientations, thus reducing the stress and permitting gradual creep.

Impact Strength Impact strength is probably related to crystallinity, but the precise relationships are far from clear. Crystallinity often decreases impact strength (ref 52,116,143,145,155) (Figure 5-48), and large crystallites particularly produce brittleness (ref 88). On the other hand, crystallinity sometimes produces moderate to high impact strengths (ref 156,157).

More detailed analysis suggests that, in polymers above their glass transition temperature, impact strength will be high even in the presence of crystallinity, but inverse to the degree of crystallinity (ref 158). When a polymer is below its glass transition temperature, at intermediate degrees ($\leq 60\%$) of crystallinity the polymer may still be tough and horn-like and have high impact strength; but at high degrees (70% to 90%) of crystallinity it is apt to be quite brittle (ref 128), because the sharp-pointed crystalline edges act as stress concentrators (ref 159).

COMPLEX MECHANICAL PROPERTIES When the complexities of crystalline stucture are superimposed upon the more complex types of mechanical properties, the results can be quite complicated. Three examples may be noted at this point:

Stress-cracking Stress-cracking is observed when a polymer is simultaneously immersed in a weak solvent or strong aqueous surfactant,

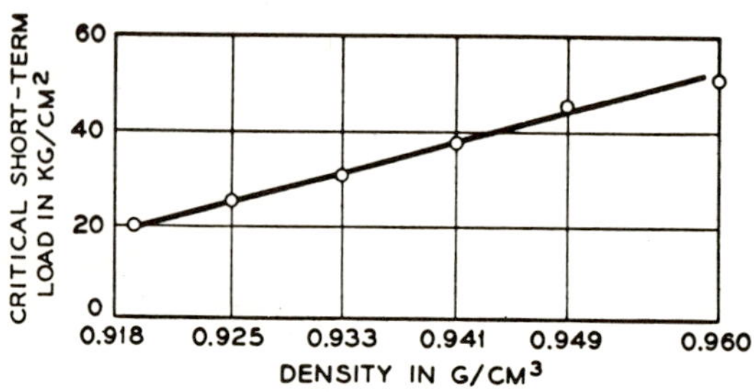

Figure 5-49. Effect of crystallinity upon critical load to stress-crack polyethylene (ref 162). Note that increasing density indicates increasing crystallinity.

subjected to a moderate mechanical stress insufficient to cause immediate failure, and held under these conditions for an extended period of time; under these combined conditions, brittle failure often occurs. The effect of crystallinity on stress-cracking depends upon the specific type and conditions of the test (ref 161). When a high constant short-term stress is applied and critical load is measured, modulus is the controlling factor, and stress-crack resistance increases with crystallinity (Figure 5-49); whereas when a constant long-term strain is applied and failure time is measured, flexibility is the controlling factor, and stress-crack resistance is generally inverse to crystallinity (Table 5-7, Figure 5-50). Thus the first type of test would be more useful for applications like water pipe, the latter for applications like power cable.

When a molten polymer is quenched to low crystallinity and then annealed to cause crystallization, increasing the annealing time increases crystallinity and decreases stress-crack resistance (Figure 5-51); but stress-crack resistance continues to decrease long after crystallinity has

Table 5-7. Effect of Crystallinity upon Failure Time in Stress-cracking of Polyethylene (ref 163)

Density at 23°C	Cracking Time, F_{50} (hr)
0.9218	>1000
0.9320	7
0.9415	>1000
0.9472	130
0.9492	200
0.9502	12
0.9573	72
0.9613	12

Note that increasing density indicates increasing crystallinity.

levelled off, suggesting that further changes in the texture of crystallinity are also important (ref 166).

Cyclic Stress Fatigue Cyclic stress fatigue can result from repeated applications of even rather low unit stresses, until cracking occurs (ref 167). One possible mechanism to explain such failure may start with the ability of mechanical stress to induce crystallization, and the effects of crystallinity on flexibility and on ultimate mechanical properties (all covered in Chapter 5) (ref 137).

Lubricity Mechanical theories of adhesion and friction often depend upon deformation of adjacent surfaces to make them conform more closely to each other's microscopic irregularities. Since crystallinity restrains molecular mobility and creep, it is not surprising that amorphous elastomers have high friction, while highly crystalline nylon and polyacetal resins have high lubricity. In polyethylene film, resistance to blocking, and improved slip during handling, increase with increasing crystallinity (ref 116).

Balance of Mechanical Properties Since crystallinity produces such a range of effects upon individual mechanical properties, it is useful

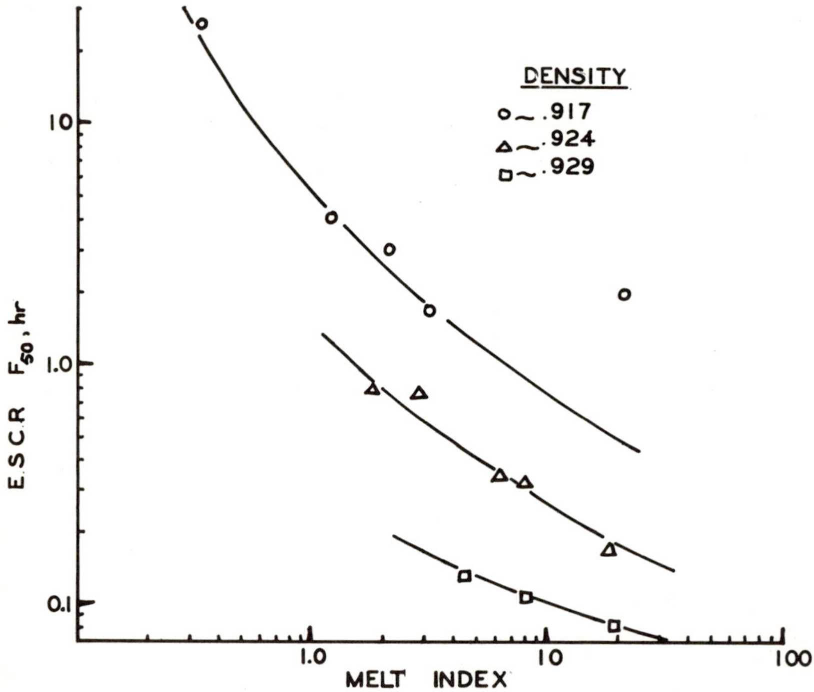

Figure 5-50. Effect of crystallinity upon failure time in stress-cracking of polyethylene (ref 164). Note that increasing density indicates increasing crystallinity.

in summary to consider the balance of these effects in specific applications. In fibers, where the fine diameter and monoaxial construction already produce considerable flexibility, the main requirement is maximum axial modulus and strength, which are conveniently produced by high crystallinity; and thus all fiber-forming polymers rely upon maximum development of crystallinity to produce optimum properties. In stereoregular rubbers, temporary crystallization on stretching combines soft flexibility with high strength and elasticity (ref 168). Polymers of moderate crystallinity, between their glass transition and melting temperatures, combine the strength of the crystalline phase plus the impact resistance of the flexible amorphous phase, and are generally hard tough strong micro-composites producing optimum properties for plastics applications (ref 169,170) (Chapter 7).

Such structures are particularly conveniently produced by block copolymerization, thus combining a balance of rigidity and impact strength superior to what could be obtained in homopolymers or blends;

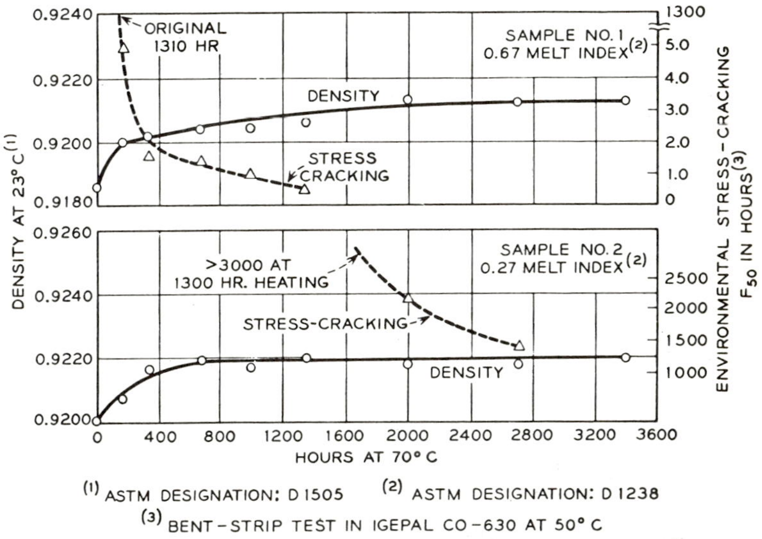

Figure 5-51. Effects of annealing upon density (crystallinity) and stress-cracking of polyethylene (ref 165).

for example, the impact strength of ethylene/propylene block copolymers is far superior to either homopolymer (ref 63) (Figure 5-52). Similarly, polyurethane block copolymer fibers combine elasticity, strength, and tear resistance in this way (ref 171).

Acoustic Properties

Crystallinity inhibits short rapid displacements of molecular segments, retarding them to slower longer relaxation times, and thus decreases the velocity of sound in polyethylene at higher density (ref 136).

Thermal Properties

Crystallinity has a number of important effects upon the thermal properties of a polymer. Its most general effects are the introduction of a sharp melting point and the stiffening of thermal mechanical properties. It also has specialized effects upon thermal conductivity, heat capacity, and thermal expansion.

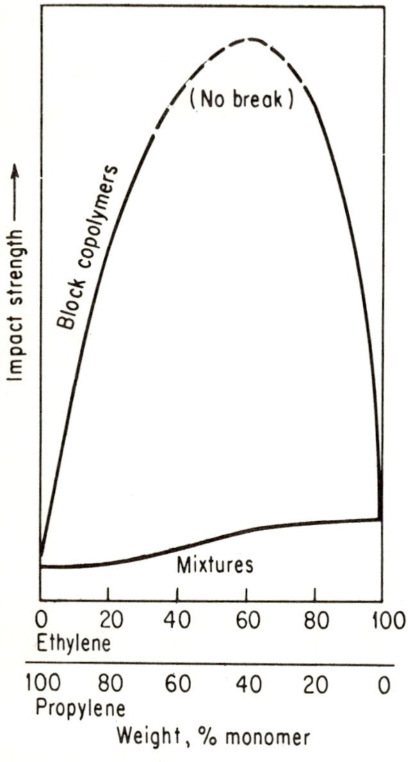

Figure 5-52. Impact strength of ethylene/propylene block copolymers (ref 63). Reprinted by permission of Modern Plastics Encyclopedia, McGraw-Hill, Inc.

Melting Behavior When the temperature of a solid amorphous polymer is raised, larger segments become mobile with greater frequency, and the polymer gradually softens through a rubbery liquid stage to a fluid state. When a crystalline polymer is heated, on the other hand, the molecules in the tightly-packed regular crystalline lattice remain firmly fixed until they reach a temperature at which their thermal energy exceeds the forces holding the crystal together, and at this point the crystal melts sharply, releasing the molecules directly into the fluid state (ref 172). Thus each crystalline polymer has a characteristic melting point (Table 5-4). In fact the melting point is so characteristic of polymer crystallinity that it is often taken as a measure of crystallinity, and particularly of the factors and forces in the polymer structure that favor and control crystallinity (Section on Factors Determining Crystallinity).

The melting point T_m of a crystalline polymer, shown in the equation $T_m = \Delta H/\Delta S$, is a direct function of the attractive forces holding the crystal together, such as regularity and polarity, as measured by the heat of fusion ΔH; and an inverse function of the entropy ΔS gained by converting *(a)* the statistically improbable regularity of the crystalline lattice into *(b)* the statistically much more probable random coil of the molten polymer.

Factors which Increase the Melting Point Structural factors which function to hold the crystal lattice together have already been discussed in detail; these all tend to raise ΔH and thus T_m.

Other factors controlling melting point are the perfection of crystallinity, the percentage of crystallinity, and the size of the crystals formed. As these increase, the melting point increases somewhat and occupies a narrower temperature range (ref 173–175).

Factors which Decrease the Melting Point Structural factors which increase the change in regularity and entropy during melting, and thus inversely affect the melting point, are not as well understood. It is clear that polyethylene crystallizes in a zigzag, highly regular conformation, and melts to a flexible, highly random coil, and that the resulting large increase in entropy contributes to its relatively low melting point (ref 176). Conversely, the high melting point of polytetrafluoroethylene is due to the stiffness of the rod-like molecule, which permits little random coiling even in the molten state, and thus provides little gain in entropy. When this reasoning is extended to the high melting points of polyamides and polyureas, and used to imply some degree of organization of the molecules in the molten polymer (ref 177), however, it is obvious that this theory requires further study of molecular conformation and flexibility in the molten state.

THERMAL MECHANICAL PROPERTIES *Modulus* Amorphous polymers are rigid and glassy at low temperatures because the individual segments of the polymer molecules have too little thermal energy and are too closely packed to be able to move past each other. As the temperature is raised and their thermal vibration increases, they expand gradually until there is enough free volume to permit segments of molecules to move past each other. As they pass the critical glass transition temperature range they become stiffly flexible and leathery, then softly flexible and rubbery. This rubbery plateau continues up to the point at which large segments of the polymer molecules slip past each other easily enough to permit whole molecules to disentangle occasionally, producing rubbery flow; and at still higher temperatures whole molecules disentangle and flow past each other easily, producing true liquid flow.

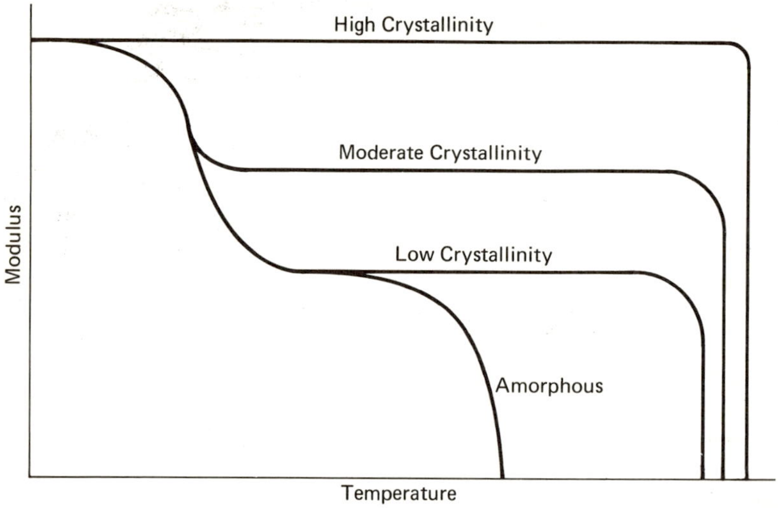

Figure 5-53. Effect of crystallinity upon modulus vs temperature curve (ref 125,128,178).

This characteristic thermal mechanical behavior of amorphous polymers is sharply modified by the introduction of crystallinity (Figure 5-53). Low degrees of crystallinity, as in polyvinyl chloride, do not affect the glassy, leathery, or rubbery regions, but introduce occasional points of crystalline bonding which prevent the molecules from disentangling and flowing past each other; this extends the rubbery region up to much higher temperatures until crystalline melting of these points occurs, at which time the rubbery state shows a sharp transition to a fluid liquid state. Moderate degrees of crystallinity, as in low-density polyethylene, do not affect the rigid glassy state or the glass transition, but introduce considerable stiffening into the flexible polymer, making it stiffer and more leathery; they also extend this leathery plateau out to much higher temperatures until crystalline melting occurs, at which point there is a sharp transition to a fluid liquid state.

High degrees of crystallinity—as in high-density polyethylene, polypropylene, polyoxymethylene, and nylons—still have little or no effect upon the rigidity of the glassy amorphous state at low temperatures, and only minor effects upon the phenomenon of the glass transition, but they introduce so much stiffening into the flexible amorphous state that it becomes quite tough and rigid; they also extend this tough rigid state out to much higher temperatures, and thus greatly broaden the

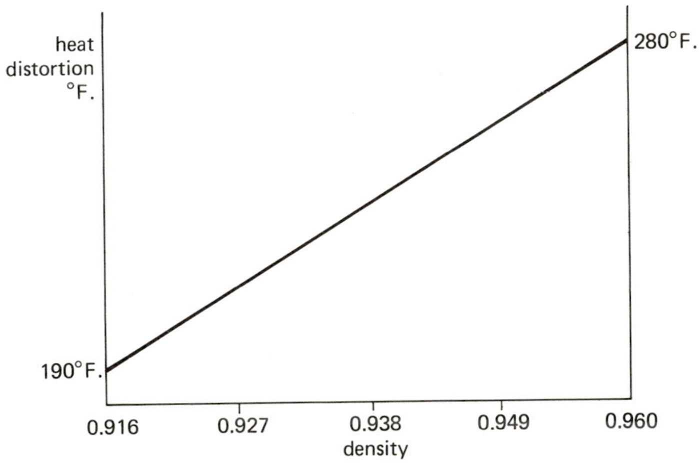

Figure 5-54. Effect of crystallinity upon heat distortion temperature of polyethylene (ref 121). Note that increasing density indicates increasing crystallinity.

range of useful rigid properties, until true crystalline melting occurs quite sharply to a fluid liquid state. At very high degrees of crystallinity even the completely rigid state may become somewhat higher in modulus; and the glass transition has very little effect upon the rigidity of the polymer, which remains fairly constant all the way up to the sharp crystalline melting point, at which a rapid transition occurs into the fluid liquid state.

To some extent crystallinity may reduce the mobility of segments of the adjacent amorphous polymer phase, and increase the temperature at which they have enough mobility to produce a glass transition. This may raise the glass transition temperature, heat distortion temperature (Figure 5-54), Vicat softening temperature (Figures 5-55 and 5-56),

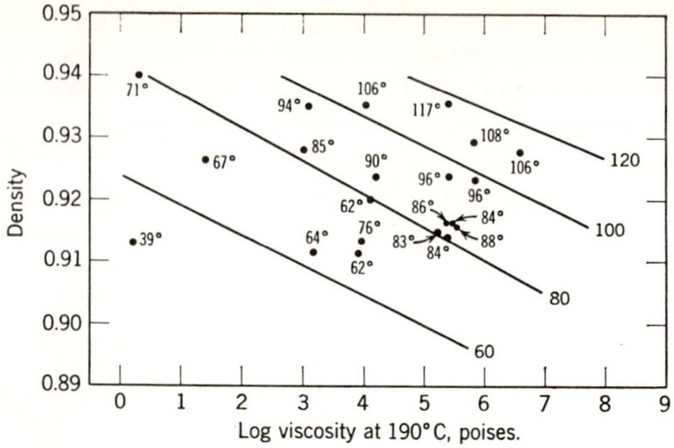

Figure 5-55. Effect of crystallinity upon Vicat softening temperature of low-density polyethylene (ref 187). Note that increasing density indicates increasing crystallinity.

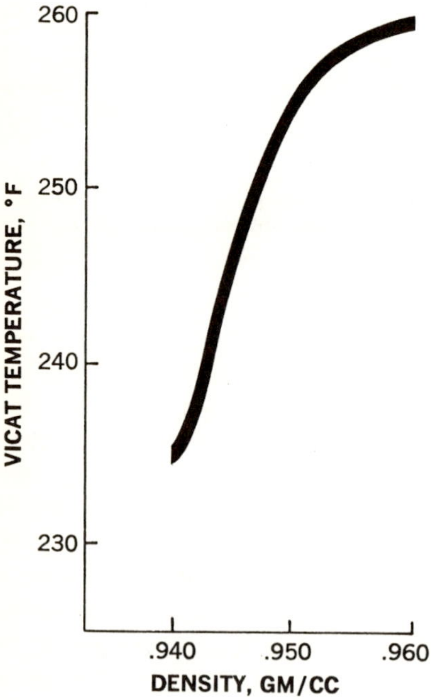

Figure 5-56. Effect of crystallinity upon Vicat softening temperature of high-density polyethylene (ref 122). Note that increasing density indicates increasing crystallinity.

brittleness temperature (Figure 5-57), and other similar measures of the transition from rigid to flexible properties (ref 179,180).

Impact Strength In addition to the interacting effects of crystallinity and temperature on modulus, this interaction also has interesting and useful effects on resistance to brittle failure under impact. In amorphous

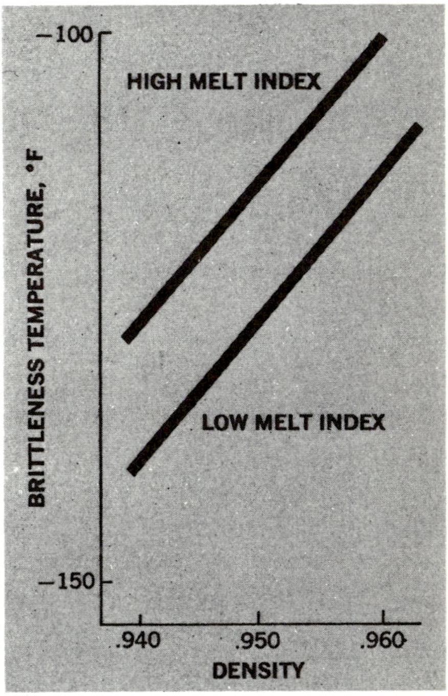

Figure 5-57. Effect of crystallinity upon brittleness temperature of polyethylene (ref 122). Note that increasing density indicates increasing crystallinity.

polymers the optimum combination of rigidity and impact strength generally occurs in a temperature range between the glass transition temperature and about 20°C above this transition temperature (ref 181,182). In partially crystalline polymers the optimum combination of rigidity and impact strength generally occurs in the much broader range between the glass transition temperature and the melting point (ref 181,183). In general this range is large enough to be of considerable practical value. In symmetrical crystalline polymers like polyvinylidene chloride, the glass transition temperature may be about one half of the melting temperature in °K; while in unsymmetrical crystalline polymers like polychlorotrifluoroethylene and polyethylene terephthalate, the ratio is about two thirds (ref 100,184). The range also increases with increasing molecular weight (ref 185) (Figure 5-58).

All-cis and all-trans polydiene elastomers can be cooled to a true crystalline melting point, and become hard and tough below this melting point (ref 188). Some crystalline polymers may be brittle at very low temperatures, but have polycrystalline secondary transitions at moderate temperatures, above which they become tough and still retain their rigidity up to the primary melting point at higher temperatures (ref 189).

Block copolymers can have two separate melting points, one for each of the individual ingredients, and the range between these two melting points can combine the rigidity of the high-melting blocks with the tough impact strength of the low-melting blocks (ref 63,67) (Figures

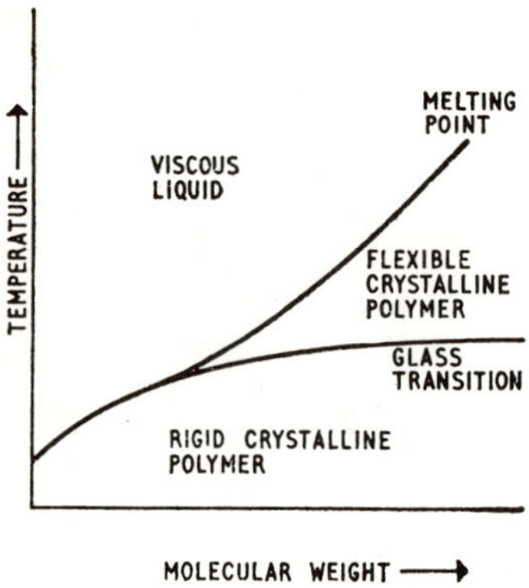

Figure 5-58. Effect of molecular weight upon range between glass transition and melting points (ref 186).

5-59, 5-60). As indicated earlier, such structures offer great promise of practical utility.

Thermal Conductivity Conduction of heat through crystalline solids is facilitated by coupled vibrations of the atoms in the lattice structure. For this reason, thermal conductivity of crystalline polymers is about twice as high as that of amorphous polymers (ref 190).

Heat Capacity The amount of heat needed to raise the temperature of a solid polymer increases gradually with temperature (Figure 5-61). In amorphous polymers it levels off above T_g; whereas in highly crystalline polymers, it shows a sharp discontinuity at T_m representing the tremendous amount of thermal energy used in melting the crystalline structure (ref 192).

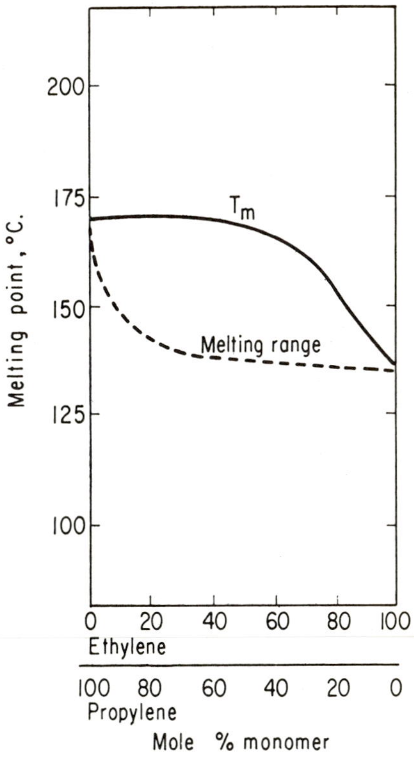

Figure 5-59. Broad melting range of ethylene/propylene block copolymers (ref 63). Reprinted by permission of Modern Plastics Encyclopedia, McGraw-Hill, Inc.

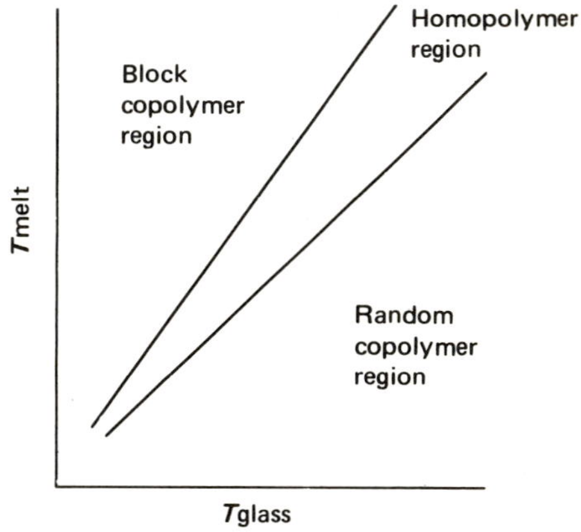

Figure 5-60. Melting vs glass transition temperatures of copolymers (ref 67). Turner Alfrey and Edward F. Gurnee, *Organic Polymers*, © 1967. Reprinted by permission of Prentice-Hall, Inc., Englewood Cliffs, New Jersey.

Coefficient of Thermal Expansion In a partly crystalline polymer below T_g, crystallinity has little effect on thermal expansion (ref 192) (Figure 5-62). Between T_g and T_m, increasing crystallinity decreases the rate of thermal expansion; and approaching T_m produces sharp melting to the less dense liquid phase. These considerations are important primarily in process engineering thermodynamics.

Electrical Properties

Crystallinity restricts the mobility of polar groups in the polymer molecule, and thus affects dielectric properties of the polymer. In the

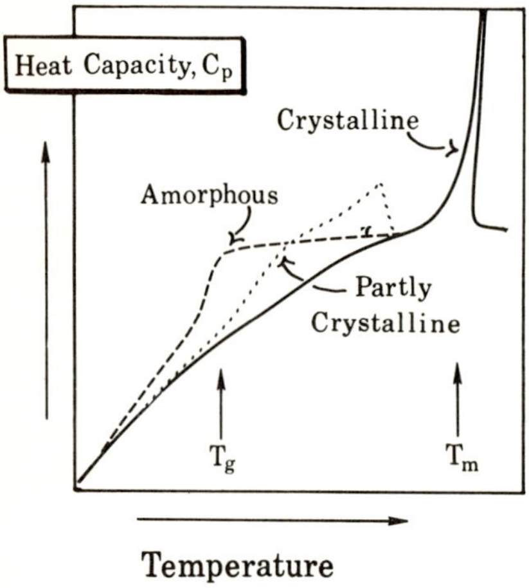

Figure 5-61. Heat capacity of amorphous and crystalline polymers (ref 191).

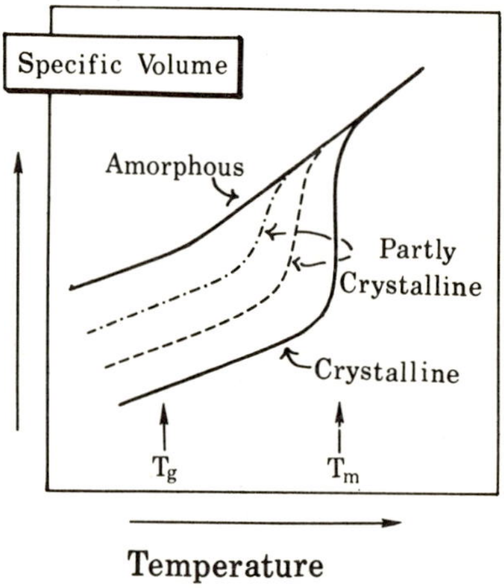

Figure 5-62. Thermal expansion of amorphous and crystalline polymers (ref 191).

molten state, thermal randomization of polar groups is greater than the orienting effect of an electric field, and dielectric constant is rather low (Figure 5-63). With decreasing temperature, thermal randomization decreases, the orienting and polarizing effect of the electric field remains, and thus dielectric constant increases. At the melting point, the molecules are frozen into the lattice structure, further reducing mobility and increasing polarization and dielectric constant; the sudden increase in density also plays a role here by increasing the concentration of polar groups per unit volume. Below the melting point, continued decrease in temperature can have either of two effects. If polar side groups still have considerable freedom of rotation, as in polymethyl methacrylate, these still

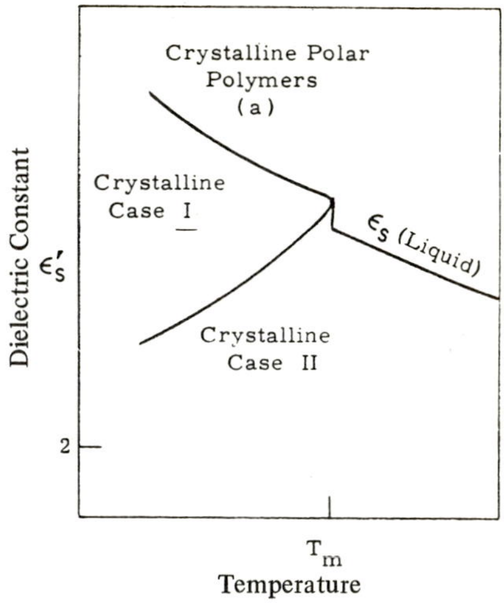

Figure 5-63. Effect of temperature upon dielectric constant of polar crystalline polymers (ref 193).

randomize and counteract the polarizing effect of the electric field; as temperature continues to decrease, the mobility of these side groups decreases, the orienting and polarizing effect of the electric field takes control, and dielectric constant continues to rise (Case I). If the polar groups are locked into fixed positions in the crystalline lattice, as in polychlorotrifluoroethylene, they can no longer respond to the orienting effect of the electric field, and dielectric constant decreases; as temperature decreases, these polar groups become less and less able to respond, and dielectric constant continues to decrease (Case II) (ref 194). Similarly dielectric loss index behavior would parallel the effects on dielectric constant (Figure 5-64).

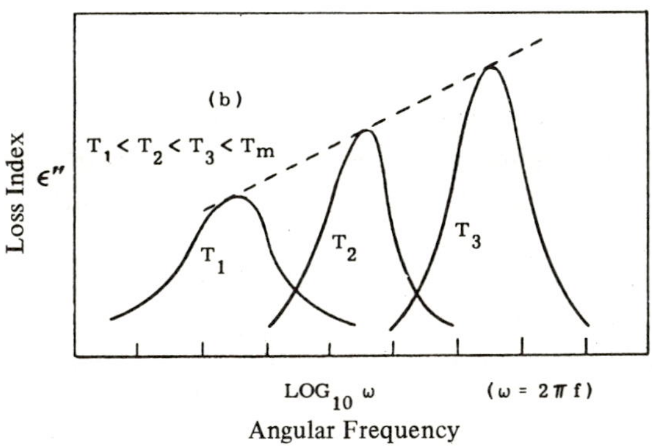

Figure 5-64. Effect of temperature upon loss index vs frequency of polar crystalline polymers (ref 195).

This is still only a simplified schematic description, and actual polymer structures would be expected to display much more complicated behavior. Thus for example in polyethylene terephthalate, the cis and trans forms of the terephthalic group affect dipole activity, and crystallization lowers dissipation factor and shifts the peak to lower temperatures (ref 196) (Figure 5-65).

Since polymers are used as electrical insulators over a considerable range of temperatures, and dielectric losses often generate further heat during use, such interesting effects of crystallinity and temperature upon

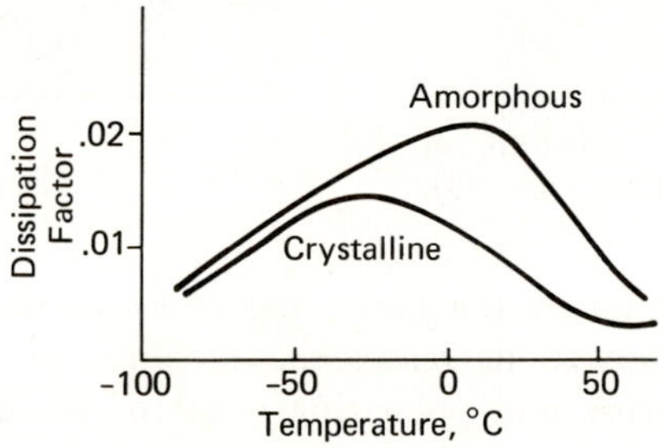

Figure 5-65. Effect of crystallinity upon dissipation factor vs temperature in polyethylene terephthalate film (ref 197).

dielectric properties are important considerations in the design and performance of electrical and electronic equipment.

Optical Properties

Crystallinity increases the density of the polymer, which decreases the speed of light passing through it and thus increases the refractive index. When crystals are larger than the wavelength of visible light—about 400–700 $m\mu$ (ref 198)—light passing through many successive crystalline and amorphous areas is scattered, and clarity of the polymer decreased (ref 80,199,200). Large single crystals scatter light at wide angles and thus cause haze (ref 201).

Clarity of crystalline polymers can be improved by quenching or using nucleating agents to produce a larger number of finer crystals, smaller than the wavelength of visible light; or by random copolymerization to reduce both the extent and the size of crystallinity (ref 52). Thus for example, the clearest polyethylene film is produced by melt extrusion followed immediately by casting the molten film onto chilled steel rolls to accelerate nucleation and form many fine crystals as rapidly as possible (ref 202). Conversely, slow cooling and annealing produce a small number of large crystals and cause opacity in the crystalline polymer.

X rays, with wavelengths of 1–120 $\times$ 10^{-9} cm (ref 203), are very useful in studying the ultimate details of crystalline structure, which fall in this same range of dimensions (ref 204–206) (Figures 5-66 to 5-68).

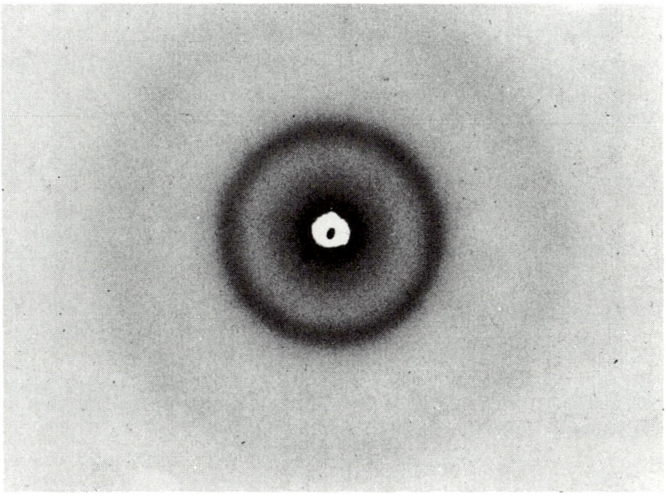

Figure 5-66. X-ray diffraction pattern of an amorphous polymer (ref 207), poly t-butyl acrylate.

Chemical Properties

The tightly-packed regular structure of crystalline polymers is so stable that it is difficult for chemical reagents to penetrate or separate the molecules from such a structure, unless considerable energy is available to help in this separation. This has a number of effects upon the chemical behavior of crystalline polymers.

PLASTICIZATION High crystallinity prevents compatibility with plasticizers, because they are unable to separate the polymer molecules sufficiently to move in between them (ref 80). Thus plasticization of polymers like polyethylene and nylon is quite difficult and limited in

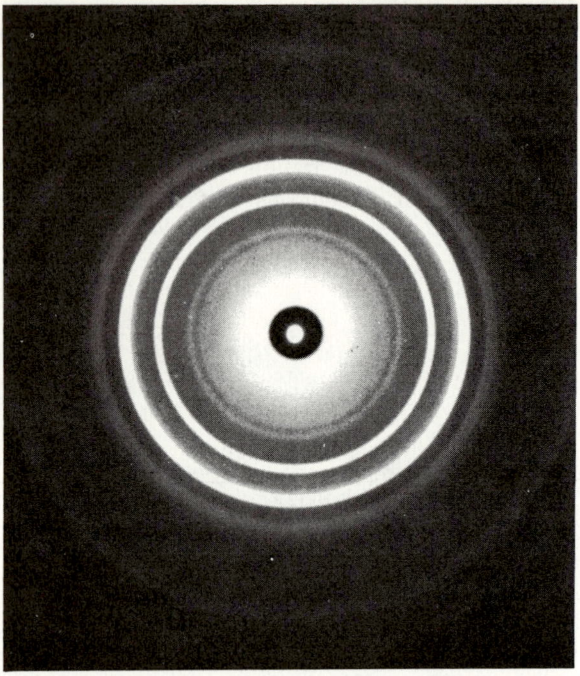

Figure 5-67. X-ray diffraction pattern of a crystalline polymer (ref 208), polyethylene oxide.

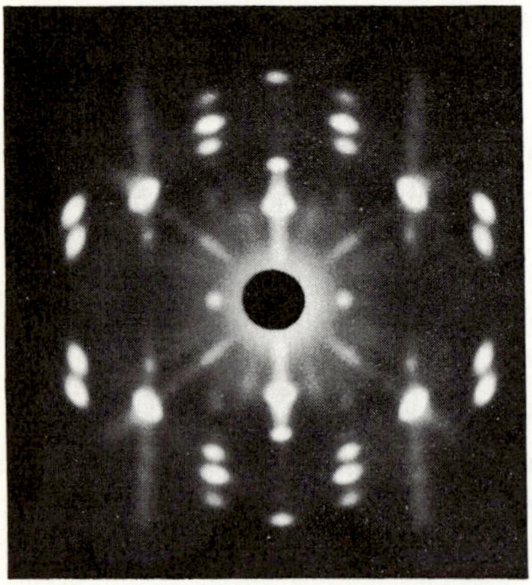

Figure 5-68. X-ray diffraction pattern of a polymeric fiber (ref 209), poly-4-methyl-1-pentene.

its application. On the other hand, low crystallinity may be decreased or destroyed by the addition of plasticizers (ref 80), as in polyvinyl chloride.

SOLUBILITY Whereas linear amorphous polymers dissolve easily in a range of organic solvents, crystalline polymers are usually insoluble, and can be dissolved only under limited conditions in a few specific solvents (ref 80,170,181,210,211). This is because the attractive forces and stability of the crystal structure are greater than those that result from interaction between polymer and solvent. Thus crystallinity is a stronger force than the solubility parameter alone (ref 212).

Crystalline polymers can be dissolved above their melting point, where sufficient thermal energy has already been supplied to separate the polymer molecules from the tight crystalline structure (ref 213). Nonpolar polymers can be dissolved just below the melting point, where the thermal energy alone is almost sufficient to separate the molecules and the polymer-solvent interactions supply what little additional energy is required (ref 214). The interaction between polar crystalline polymers and polar solvents can provide a greater proportion of the energy needed to dissolve the crystal (ref 215), particularly when strong hydrogen bonds can form between the solvent and the polymer (ref 216,217). If polar forces and hydrogen bonding between polymer and solvent are strong enough, the polymer may dissolve above the melting point and still remain in solution on cooling again (ref 218).

The solvent resistance of crystalline polymers provides useful chemical resistance in many household and mechanical applications. On the other hand, it makes solution-spinning of fibers quite difficult; and solution-casting of coatings and adhesives too impractical for wide commercial use.

SORPTION, DIFFUSION, AND PERMEABILITY Sorption, diffusion, and permeability require free volume for the small molecules of solvents or gases to dissolve and permeate through the polymer (ref 219); thus they are limited to the amorphous phase in partially crystalline polymers (ref 80,220) (Figure 5-69). Diffusion occurs mainly in the amorphous regions, and motion around the crystalline areas is a tortuous and difficult path (ref 222,223). Permeability depends upon both solubility and diffusion; since crystallinity reduces each of these, it has even a greater effect in reducing their product (ref 223). Thus crystallinity of a polymer greatly reduces its permeability (ref 80,224,225) (Tables 5-8, 5-9). This is particularly useful in packaging of foods, and may provide superior protective coatings as well. In polyethylene film, for example, increasing crystallinity reduces gas, liquid, and moisture vapor transmission, and improves resistance to grease as well (ref 116) (Figures 5-70, 5-71).

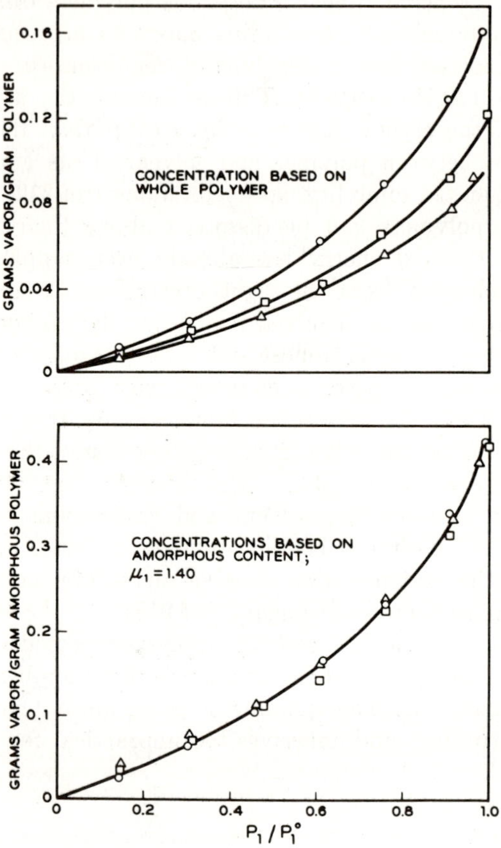

Figure 5-69. Effect of crystallinity upon sorption (ref 221).
Sorption isotherms for methyl bromide and different density polyethylenes at 0°C. ○ density 0.922 g/cc, □ 0.938 g/cc, △ 0.954 g/cc.

Dyeability Fibers need high crystallinity for mechanical strength, but they also need amorphous areas for dyeability. For example, the high crystallinity of polyacrylonitrile fibers makes dyeing very difficult. Fortunately, grafting them with 10 percent of polyvinyl alcohol can triple their dyeability without affecting tenacity, extensibility, or water

Table 5-8. Effect of Crystallinity upon Permeability of Polymers: Values for $P \times 10^{10}$ cm^3 sec^{-1} mm cm^{-2} cmHg^{-1} (ref 226)

Polymer	Permeability				Ratios to N_2 per $m = G$ value			Nature of polymer
	N_2(30°C)	O_2(30°C)	CO_2(30°C)	90 % R.H. H_2O(25°C)	P_{O_2}/P_{N_2}	P_{CO_2}/P_{N_2}	P_{H_2O}/P_{N_2}	
Poly(vinylidene chloride)	0.0094	0.053	0.29	14	5.6	31	1,400	Crystalline
PCTFE	0.03	0.10	0.72	2.9	3.3	24	97	Crystalline
Poly(ethylene terephthalate)	0.05	0.22	1.53	1,300	4.4	31	26,000	Crystalline
Rubber hydrochloride (Pliofilm ND	0.08	0.30	1.7	240	3.8	21	3,000	Crystalline
Nylon 6	0.10	0.38	1.6	7,000	3.8	16	70,000	Crystalline
PVC (unplasticised)	0.40	1.20	10	1,560	3.0	25	3,900	Slight crystalline
Cellulose acetate	2.8	7.8	68	75,000	2.8	24	2,680	Glassy
Polyethylene (d = 0.954–0.960	2.7	10.6	35	130	3.9	13	48	Crystalline
Polyethylene (d = 0.922)	19	55	352	800	2.9	19	42	Some crystalline
Polystyrene	2.9	11	88	1,200	3.8	30	4,100	Glassy
Polypropylene	—	23	92	680	—	—	—	Crystalline
Butyl rubber	3.12	13.0	51.8	—	4.1	16.2	—	Rubbery
Methyl rubber	4.8	21.1	75	—	4.4	15.6	—	Rubbery
Polybutadiene	64.5	191	1,380	—	3.0	21.4	—	Rubbery
Natural rubber	80.8	233	1,310	—	2.9	16.2	—	Rubbery

Table 5-9. Effect of Density (Crystallinity) upon Permeability (ref 227)

Polymer	Density, g/cc	% crystallinity (approx.)	N_2			O_2			CO_2		
			P, 30°C	P_o	E_P	P, 30°C	P_o	E_P	P, 30°C	P_o	E_P
Polychlorotrifluoroethylene	—	30	0.008	—	—	0.05	—	—	0.11	—	—
	—	80	0.004	—	—	0.013	—	—	0.03	—	—
Polymethylene	0.933	—	0.15	0.52	10.4	—	—	—	—	—	—
Polyethylene	*	0	16	0.021	5.7	—	—	—	—	—	—
	0.922	60	1.9	47	11.7	5.5	14	10.3	25.2	1.5	8.2
	0.938	69	0.66	2.3	10.5	2.1	1.2	9.4	7.4	0.16	7.4
	0.954	78	0.33	1.5	10.6	1.1	0.23	8.8	4.3	0.089	7.4
	0.960	81	0.27	0.7	10.3	1.06	—	—	3.5	—	—
	0.965	83	—	—	—	0.5	0.11	8.8	2.5	0.041	7.3
Polypropylene	0.907	—	0.42	540	14.0	2.1	37	11.4	8.4	3.2	9.1

P in units of cc (STP)/mm/cm^2/sec/cm Hg $\times 10^9$.
P_o in units of cc (STP)/mm/cm^2/sec/cm Hg $\times 10^2$.
E_P in units of kcal/mole.
* Extrapolated from the melt.

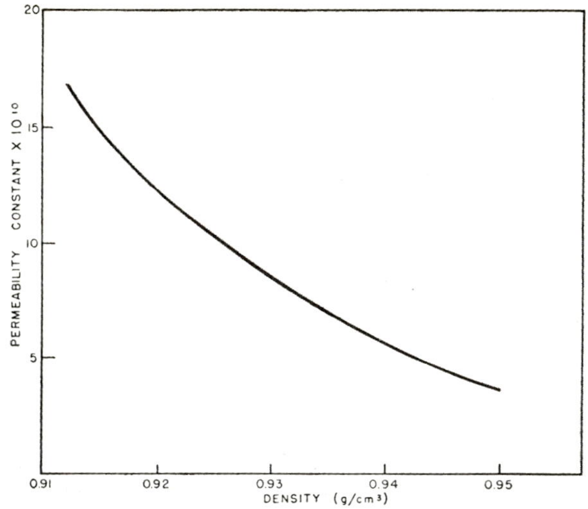

Figure 5-70. Effect of density (crystallinity) upon nitrogen permeability of polyethylene film (ref 234).

sensitivity (ref 228) (Table 5-10). Similar problems and techniques are found in polypropylene fibers.

Stress Cracking Stress cracking probably starts at weak amorphous interfaces between crystalline regions (ref 230). Thermal stress cracking is worst at high crystallinity (ref 231) (Figure 5-72). Environmental stress-cracking also depends upon crystallinity, but the exact relationship is complex and depends upon the tests used and applications

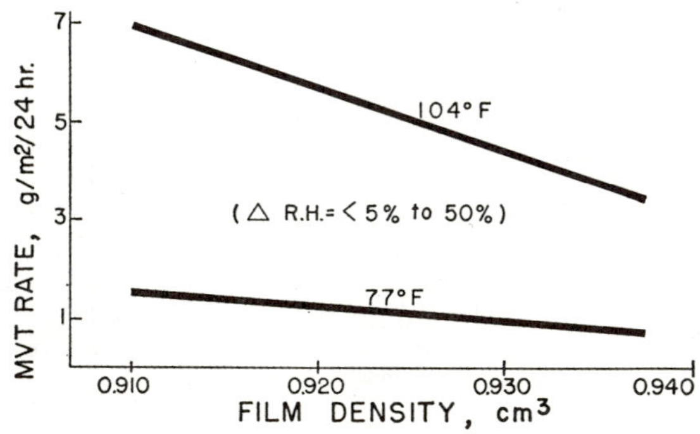

Figure 5-71. Effect of density (crystallinity) upon moisture vapor transmission through polyethylene film (ref 234).

Table 5-10. Mechanical Properties of Fibers Made from Polyacrylonitrile and from a Graft Copolymer of Acrylonitrile and Vinyl Alcohol Containing Ten Percent, by Weight, of Poly(vinyl Alcohol) (ref 229)

Property	Polyacrylonitrile	Graft Copolymer
Denier	285.3	192.4
Tenacity of dry fiber, *g* per denier	3.3	3.1
Ratio of tenacities of wet and dry fibers	0.97	0.94
Percent extensibility of dry fiber	14.6	15.6
Ratio of extensibilities of wet and dry fibers	1.10	1.03
Percent shrinkage in boiling water	8.1	6.8
Percentage of dye removed from dye bath by fiber in 60 minutes at 100°C	13.2	40

previously considered. These problems arise in many areas, particularly in high-density polyethylene bottles and pipe, and are most readily solved by copolymerization to reduce crystallinity slightly (ref 233).

Price

Historically amorphous polymers were first synthesized easily from low-cost raw materials. The first crystalline polymers were linear condensation polymers which required more complex synthetic routes, and were also more difficult to polymerize, making them much higher in

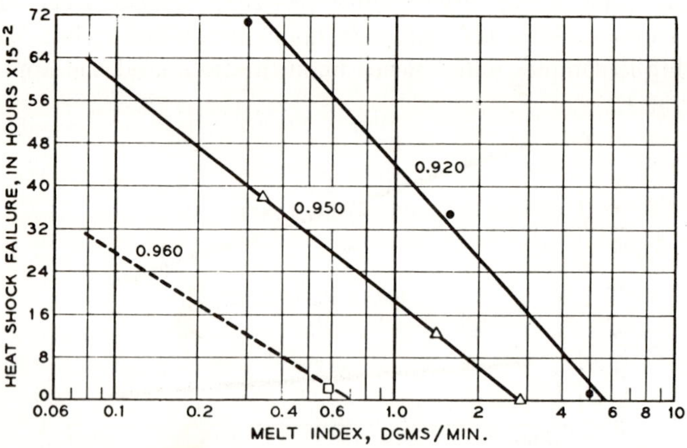

Figure 5-72. Effect of density (crystallinity) upon thermal stress-cracking (ref 232). Thermal stress-cracking versus melt index in polyethylene wire insulations at 70°C.

cost. More recently polymerization of low-cost olefins has made crystalline polymers among the most inexpensive polymers known. Thus cost at present depends primarily upon chemical composition and functional groups, hardly at all upon crystallinity itself.

It is generally recognized that crystalline polymers have somewhat higher density than amorphous polymers, which increases their cost per unit volume somewhat, and is thus a factor in ultimate economic considerations (ref 235).

Applications

In general low crystallinity is desirable in elastomers (ref 168,236), medium crystallinity in engineering plastics (ref 237), and high crystallinity in fibers (ref 117,236,238,239).

5-3 ORIENTATION

The amorphous and crystalline states of polymers have been thoroughly studied and are quite well understood. The oriented state—intermediate between these two in regularity, but quite different in its morphological phenomena—may be equally important, particularly in practical applications such as fiber, films, and probably also foams; but it has been much less studied and is much less understood. This is partly because sufficient analytical techniques have not yet been developed for such studies, and partly because of a popular confusion between crystallinity and orientation. These phenomena should be clearly distinguished from each other; and it may be hoped that present and future development of more appropriate analytical techniques will eventually lead to a clearer understanding and utilization of the oriented state. For a review of what has already been accomplished, it is convenient to consider first the processing techniques that lead to orientation, and then the effects of orientation upon properties and applications.

Processing → Orientation

When a molten polymer is cooled from the liquid to the solid state the molecules, whether amorphous or crystalline, lie in a random orientation in the total sample. Under directional mechanical stress the axes of the polymer segments, or of the crystallites, tend to rotate into a position parallel to the direction of stress (Figure 5-73). This is the mechanism by which orientation occurs.

Relations between Orientation and Crystallization

It is best to begin by examining the relationships between orientation and crystallinity.

When crystallizable polymers have been melted and quenched into the amorphous state, and then are oriented by mechanical stress, such orientation may make their molecules parallel enough to induce crystallization, and this is the main reason for the popular confusion between crystallinity and orientation. When crystallites already exist in the amorphous matrix, orientation will make these crystallites parallel (ref 240). If a polymer crystallizes too well from the melt, it may not contain enough amorphous matrix to permit orientation, and will break during stretching (ref 241). Most partially crystalline polymers can be drawn 4 to 5 fold (ref 108). Whereas a polymer may range from completely amorphous to highly crystalline, the maximum degree of orientation is probably only partial, and only includes a small fraction of the total polymer molecules (ref 242).

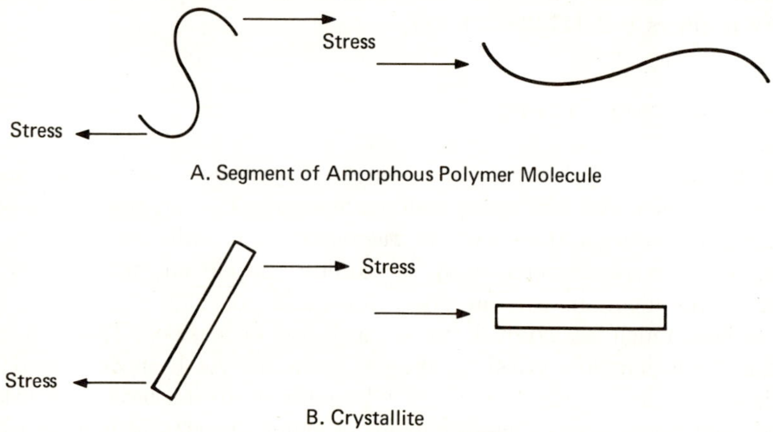

Figure 5-73. Orienting effect of directional mechanical stress.

Importance of Mobility during Orientation

Orientation requires considerable mobility of large segments of the polymer molecules, and thus cannot occur below the glass transition temperature. It is achieved by bringing the polymer temporarily above its glass transition temperature. This can be accomplished in a variety of ways.

Cold Stretching A polymer may be oriented by cold stretching, apparently below its glass transition temperature, if there is sufficient

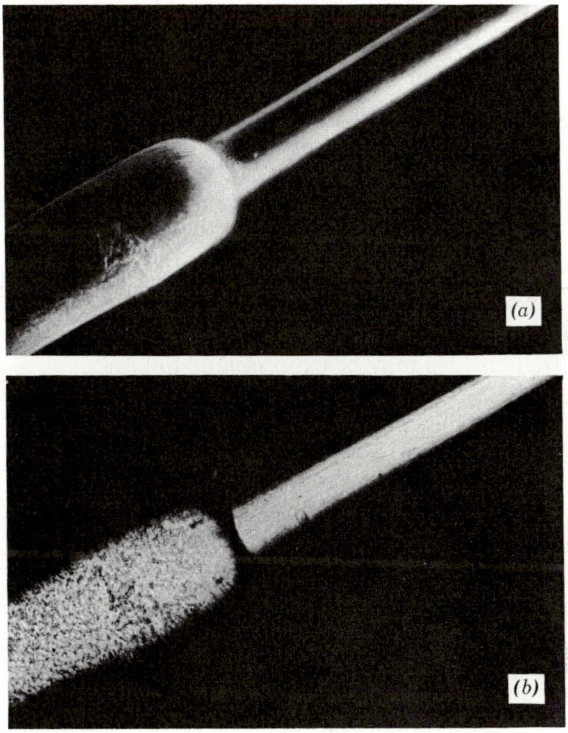

Figure 5-74. Cold-drawing of polyethylene fiber (ref 243).
(*a*) Cold drawn fiber of polyethylene, showing sections of drawn and undrawn material. (*b*) Same, observed between crossed polarizers, showing evidence of random orientation in the undrawn section and orientation parallel to the direction of stress in the drawn section.

internal friction to convert mechanical into thermal energy and thus produce local heating above the glass transition temperature. This occurs characteristically in the necking of fibers during cold-drawing (Figure 5-74). It has been suggested that in moderately crystalline polymers (35% to 75% crystallinity), cold-drawing may occur by local stress-melting of their crystalline regions (ref 244) (Figure 5-75).

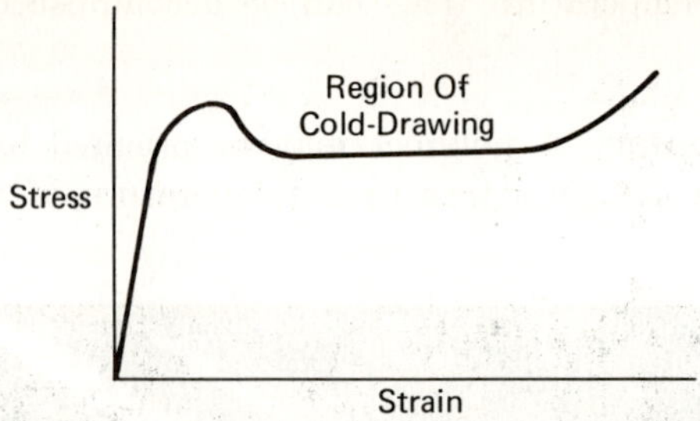

Figure 5-75. Stress-strain relationship in cold-drawing (ref 244,245).

WARM STRETCHING Most commercial orientation is produced by warming the polymer slightly above its glass transition temperature to produce rapid and easy mobility and processing. At still higher temperatures, excessive thermal motion and mobility may be sufficient to produce rerandomization to higher entropy levels and thus destroy orientation (Figure 5-76). This also may be expressed as follows:

$$\text{Amorphous} \underset{\text{Thermal Rerandomization}}{\overset{\text{Mechanical stress}}{\rightleftharpoons}} \text{Oriented}$$

Whereas thermal rerandomization depends entirely upon the thermal motion of the atoms and segments in the polymer molecule, orientation

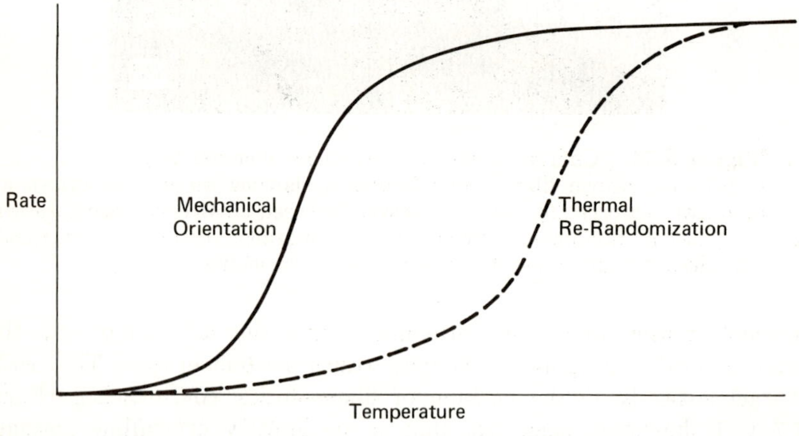

Figure 5-76. Mechanical orientation versus thermal rerandomization in warm stretching.

benefits from the added external effect of the mechanical stress applied by stretching. This produces the margin of advantage for producing orientation during processing. For most effective orientation, it is best to stretch rapidly somewhat above the glass transition temperature and then rechill immediately to below the glass transition temperature.

To prevent later thermal rerandomization and loss of orientation, which might occur during use at temperatures above the glass transition temperature, it may be best to follow the orientation step by a quick high-temperature heat-setting by crystallization or possibly vulcanization. It has been noted that stretched rubber can be vulcanized and retain its stretched form afterwards in this way (ref 238).

Wet Stretching For polymers whose glass transition temperature is above their decomposition temperature, orientation can be accomplished by swelling them temporarily with liquids to lower the glass transition temperature of the total mass, particularly in solution processing. Thus polyacrylonitrile fibers can be drawn during solution spinning, and cellulose viscose fibers and films can be drawn during coagulation, while they still contain enough fugitive solvent to plasticize them and lower their glass transition temperature below the decomposition temperature (ref 90,246). Final removal of the plasticizing solvent makes the orientation permanent.

Axes of Orientation

Monoaxial stretching produces bundles or rods of parallel molecules and is of particular practical importance in all fibers and in some film, particularly for tape applications (Figure 5-77).

Two-dimensional planar orientation may be accomplished by biaxial stretching in two separate steps, using higher speed take-off equipment for the initial stretch in the machine direction, and tenter frame (Figure 5-78) or blown bubble (Figure 5-79) for the transverse stretch. Two-dimensional planar orientation may also be accomplished by a single multiaxial stretch in all directions in the plane, but there is considerable belief that there is a distinct difference between the structures produced by biaxial and multiaxial stretching. Such planar two-dimensional orientation is less clearly understood, but is extremely important in most commercial films.

Plastic foams are produced by expanding tiny gas bubbles in a semimolten polymer, stretching the interfacial walls between the gas bubbles, and thus orienting them in a single-step multiaxial stretch. It is even conceivable that at the corner where three bubbles intersect, the interfacial walls between them may be subjected to some type of three-dimensional orientation. This mechanism is still very little understood.

Orientation during Other Process Techniques

Orientation is also of some significance in a number of other conventional plastic processes. During molten flow, gradual disentanglement

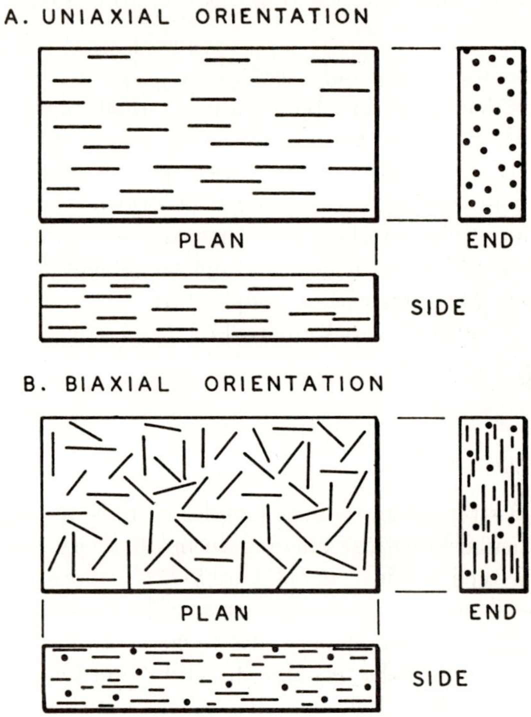

Figure 5-77. Schematic diagram of mono and biaxial orientation (ref 247).

and orientation of polymer molecules parallel to each other produces the non-Newtonian viscosity phenomena which are commonly observed in polymer processing (ref 248–250). In thermoforming of plastic sheet, stretching and orientation are major considerations in both processing and final properties (ref 251). During calendering of film and sheet

Figure 5-78. Biaxial orientation of film on a tenter frame (ref 246).
Biaxial orientation of plastic film by a tentering device. Grips moving along diverging tracks provide transverse stretching (schematic). Turner Alfrey and Edward F. Gurnee, *Organic Polymers*, © 1967. Reprinted by permission of Prentice-Hall, Inc., Englewood Cliffs, New Jersey.

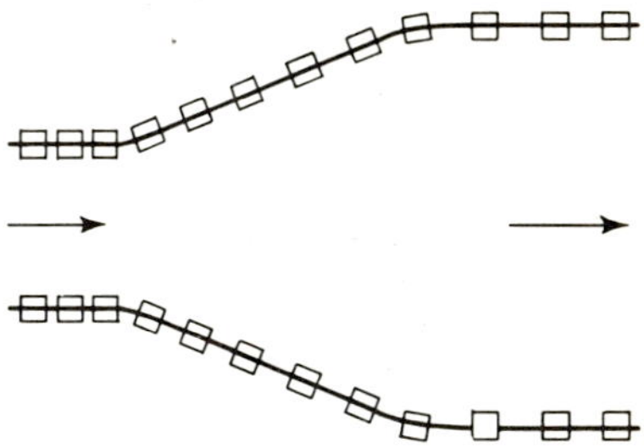

there is often some orientation of molecules parallel to the machine direction, and thus some anisotropy in resulting properties (ref 252). Even in injection molding there is some orientation due to the linear flow of the molten polymer molecules through the narrow orifices in the runner and gate, and even during the filling of long molds (ref 247,253). In the application of coatings from solution, it is common to use a mixture of good solvent and poor solvent, chosen so that the good solvent will evaporate somewhat more rapidly, gradually producing a gel of polymer and poor solvent, after which evaporation of poor solvent from the gel produces biaxial orientation, which improves the properties of the finished coating (ref 254). In all of these process techniques, orientation occurs incidentally in addition to the primary process but has important effects upon the properties of the resulting product.

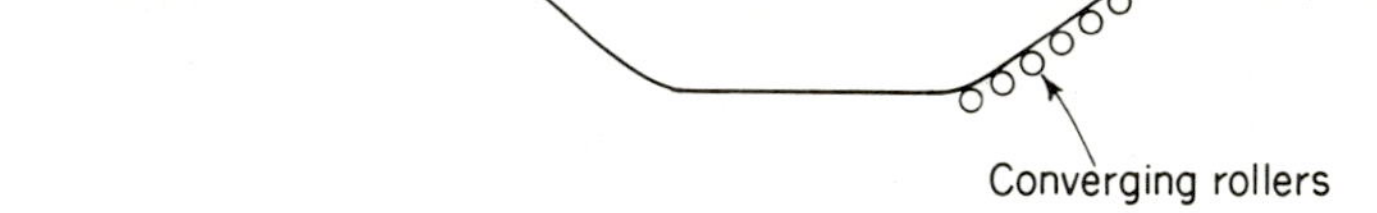

Figure 5-79. Biaxial orientation of film by a blown bubble (ref 248).
Biaxial orientation of plastic tube by a trapped air bubble. Machine-direction stretching accomplished by differential roll-speeds; transverse stretching by bubble. Turner Alfrey and Edward F. Gurnee, *Organic Polymers*, © 1967. Reprinted by permission of Prentice-Hall, Inc., Englewood Cliffs, New Jersey.

Orientation → Properties

Orientation introduces added complexity into processing and generally produces anisotropic properties which are of considerable importance, particularly in fiber, film, and foam applications.

Processability

Orientation complicates processing in a variety of ways. Some of these have already been discussed. Several more may be noted at this point.

Pseudoplastic and thixotropic viscosity behavior, during the flow of polymers in molten or dissolved form, is due to their increasing disentanglement at increasing rates of flow, producing parallel alignment of molecules and easier slip motion past each other at high rates of shear (ref 248–250).

Oriented films and fibers may lose some of their orientation because of excessive thermal motion during later thermal processing such as post-forming and heat sealing, and also in the use of solvents for sealing and printing. On the other hand, heat-shrinkable films and fibers are conveniently made by temporary orientation, which can be lost when desired by warming above the glass transition temperature to produce disorientation and shrinkage back to their original dimensions.

In thermoforming of films, if the unoriented film tends to sag excessively during warming, preorientation may give it sufficient hot strength for more suitable thermoformability.

These examples illustrate some of the complex effects of orientation upon processability.

Orientation generally aligns polymer molecules in parallel bundles or plates, along one or two of their three axes respectively. Many polymer properties depend upon the direction of such alignment. If a particular property increases in the direction of alignment, it generally decreases perpendicular to this direction. These effects are illustrated in most of the following properties.

Mechanical Properties

Mechanical properties depend directly upon the relationship between the axes of orientation of the polymer molecules and the axes of mechanical stress upon these molecules.

REVERSIBLE MECHANICAL PROPERTIES Reversible mechanical properties, such as modulus and stiffness, increase in the direction of orientation (ref 255–257) (Figure 5-80). Conversely, in a direction perpendicular to the axis of orientation, modulus decreases and flexibility increases (ref 63,257,259) (Table 5-11), because stress along the axis of orientation of the polymer molecules is applied against the strong covalent bonds

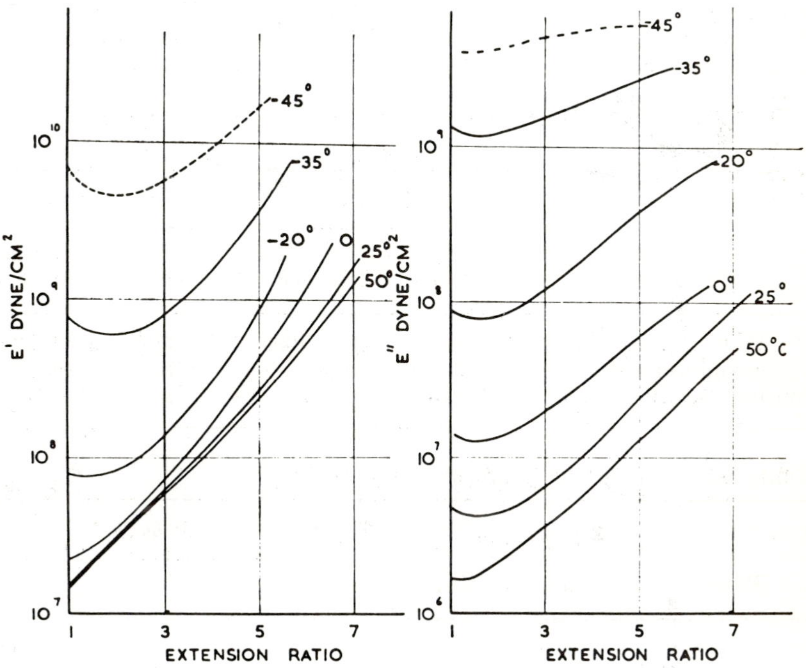

Figure 5-80. Effect of stretching upon modulus of rubber (ref 258). Dynamic Young's moduli of rubber as a function of the elongation at 1 kc/sec.

within the polymer molecules, whereas stress perpendicular to the axis of orientation is applied only against the weak secondary forces between the polymer molecules. These effects are important to the toughness and flexibility of most films and all fibers.

ULTIMATE TENSILE STRENGTH Ultimate tensile strength generally increases significantly in the direction of orientation and decreases per-

Table 5-11. Relative Changes in Young's Moduli for Highly Oriented Polymers (ref 260)

Polymer	Modulus Parallel to Chains/Modulus of Unoriented Polymer	Modulus Perpendicular to Chains/Modulus of Unoriented Polymer
Polyethylene terephthalate	17.8	0.67
Nylon 66	19.2	0.57
Viscose rayon	9.3	0.67
Polyacrylonitrile	5.8	0.67

Table 5-12. Biaxial Orientation of Polystyrene and Polymethyl Methacrylate Sheets (ref 265)

	Polystyrene		Polymethyl Methacrylate	
Property	Unoriented	Biaxial Oriented	Unoriented	Biaxial Oriented
Tensile strength (psi)	5000–9000	7000–12000	7500–10000	8000–11000
Elongation to break (%)	1–3.6	8–18	5–10	25–50
Impact strength	0.25–0.5	>3	4	15

Table 5-13. Orientation of Rayon Fibers (ref 266)

		Tenacity	
Fiber	Condition	*g*/denier	psi
Viscose rayon	Poorly oriented	1.7 to 1.9	33,000 to 37,000
	Moderately oriented	2.0 to 2.6	38,000 to 50,000
	Highly oriented	3.0 to 3.5	58,000 to 67,000
	Extremely oriented	Up to 5.8	Up to 111,000
Acetate rayon	Poorly oriented	1.3 to 1.6	22,000 to 27,000
	Moderately oriented	1.8 to 2.3	31,000 to 39,000
	Highly oriented	3.5 to 4.0	60,000 to 68,000

pendicular to it, for the same reason (ref 261–264) (Tables 5-12 to 5-15, Figure 5-81). The changes in tensile strength may also be due to the fact that stress-concentrators, such as microscopic or submicroscopic flaws, become parallel to the axis of orientation; when stress is applied along the same axis, it no longer tends to pull them apart, whereas perpendicular to this axis it pulls them apart very effectively (ref

Table 5-14. Monoaxial Orientation of Cellophane Film (ref 266)

Angle between the direction of stretching and a given direction in the film, deg	Tenacity			
	Before stretching		After stretching	
	kg/sq mm	psi	kg/sq mm	psi
0	11.6	16,500	26.5	37,500
30	11.2	15,900	15.7	22,300
60	11.4	16,200	10.2	14,500
90	11.5	16,400	4.5	6,400
120	11.9	16,900	9.3	13,200
150	11.2	15,900	14.7	20,900
180	11.1	15,800	25.9	36,900

269,270) (Figure 5-82). The improvement of tensile strength has great practical value in most applications of films and fibers.

ULTIMATE ELONGATION The effects of orientation upon ultimate elongation are more complex. Moderate orientation, particularly in rigid amorphous polymers like polystyrene, increases ductility and ultimate elongation markedly in the direction of orientation and decreases it in the transverse direction (ref 271–274) (Tables 5-12, 5-15). High degrees of orientation, and orientation of ductile polymers, can have the opposite

Table 5-15. Monoaxial Orientation of Polystyrene Sheet (ref 267)

Birefringence $\Delta n \times 10^4$	Tensile Strength (psi)		Elongation to Break (%)		Izod Impact Strength	
	$\parallel$	$\perp$	$\parallel$	$\perp$	$\parallel$	$\perp$
1.1	3940	3440	2.4	1.8	0.24	0.22
4.3	5340	4240	3.1	2.1	0.32	0.21
9.1	6320	4110	3.9	2.6	0.36	0.20
16.3	7550	4140	4.2	1.9	0.65	0.21
25.4	7640	3710	5.0	1.8	1.36	0.18
30.9	7590	2630	6.8	1.3	—	—
41.4	8660	3440	5.2	2.1	1.54	0.23
51.8	10,170	4550	4.4	2.5	1.58	0.18
53.7	8440	1290	7.0	0.7	—	—

$\parallel$ Stress applied parallel to chain orientation.
$\perp$ Stress applied perpendicular to direction of orientation.

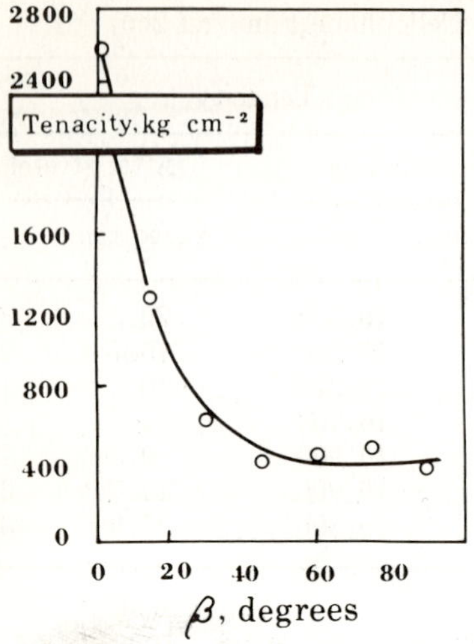

Figure 5-81. Monoaxial orientation of poly(ethylene terephthalate) film (ref 268).

Relationship between the tensile strength of uniaxially oriented films of poly(ethylene terephthalate) and the angle, β, between the direction of orientation and the direction of the applied force.

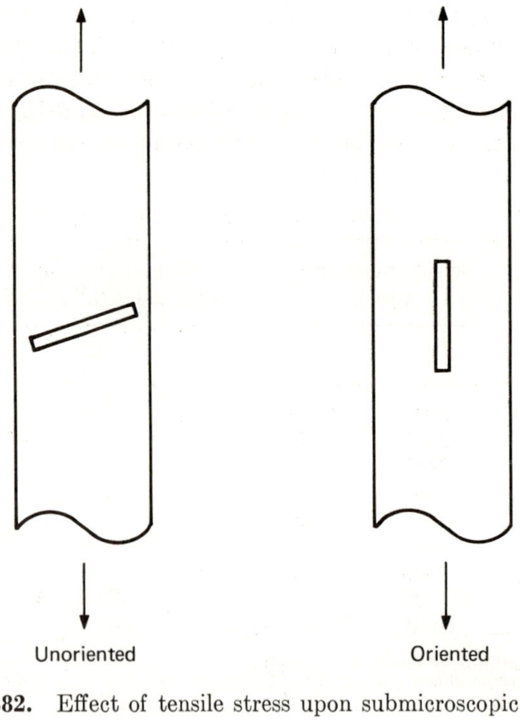

Figure 5-82. Effect of tensile stress upon submicroscopic flaws in fibers.

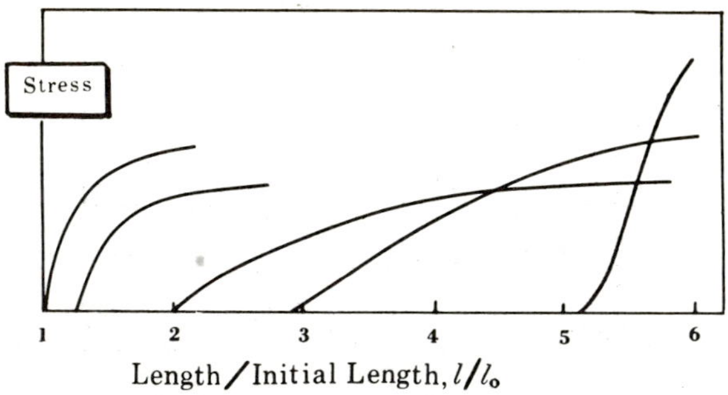

Figure 5-83. Effects of orientation upon stress-strain behavior of polyethylene fiber (ref 273). Load-extension plots for polyethylene fibers oriented to various extents, then restretched under conditions where necking did not occur.

effect (ref 273, 275–277) (Figure 5-83, Table 5-16). In these latter effects, orientation has already used up much of the inherent extensibility of the polymer, whereas transverse elongation can still occur by deorientation and reorientation in the direction of transverse stress (ref 279).

IMPACT STRENGTH Biaxial orientation increases impact strength significantly (ref 63,265,280,281) (Tables 5-12, 5-15, Figures 5-84 to 5-86), and is thus very useful in most film applications. The mechanism

Table 5-16. Effect of Orientation upon Ultimate Elongation (ref 278)

Linear Stretch during Orientation at 400°F	Ultimate Elongation* of Stretched Sheet	
	High-impact Styrene	ABS Polyblend
10%	21%	62%
20	19	58
30	18	56
40	17	53
50	16	48
60	15	46
70	15	42
80	14	39

* Rate of test 20 inches/minute

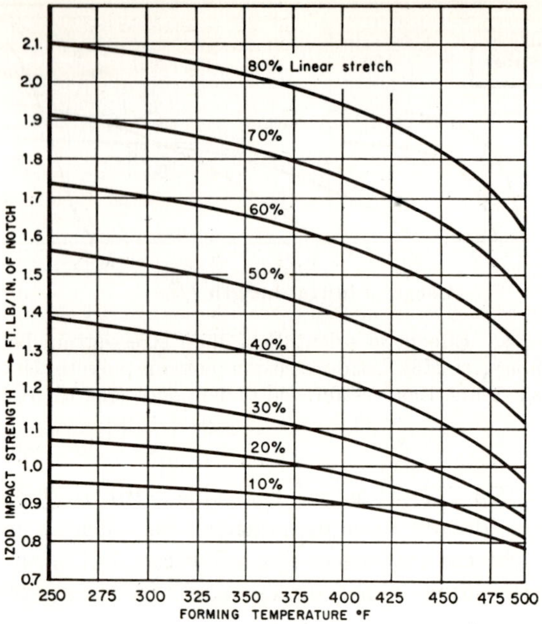

Figure 5-84. Effect of biaxial orientation upon Izod impact strength of high-impact polystyrene sheets (ref 282).

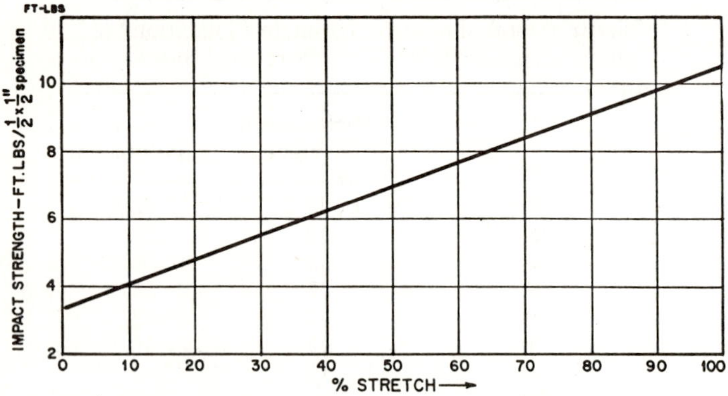

Figure 5-85. Effect of orientation upon Charpy impact strength of acrylic sheet (ref 282).

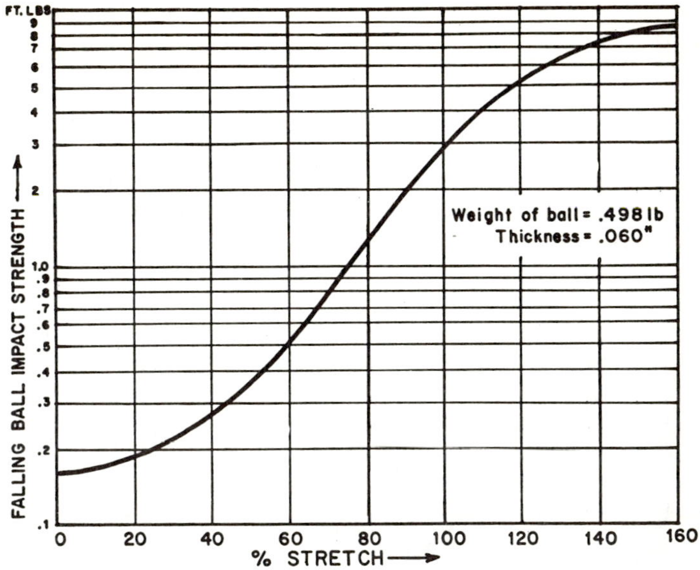

Figure 5-86. Effect of orientation upon falling ball impact strength of acrylic sheet (ref 283).

of this improvement is indicated by the tremendous increase in the area under the stress-strain curve, which is often used as a measure of toughness (ref 267,275). In monoaxial orientation, again, impact strength increases in the direction of orientation but decreases in the transverse direction (ref 265,284) (Figure 5-87); in fact, transverse impact on a

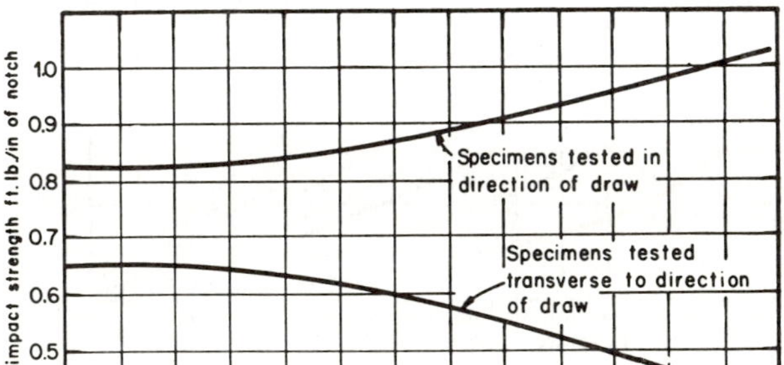

Figure 5-87. Effect of monoaxial orientation upon impact strength of high-impact polystyrene sheet (ref 284).

monoaxially oriented film will actually break it into bundles of fibers (ref 265).

COMPLEX MECHANICAL PROPERTIES These improvements in flexibility and toughness also produce consequent and useful improvements in ultimate flex life on repeated flexure (ref 268), and in resistance to creep, crazing, and stress-cracking (ref 281,285–288). On the other hand, properties which depend upon stresses perpendicular to the axis

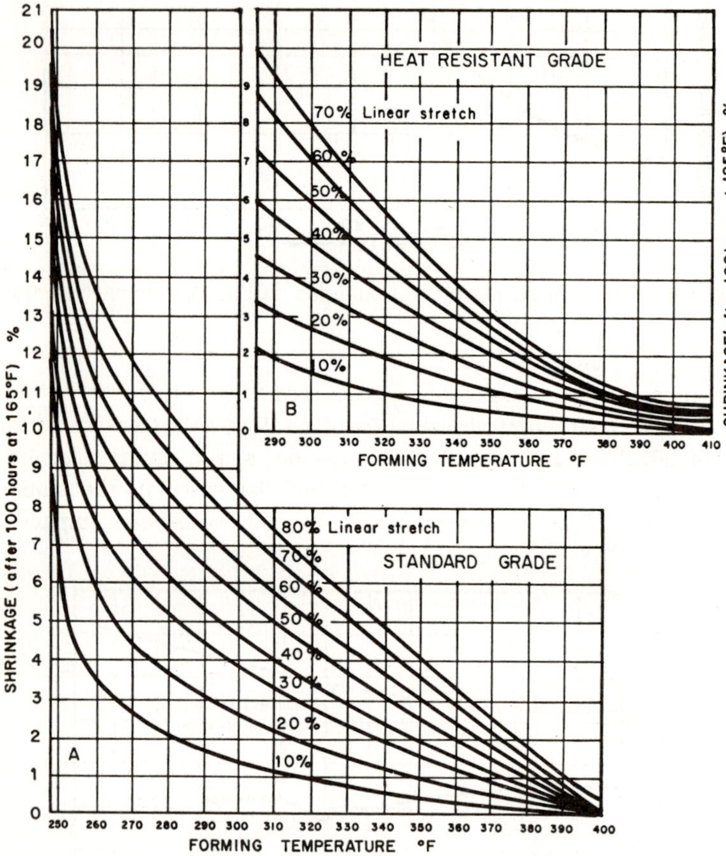

Figure 5-88. Hot shrinkage of oriented impact styrene sheet (ref 289). The effect of biaxial stretching at different forming temperatures on the dimensional stability of high-impact polystyrene sheetings.

A. Shrinkage of standard sheeting after 100 hours exposure at 165°F.

B. Shrinkage of heat-resistant sheeting after 100 hours exposure at 185°F.

of orientation in fibers and films, such as abrasion resistance, and fibrillation and delamination during repeated flexure, generally become worse due to orientation (ref 265,268,289). All of these effects are in accordance with the same fundamental mechanisms already discussed.

Thermal Properties

DIMENSIONAL STABILITY Orientation decreases dimensional stability at higher temperatures. Because the molecules have been stretched

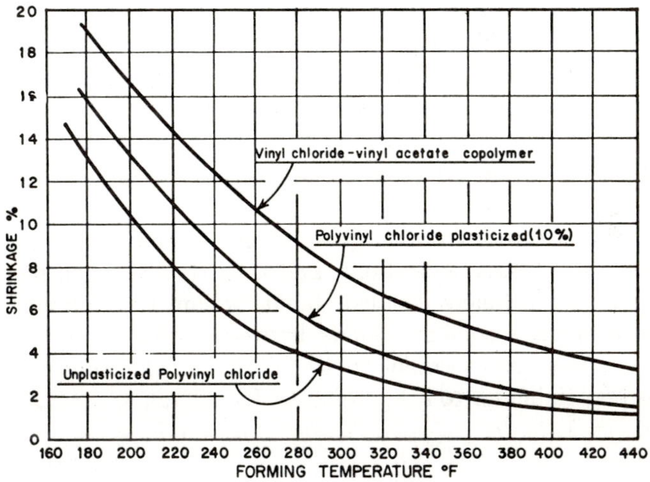

Figure 5-89. Hot shrinkage of oriented rigid vinyl sheet (ref 293). The effect of biaxial stretching (20 to 25 percent) on the dimensional stability of rigid vinyl sheetings. Shrinkage after 100 hours exposure at 165°F.

into a statistically improbable conformation of thermodynamically high energy level, on rewarming they tend to retract to their original randomly-coiled conformation as soon as they acquire sufficient mobility, disorienting to a lower energy and higher entropy level. This makes them deform and distort at lower temperatures than unoriented materials, and causes them to shrink back toward their original size and shape on rewarming (ref 290–292) (Figures 5-88 and 5-89). On the other hand, it can be useful in the production of shrink film, which is first stretched and oriented, then chilled below the glass transition temperature, used for packaging, and finally reheated above the glass transition temperature, permitting it to shrink tightly around the object wrapped in it.

Glass Transition Temperature Effects of orientation on glass transition temperature vary with the type of test and test conditions (ref 294). Generally orientation restrains molecular mobility and increases glass transition temperature somewhat (ref 294,295). On the other hand, disorientation above the glass transition temperature produces more complex effects on properties than are observed in simple amorphous polymers.

Coefficient of Thermal Expansion Coefficient of linear thermal expansion decreases parallel to the axis of orientation α_{11} and increases perpendicular to it $\alpha_{\perp}$. Consequently their resulting coefficient of volumetric thermal expansion α hardly changes at all (ref 296,297). For an incompressible material, $\alpha = \alpha_{11} + 2\alpha_{\perp}$.

Thermal Conductivity Thermal conductivity λ depends upon transmission of thermal vibrations, and thus travels much more readily along the strong primary covalent bonds within the polymer molecule than across weak secondary bonds between polymer molecules. Consequently it increases parallel to the direction of orientation λ_{11} and decreases perpendicular to it $\lambda_{\perp}$ (ref 298). In general, $3/\lambda = 1/\lambda_{11} + 2/\lambda_{\perp}$ or $\lambda^3/\lambda_{11}\lambda_{\perp} = 1$.

Electrical Properties

Orientation decreases dissipation factor in the direction of orientation, and increases it perpendicular to orientation (Figure 5-90). Since modulus changes in the opposite way, as previously mentioned, this indicates that polar vibrations along a stretched polymer molecule are decreased, while transverse vibrations between the stretched molecules become easier. Thus when polyethylene terephthalate is extruded and stretched to make film for capacitors, orientation is necessary to produce tough flexible mechanical properties, but the concommitant increase in dielectric loss across the film is at best a necessary evil.

Optical Properties

Biaxial orientation of crystalline polymers generally improves clarity of films (ref 63,115) (Figure 5-91). This is because stretching breaks up large crystalline structures into units smaller than the wavelength of visible light (ref 106, 237). This is very beneficial in packaging and other visual applications.

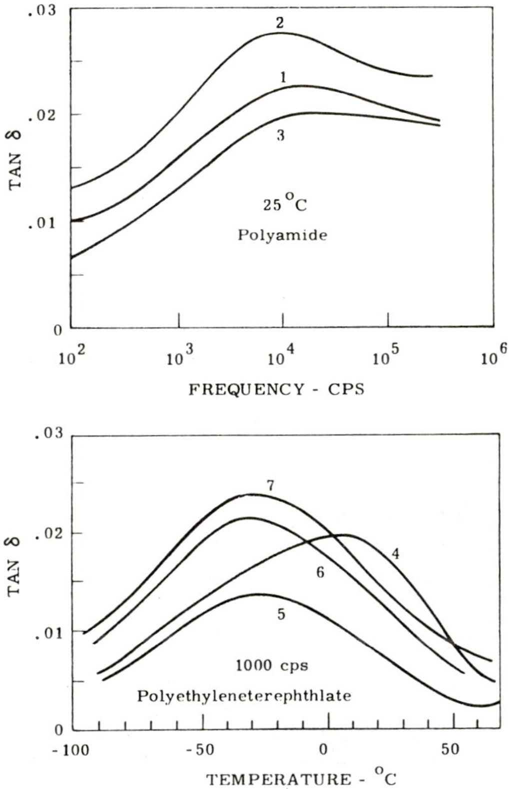

Figure 5-90. Effect of orientation upon dissipation factor tan δ of polymers (ref 197).

1—Unoriented polyamide
2—Oriented polyamide, electrical stress perpendicular to the orientation
3—Oriented polyamide, electrical stress parallel to the orientation
4—Amorphous polyethylene terephthalate
5—Crystalline polyethylene terephthalate
6—Oriented (one direction) polyethylene terephthalate
7—Doubly oriented polyethylene terephthalate

Monoaxial orientation produces anisotropic refractive index and thus birefringence, especially in crystalline polymers (Figure 5-92). This provides a very useful analytical tool for structural studies (ref 299,301).

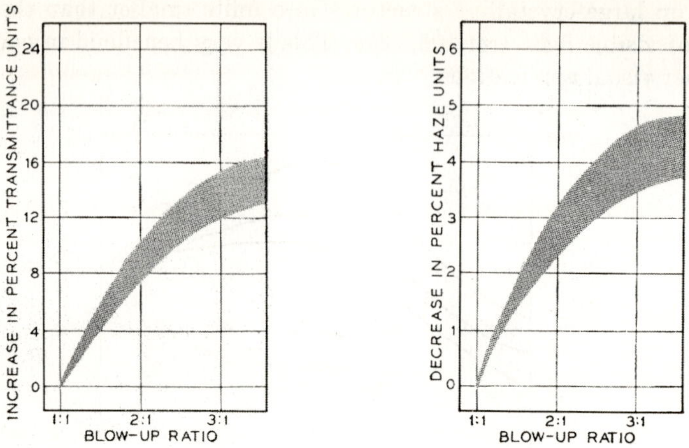

Figure 5-91. Improvement in clarity of polyethylene film by orientation (ref 234).

Chemical Properties

Simple orientation alone increases sorption and solubility; but when it induces crystallization, the overall net effect is a decrease in these properties (ref 302). Similarly, residual stresses in the structure of foamed plastics make them more susceptible to chemical attack (ref 303).

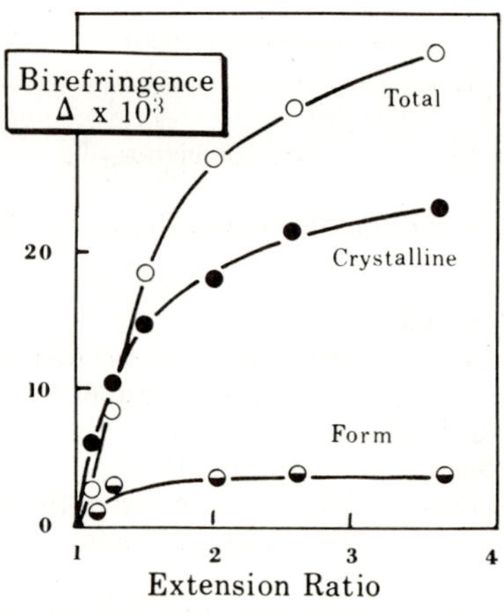

Figure 5-92. Birefringence in monoaxially oriented polyethylene (ref 300).

Birefringence of a sample of partly crystalline polyethylene: ○ total birefringence, ● birefringence due to crystallites and ◒ birefringence due to form effects.

Orientation of amorphous polymers affects molecular mobility and permeability in similar ways: permeability decreases in the direction of orientation and increases perpendicular to it (ref 304). Orientation of crystallizable and crystalline polymers decreases permeability still further (ref 305). These effects are particularly important in packaging films.

Optimum application of adhesives to substrates may benefit from orientation of dipoles at the interface (ref 306). This of course depends upon the type of adhesion and involves complex questions of adhesion mechanism theory.

Price and Balance of Properties

The process of orientation is fairly expensive, and increases the cost per unit weight of the polymer product. On the other hand, the yield of fiber or film increases considerably on stretching and the quality improves greatly.

Fibers in the unoriented state are generally weak, brittle, and useless; orientation makes them strong, flexible, and highly valuable (ref 259). The same is true of many films, notably polystyrene and polyethylene terephthalate.

Optimum application of coatings from a mixture of low-boiling good solvent and higher-boiling poor solvent produces intermediate gelation followed by biaxial orientation, which produces strength and toughness in the finished coating (ref 254).

Thus on ultimate balance, maximum economic value depends upon the relative cost of stretching vs the increase in yield and properties.

REFERENCES

1. Brydson, J. A., *Plastics Materials,* Iliffe, 1966, p. 55.
2. Miller, M. L., *The Structure of Polymers,* Reinhold, 1966, p 527.
3. Morton, W. E., & Hearle, J. W. S., *Physical Properties of Textile Fibers,* Butterworths, 1962, p 43.
4. Baer, Eric, *Engineering Design for Plastics,* Reinhold, 1964, p 89.
5. Billmeyer, F. W., Jr., *Textbook of Polymer Science,* Interscience, 1962, p 135.
6. Miller, *op. cit.,* pp 257–59.
7. *Ibid.,* pp 530–7.
8. *Ibid.,* pp 505–6.
9. Alfrey, Turner, & Gurnee, E. F., *Organic Polymers,* Prentice-Hall, 1967, pp 7–8.

10. Billmeyer, *op. cit.*, pp 145–47.
11. Mark, H. F., Atlas, S. M., & Cernia, Enrico, *Man-Made Fibers*, Interscience, 1967, pp 303–8.
12. *Ibid.*, p 343.
13. Nielsen, L. E., *Mechanical Properties of Polymers*, Reinhold, 1962, pp 41–43.
14. Baer, *op. cit.*, p 94.
15. Billmeyer, *op. cit.*, pp 140–43.
16. Mark, *op. cit.*, pp 308–13.
17. *Ibid.*, pp 348–52.
18. Miller, *op. cit.*, pp. 508–11.
19. Alfrey, *op. cit.*, p 8.
20. Billmeyer, *op. cit.*, p 142.
21. *Ibid.*, pp 147–48.
22. Mark, *op. cit.*, pp 334–37.
23. *Ibid.*, pp 352–53.
24. Nielsen, *op. cit.*, pp 42–43.
25. *Ibid.*, p 43.
26. Mark, *op. cit.*, p 352.
27. Nielsen, *op. cit.*, p 34.
28. *Ibid.*, p 38.
29. Alfrey, *op. cit.*, p 39.
30. Billmeyer, *op. cit.*, pp 216–17.
31. *Ibid.*, p 217.
32. Meares, Patrick, *Polymers: Structure and Bulk Properties*, Van Nostrand, 1965, p 105.
33. *Ibid.*, p 92.
34. Billmeyer, *op. cit.*, p 118.
35. *Ibid.*, p 125.
36. Brydson, *op. cit.*, pp 53–54.
37. Miller, *op. cit.*, pp 410–12.
38. *Ibid.*, p 410.
39. Alfrey, *op. cit.*, p 36.
40. Billmeyer, *op. cit.*, pp 120–22.
41. Miller, *op. cit.*, pp 413–16.
42. Billmeyer, *op. cit.*, p 120.
43. Miller, *op. cit.*, p 414.
44. *Ibid.*, p 415.
45. Billmeyer, *op. cit.*, pp 219–20.
46. Miller, *op. cit.*, p 300.
47. *Ibid.*, p 534.
48. *Ibid.*, p 131.
49. Billmeyer, *op. cit.*, p 220.
50. *Ibid.*, pp 225–26.
51. Alfrey, *op. cit.*, pp 41–43.

52. Modern Plastics Encyclopedia 1968, McGraw-Hill, 1967, p 19.
53. Nielsen, *op. cit.*, pp 30–37.
54. Tobolsky, A. V., *Properties and Structure of Polymers,* Wiley, 1960, p 205.
55. Baer, *op. cit.*, p. 105.
56. Alfrey, *op. cit.*, p 42.
57. Billmeyer, *op. cit.*, pp 213–14.
58. Nielsen, *op. cit.*, p 31.
59. *Ibid.*, p 37.
60. *Ibid.*, pp 32–37.
61. Alfrey, *op. cit.*, pp 41–42.
62. Billmeyer, *op. cit*., p 214.
63. Modern Plastics Encyclopedia, *op. cit.*, p 20.
64. *Ibid.*, p 21.
65. Ham, G. E., *Copolymerization,* Interscience, 1964, pp 204–9.
66. Meares, *op. cit.*, p 93.
67. Alfrey, *op. cit.*, p 46.
68. Eastman Technical Bulletins on Polyallomer.
69. Belding-Heminway Technical Bulletins.
70. Miller, *op. cit.*, p 533.
71. Nielsen, *op. cit.*, p 32.
72. Baer, *op. cit.*, pp 773–75.
73. Miller, *op. cit.*, pp 493–94.
74. Tobolsky, *op. cit.*, pp 49–52.
75. Brydson, *op. cit.*, p 45.
76. Miller, *op. cit.*, pp 527–28.
77. Modern Plastics Encyclopedia, *op. cit.*, p 15.
78. Morton, Maurice, *Introduction to Rubber Technology,* Reinhold, 1959, p 54.
79. Nielsen, *op. cit.*, p 39.
80. Billmeyer, *op. cit.*, p 227.
81. Baer, *op. cit.*, pp 75–80.
82. Miller, *op. cit.*, pp 527–31.
83. Nielsen, *op. cit.*, pp 33–5.
84. Brydson, *op. cit.*, p 40.
85. Meares, *op. cit.*, pp 116–17.
86. Modern Plastics Encyclopedia, *op. cit.*, pp 14–15.
87. Morton, *op. cit.* (ref 3), p 44.
88. Brydson, *op. cit.*, pp 40–42.
89. Nielsen, *op. cit.*, pp 254–55.
90. Golding, Brage, *Polymers and Resins,* Van Nostrand, 1959, p 550.
91. Baer, *op. cit.*, 113–18.
92. Billmeyer, *op. cit.*, pp 154–57.
93. Miller, *op. cit.*, pp 515–18.
94. Billmeyer, *op. cit.*, pp 155–56.

95. Miller, *op. cit.*, p 516.
96. Baer, *op. cit.*, p 116.
97. Meares, *op. cit.*, pp 132–34.
98. Miller, *op. cit.*, p 517.
99. Nielsen, *op. cit.*, pp 184–85.
100. Nielsen, *op. cit.*, p 29.
101. Baer, *op. cit.*, pp 113–15.
102. Miller, *op. cit.*, pp 516–17.
103. Nielsen, *op. cit.*, pp 251–53.
104. Baer, *op. cit.*, p 115.
105. Alfrey, *op. cit.*, p 84.
106. Billmeyer, *op. cit.*, p 166.
107. Miller, *op. cit.*, pp 559–60.
108. Golding, *op. cit.*, p 549.
109. Alfrey, *op. cit.*, pp 83–84.
110. Kinney, G. F., *Engineering Properties and Applications of Plastics,* Wiley, 1957, p 147.
111. Meares, *op. cit.*, p 227.
112. *Ibid.*, p 229.
113. Schildknecht, C. E., *Vinyl and Related Polymers,* Interscience, 1952, p 581.
114. Meares, *op. cit.*, pp 227–29.
115. Brydson, *op. cit.*, p 43.
116. Modern Plastics Encyclopedia 1967, McGraw-Hill, 1966, p 552.
117. Billmeyer, *op. cit.*, p 231.
118. Nielsen, *op. cit.*, p 183.
119. Boenig, H. V., *Polyolefins: Structure and Properties,* Elsevier, 1966, p 119.
120. Baer, *op. cit.*, p 109.
121. Celanese Technical Bulletin on Fortiflex Polyethylene.
122. DuPont Technical Bulletin on Alathon Polyethylene.
123. Billmeyer, *op. cit.*, p 206.
124. Nielsen, *op. cit.*, pp 62–63.
125. Tobolsky, *op. cit.*, pp 198–99.
126. Alfrey, *op. cit.*, p 51.
127. Meares, *op. cit.*, pp 227–28.
128. Billmeyer, *op. cit.*, pp 206–7.
129. Miller, *op. cit.*, p 538.
130. Billmeyer, *op. cit.*, p 207.
131. Miller, *op. cit.*, p 537.
132. *Ibid.*, pp 537–38.
133. Baer, *op. cit.*, p 110.
134. Billmeyer, *op. cit.*, pp 208–9.
135. *Ibid.*, p 221.
136. Miller, *op. cit.*, p 539.

137. Nielsen, *op. cit.*, p 232.
138. Boenig, *op. cit.*, p 83.
139. *Ibid.*, p 92.
140. Mark, *op. cit.*, p 331.
141. Billmeyer, *op. cit.*, p 9.
142. Golding, *op. cit.*, p 573.
143. McFarlane, S. B., *Technology of Synthetic Fibers,* Fairchild Publications, 1953, p 93.
144. Boenig, *op. cit.*, p 57.
145. Nielsen, *op. cit.*, p 252.
146. Billmeyer, *op. cit.*, pp 210–11.
147. Miller, *op. cit.*, p 515.
148. Nielsen, *op. cit.*, p 129.
149. Billmeyer, *op. cit.*, pp 528–29.
150. Meals, R. N., & Lewis, F. M., *Silicones,* Reinhold, 1959, pp 36–37.
151. Billmeyer, *op. cit.*, p 529.
152. Alfrey, *op. cit.*, p 91.
153. Nielsen, *op. cit.*, pp 62–64.
154. *Ibid.*, p 87.
155. Billmeyer, *op. cit.*, p 221–22.
156. Brydson, *op. cit.*, p 59.
157. Tobolsky, *op. cit.*, pp 80–83.
158. Nielsen, *op. cit.*, p 125.
159. *Ibid.*, pp 125–26.
160. Wild, Leslie, Woldering, J. F., & Guliana, R. T., SPE Annual Technical Conference, *13,* 93 (May, 1967).
161. Baer, *op. cit.*, pp 753–54.
162. *Ibid.*, p 753.
163. *Ibid.*, p 754.
164. Wild, *op. cit.*, p 95.
165. Baer, *op. cit.*, p 756.
166. *Ibid.*, pp 754–56.
167. Nielsen, *op. cit.*, p 230.
168. Billmeyer, *op. cit.*, pp 230–31.
169. *Ibid.*, pp 232–33.
170. Golding, *op. cit.*, pp 543–45.
171. Miller, *op. cit.*, pp 482–83.
172. *Ibid.*, p 524.
173. Meares, *op. cit.*, pp 132–34.
174. Miller, *op. cit.*, pp 524–25.
175. Nielsen, *op. cit.*, pp 30–31.
176. Miller, *op. cit.*, p 529.
177. Baer, *op. cit.*, p 77.
178. Tobolsky, *op. cit.*, p 75.
179. Billmeyer, *op. cit.*, pp 221–23.

180. Miller, *op. cit.*, pp 540–42.
181. *Ibid.*, p 542.
182. Tobolsky, *op. cit.*, p 81.
183. *Ibid.*, pp 81–82.
184. *Ibid.*, p 69.
185. Brydson, *op. cit.*, pp 33–34.
186. *Ibid.*, p 34.
187. Billmeyer, *op. cit.*, p 223.
188. Miller, *op. cit.*, pp 413–14.
189. Nielsen, *op. cit.*, pp 44–45.
190. Miller, *op. cit.*, p 563.
191. *Ibid.*, p 541.
192. *Ibid.*, pp 541–42.
193. Baer, *op. cit.*, p 540.
194. *Ibid.*, pp 539–40.
195. *Ibid.*, p 541.
196. *Ibid.*, pp 561–62.
197. *Ibid.*, p 562.
198. *Ibid.*, p 590.
199. Alfrey, *op. cit.*, p 126.
200. Brydson, *op. cit.*, pp 41–42.
201. Miller, *op. cit.*, p 550.
202. Modern Plastics Encyclopedia 1968, p 544.
203. Duff, A. W., *College Physics*, Longmans Green, 1937, p 440.
204. Baer, *op. cit.*, pp 97–98.
205. Billmeyer, *op. cit.*, pp 97–99.
206. Miller, *op. cit.*, pp 495–507.
207. *Ibid.*, p 496.
208. Baer, *op. cit.*, p 97.
209. Miller, *op. cit.*, p 500.
210. Billmeyer, *op. cit.*, p 25.
211. Golding, *op. cit.*, pp 19–20.
212. Baer, *op. cit.*, p 1042.
213. Brydson, *op. cit.*, p 65.
214. Alfrey, *op. cit.*, p 16.
215. *Ibid.*, p 17.
216. Brydson, *op. cit.*, p 67.
217. Miller, *op. cit.*, p 543.
218. Brydson, *op. cit.*, pp 65–67.
219. Meares, *op. cit.*, pp 327–31.
220. Baer, *op. cit.*, pp 636–37.
221. *Ibid.*, p 637.
222. *Ibid.*, pp 667–69.
223. Miller, *op. cit.*, p 548.
224. Baer, *op. cit.*, pp 665–69.

225. Brydson, *op. cit.*, pp 81–82.
226. *Ibid.*, p 82.
227. Baer, *op. cit.*, p 666.
228. Miller, *op. cit.*, pp 481–82.
229. *Ibid.*, p 482.
230. Baer, *op. cit.*, pp 774–75.
231. *Ibid.*, pp 779–82.
232. *Ibid.*, p 781.
233. Phillips Petroleum Technical Bulletins on Marlex Polyethylene.
234. US Industrial Chemicals Technical Bulletins on Polyethylene.
235. Modern Plastics Encyclopedia 1968, *op. cit.*, p 27.
236. Golding, *op. cit.*, pp 21–22.
237. Billmeyer, *op. cit.*, p 232.
238. Golding, *op. cit.*, pp 548–49.
239. Morton, *op. cit.* (ref 3), pp 21–22.
240. Billmeyer, *op. cit.*, pp 164–70.
241. Meares, *op. cit.*, p 155.
242. Nielsen, *op. cit.*, p 235.
243. Billmeyer, *op. cit.*, p 165.
244. Nielsen, *op. cit.*, pp 105–08.
245. *Ibid.*, p 99.
246. Alfrey, *op. cit.*, p 116.
247. Nielsen, *op. cit.*, p 236.
248. Alfrey, *op. cit.*, p 74.
249. Bernhardt, E. C., *Processing of Thermoplastic Materials,* Reinhold, 1959, p 22.
250. Miller, *op. cit.*, p 229.
251. Bernhardt, *op. cit.*, pp 488–96.
252. Golding, *op. cit.*, p 547.
253. Bernhardt, *op. cit.*, p 348.
254. Alfrey, *op. cit.*, p 115.
255. Billmeyer, *op. cit.*, p 224.
256. Nielsen, *op. cit.*, pp 240–41.
257. *Ibid.*, pp 245–47.
258. *Ibid.*, p 246.
259. Miller, *op. cit.*, p 557.
260. Nielsen, *op. cit.*, p 247.
261. Bernhardt, *op. cit.*, pp 490–91.
262. Brydson, *op. cit.*, pp 37–38.
263. Miller, *op. cit.*, pp 560–62.
264. Nielsen, *op. cit.*, pp 241–44.
265. *Ibid.*, p 244.
266. Schmidt, A. X., & Marlies, C. A., *Principles of High Polymer Theory and Practice,* McGraw-Hill, 1948, p 365.
267. Nielsen, *op. cit.*, p 242.

268. Miller, *op. cit.*, p 562.
269. Dietz, A. G. H., *International Science & Technology*, #32, 58–69 (August, 1964).
270. Nielsen, *op. cit.*, p 241.
271. Alfrey, *op. cit.*, pp 92–93.
272. Bernhardt, *op. cit.*, pp 490–92.
273. Miller, *op. cit.*, p 561.
274. Nielsen, *op. cit.*, pp 242–44.
275. Alfrey, *op. cit.*, p 93.
276. Bernhardt, *op. cit.*, pp 492–94.
277. Nielsen, *op. cit.*, pp 243–44.
278. Bernhardt, *op. cit.*, p 494.
279. Nielsen, *op. cit.*, p 243.
280. Bernhardt, *op. cit.*, pp 489–93.
281. Brydson, *op. cit.*, pp 38–39.
282. Bernhardt, *op. cit.*, p 492.
283. *Ibid.*, p 493.
284. *Ibid.*, p 489.
285. Baer, *op. cit.*, p 778.
286. *Ibid.*, p 782.
287. Bernhardt, *op. cit.*, pp 495–96.
288. Nielsen, *op. cit.*, pp 249–51.
289. Bernhardt, *op. cit.*, p 496.
290. *Ibid.*, pp 493–95.
291. Billmeyer, *op. cit.*, pp 170–71.
292. Nielsen, *op. cit.*, pp 249–50.
293. Bernhardt, *op. cit.*, p 495.
294. Miller, *op. cit.*, pp 564–65.
295. Nielsen, *op. cit.*, p 248.
296. Miller, *op. cit.*, pp 562–63.
297. Nielsen, *op. cit.*, p 250.
298. Miller, *op. cit.*, pp 563–64.
299. *Ibid.*, pp 566–72.
300. *Ibid.*, p 572.
301. Nielsen, *op. cit.*, pp 237–38.
302. Miller, *op. cit.*, p 565.
303. Baer, *op. cit.*, p 1041.
304. *Ibid.*, p 664.
305. *Ibid.*, pp 665–67.
306. Golding, *op. cit.*, p 697.

Chapter Six

Intermolecular Bonding

The flexibility and stability of the individual polymer molecule determine how it will respond to external stress. In nearly all practical applications, however, the polymer molecule does not exist as an isolated species, but rather as a member of a mass of similar molecules. External stress is applied to the mass as a whole, and transmitted through the mass by passing from each molecule to the next adjacent molecule. To some extent the geometrical arrangement of the adjacent molecules in space affects the way in which stress is transmitted from one to the next; this has just been discussed in Chapter 5. In a more general sense, the attractive forces between adjacent molecules are the mechanism by which such stresses are transmitted, and determine the manner and extent to which they will be transmitted, from molecule to molecule throughout the mass. These attractive forces between adjacent molecules may be referred to as intermolecular bonding.

The outer shells of valence electrons in the individual atoms provide both the primary bonding forces which hold the individual molecule together, and also the secondary bonding forces which exist between adjacent molecules (ref 1). The primary covalent homopolar bond between atoms in organic molecules in general, and in organic polymers specifically, is generally 1 Å–2 Å in length and 60–100 kcal/mol in strength (ref 2,3); these are the bonds which hold the molecule together, and remain intact through all physical processes up to the point of chemical breakdown and degradation. The secondary valence or Van der Waals forces between adjacent molecules are generally more remote

(2 Å–5 Å) and weaker (1 to 20 kcal/mol) in nature; it is these secondary valence forces between molecules which change gases into liquids and solids; control volatility, miscibility, and solubility; and produce viscosity, surface tension, and friction (ref 4). The same forces operate in polymers, and become increasingly important as increasing molecular weight increases the total sum of such forces acting between two adjacent molecules.

The intermolecular forces which attract adjacent polymer molecules to each other include a number of different types of intermolecular bonding, ranging from very weak to very strong in nature: (1) The weakest and least understood secondary attractions are the London dispersion forces which result from the mobility and transient imbalance of the clouds of valence electrons in the molecule. (2) Slightly stronger are the induced dipoles produced when a polar molecule is adjacent to a non-polar molecule. (3) Permanent dipoles exist wherever two atoms of different electronegativity are bonded to each other; the existence of such dipoles on adjacent molecules produces polar attractions between them. Dipole attractions become particularly important in hydrogen bonding and crystallinity. (4) Hydrogen bonding between molecules occurs when an electronegative atom in a polymer molecule has one or more pairs of unshared electrons, and donates these toward the hydrogen atom of an adjacent polymer molecule. (5) Dipole attractions in crystalline polymers contribute greatly to the strength and stability of the crystal structures formed. (6) Ionic bonds are still stronger, and while not very common in polymer chemistry, they play an important role whenever they do occur. (7) The strongest intermolecular forces are the primary covalent bonds formed by cross-linking adjacent polymer molecules; these are actually as strong as the bonds holding the individual molecules together, and thus produce a stable 3-dimensional molecular structure of infinite size. These forces may now be analyzed in detail.

6-1 LONDON DISPERSION FORCES

In nonpolar molecules such as hydrocarbon polymers, intermolecular attraction depends mainly upon London dispersion forces. These result from the dynamic fluidity and mobility of the valence electron cloud in the molecule, producing transient unsymmetrical states of electrical unbalance and momentary polarity (ref 5,6) (Figure 6-1). They hold adjacent molecules together at an intermolecular distance $r = 3$–5 Å with a force of 1–2 kcal/mol (ref 3). When the two molecules move further apart, these dispersion forces decrease inversely $\sim 1/r^{7-8}$ (ref 6). If the molecules attempt to move any closer together, their electron clouds begin to overlap, creating repulsive forces $\sim 1/r^{12}$ (ref 3). The balance between these attractive and repulsive forces thus determines the average intermolecular distance r.

The effects of such London dispersion forces are most evident in linear nonpolar hydrocarbon polymers where no other forms of intermolecular bonding exist. Here dispersion forces are sufficient to provide additional restriction of segmental rotation (ref 7), reducing the rate and extent of simple disentanglement. Thus at low molecular weight, linear amorphous hydrocarbons like atactic polypropylene, polyisobutylene, and polybutadiene range from viscous liquids to tacky gums;

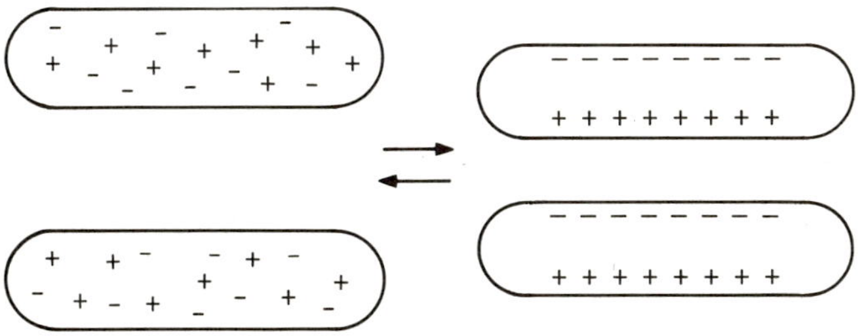

Figure 6-1. London dispersion forces.

at higher molecular weights, weak intermolecular attraction continues to permit easy rubbery extension and retraction (ref 8), while increasing numbers of chain entanglements create an internal viscosity of $\sim 10^6$ poises (ref 9) and produce a broader plateau of rubbery properties (ref 10,11). In crystalline polyolefins such as polyethylene and isotactic polypropylene, London dispersion forces undoubtedly account for the strength and fairly high melting point of these crystalline forms. In styrene polymers, the resonating electrons in the benzene ring undoubtedly contribute greatly to the dispersion forces of the electron cloud, and the resulting attractive forces help to account for the greater rigidity and fairly high softening temperature of such materials. Even in the new ultra-high-temperature polycyclic and heterocyclic polymers, the overlap of electron clouds from these resonating ring structures has been suggested as a partial explanation for their rigidity and high-temperature stability.

Thus while the mechanism of London dispersion forces remains rather vague and obscure, and their individual effects may be rather slight, their cumulative effects between adjacent high-molecular-weight polymer molecules can contribute considerably to final properties.

6-2 INDUCED DIPOLES

When a polar molecule lies adjacent to a nonpolar molecule, the polarity in the polar molecule tends to induce some electron displacement in the adjacent nonpolar molecule, creating a weak induced dipole between the two (ref 12) (Figure 6-2). While such forces have been recognized (ref 13,14), they are probably of little importance in polymer chemistry.

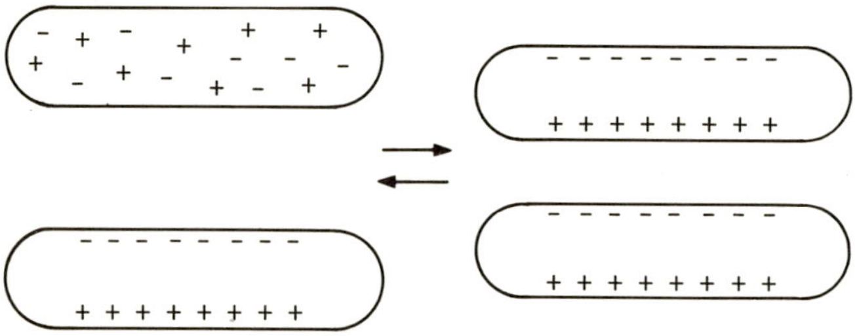

Figure 6-2. Induced dipoles.

6-3 PERMANENT DIPOLES

Whenever two atoms of different electronegativity are joined to each other, the bond between them is somewhat polar in character. Thus most organic molecules, including most polymers, contain such polar bonds; and these dipoles have many important effects on polymer properties.

Structure → Polarity

Theoretical Approach

The electronegativity and radii of individual atoms are already well established (ref 15,16) (Tables 6-1, 6-2). When any two such atoms

Table 6-1. Electronegativity Values of Some Common Elements (ref 17)

Element	Electronegativity
Caesium	0.7
Potassium	0.8
Sodium	0.9
Lithium	1.0
Aluminium	1.5
Silicon	1.8
Hydrogen	2.1
Phosphorus	2.1
Iodine	2.4
Carbon	2.5
Sulfur	2.5
Bromine	2.8
Nitrogen	3.0
Chlorine	3.0
Oxygen	3.5
Fluorine	4.0

Table 6-2. Interatomic Distances for Some Common Primary Bonds (ref 17)

Bond	Bond Length
O—O	1.32 Å
Si—Si	2.35
S—S	1.9–2.1
C—N	1.47
C—Cl	1.77
C—C	1.54
C—O	1.46
N—H	1.01
C—H	1.10
C—F	1.32–1.39
O—H	0.96
C═C	1.34
C═O	1.21
C≡N	1.15

are joined, their difference in electronegativity and the distance between them determine the polarity of the bond (ref 17,18) (Table 6-3).

Table 6-3. Dipole Moments of Common Covalent Bonds (ref 18)

Bond	Dipole Moment
C—H	0.4×10^{-18} e.s.cm
C—O	0.7
C—Cl	1.5
C=O	2.3
H—O	1.6

When three or more atoms are joined into a group or an entire molecule, the polarities of the individual bonds and the angles between these bonds determine the dipole moment of the total structure (ref

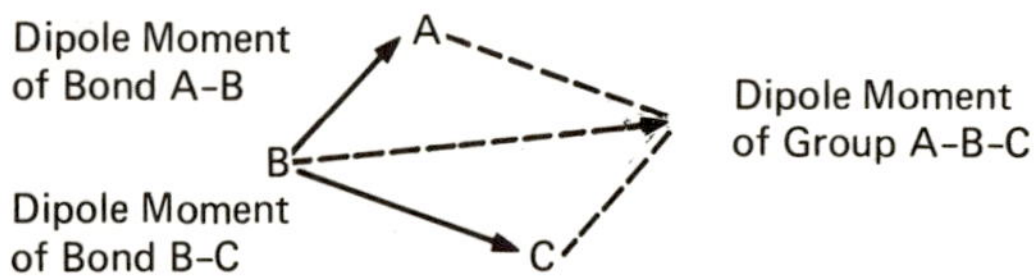

Figure 6-3. Dipole moment of a three-membered group.

19) (Figure 6-3). Thus it should be possible to start from such fundamental data, calculate the polarities in polymer structures, and correlate them with their effects on properties.

Empirical Approach

In actual practice, however, it is much more common to measure a simple overall empirical thermodynamic property of the polymer, which is clearly representative of the effect of polarity, and then use the level of this property as a relative measure of polarity in the polymer.

MELTING BEHAVIOR For example, strong intermolecular attraction favors the stability of a crystalline structure, and raises the melting point at which the crystal will finally disperse. This intermolecular attraction is measured by the enthalpy of fusion ΔH_u (Table 6-4), which

Table 6-4. Molar Cohesions of Various Linear High Polymers (ref 20)

Polymer	Structure	Molar Cohesion per 5 Å Chain Length (assuming coordination number of 4) kcal/g mole
Ethylene	$-(CH_2)_n-$	1.0
Isobutylene	$-(C(CH_3)_2-CH_2)_n-$	1.2
Butadiene	$-(CH_2-CH{=}CH-CH_2)_n-$	1.1
Natural rubber	$-(CH_2-C(CH_3){=}CH-CH_2)_n-$	1.3
Styrene	$-(CH(C_6H_5)-CH_2)_n-$	4.0
Chloroprene	$-(CH_2-CCl{=}CH-CH_2)_n-$	1.6
Vinyl chloride	$-(CHCl-CH_2)_n-$	2.6
Vinyl acetate	$-(CH(OCOCH_3)-CH_2)_n-$	3.2
Vinyl alcohol	$-(CH(OH)-CH_2)_n-$	4.2
Cellulose	$-[CH\langle CHOH-CHOH \rangle CH-O,\ CH(CH_2OH)-O \text{ ring}]_n-$	6.2
Cellulose acetate	Cellulose with acetate groups replacing alcohol groups	4.8
Synthetic polyamides	$-[OC(CH_2)_xCONH(CH_2)_y\ NH]_n-$	5.8
Silk fibroin	$-(CHR-CONH)_n-$	9.8

can be determined directly by calorimetry (ref 21,22). Thus polarity in polymer molecules produces an overall cohesive effect between these molecules, which may be measured thermodynamically and used in turn as a measure of the polar attraction between molecules.

SOLUBILITY BEHAVIOR *Cohesive Energy* Polar attraction between small molecules restricts their mobility and volatility and increases their boiling point, which is a measure of the total amount of thermal translational energy required for vaporization. Dividing the molar energy of vaporization $\Delta H - RT$ by the molar volume or molecular weight by density (M/D) produces the cohesive energy density (CED), which is a general measure of the intermolecular attraction due to polarity. Thus

$$CED = \frac{\Delta H - RT}{M/D}$$

Put another way, dividing molar energy of vaporization by molecular weight gives cohesive energy (CE) as follows: $CE = (\Delta H - RT)/M$, and dividing cohesive energy by density again gives cohesive energy density, or $CED = CE/D$.

This can be measured directly on low-molecular-weight liquids. In high-molecular-weight polymers, however, the sum of all the dipole attractions between adjacent molecules makes the cohesive energy between these molecules much greater, and supplying sufficient heat to provide the energy needed for vaporization breaks the primary bonds in the polymer molecule before it can separate the molecules from each other and permit boiling. Thus the cohesive energy density of polymers is more conveniently measured indirectly. When slightly cross-linked polymers are treated with liquids of different cohesive energy densities, liquids having cohesive energy density closest to that of the polymer will absorb into it most readily and cause the greatest degree of swelling. Cohesive energies of polymers may then be determined in this way (ref 4,23,24) (Table 6-5).

Solubility Parameter As a matter of fact, the interaction between polymers and solvents is the most common way of measuring the effect of polarity in the polymer structure upon polar attraction between polymer molecules. Similar molecules of equal polarity have uniform attraction for each other. When they are mixed with other molecules of somewhat different polarity, each species is attracted preferentially to its own kind, creating a separation of species. Thus polymer molecules and solvent molecules will only be compatible and form homogeneous solutions when they are very closely similar in polarity.

The square root of the cohesive energy density is generally known as the solubility parameter δ (ref 26,27) which is expressed thus: $\delta = (CED)^{\frac{1}{2}}$. For solvents it is readily determined by measuring the

Table 6-5. Cohesive Energy Densities of Linear Polymers (ref 25)

Polymer	Repeat Unit	Cohesive Energy Density, cal/cm^3
Polyethylene	$—CH_2CH_2—$	62
Polyisobutylene	$—CH_2C(CH_3)_2—$	65
Polyisoprene	$—CH_2C(CH_3){=}CHCH_2—$	67
Polystyrene	$—CH_2CH(C_6H_5)—$	74
Poly(methyl methacrylate)	$—CH_2C(CH_3)(COOCH_3)—$	83
Poly(vinyl acetate)	$—CH_2CH(OCOCH_3)—$	88
Poly(vinyl chloride)	$—CH_2CHCl—$	91
Poly(ethylene terephthalate)	$—CH_2CH_2OCOC_6H_4COO—$	114
Poly(hexamethylene adipamide)	$—NH(CH_2)_6NHCO(CH_2)_4CO—$	185
Polyacrylonitrile	$—CH_2CHCN—$	237

molar energy of vaporization, dividing by the molar volume to give cohesive energy density, and taking the square root to give the solubility parameter, or

$$\delta = \left(\frac{\Delta H - RT}{M/D}\right)^{\frac{1}{2}}$$

(ref 28,30) (Table 6-6). For polymers it is determined indirectly by finding solvents which produce maximum swelling or solubility (ref 24,31,33,194) (Figure 6-4) (Table 6-7).

Experimental determination of polymer solubility parameters in this way can be quite time-consuming. For a rapid approximation of solubility parameters, it is possible to use a table of molar attraction constants G for each of the individual groups in the molecule (Table 6-8), and simply take their sum ΣG as a measure of molar attraction, then divide by molar volume M/D to obtain solubility parameter (ref 35) or, $\delta = D\Sigma G/M$.

Thus cohesive energy density and solubility parameter are both

Table 6-6. Solubility Parameters of Some Common Solvents (ref 29)

	$\delta(\text{cal/sec})^{\frac{1}{2}}$		$\delta(\text{cal/sec})^{\frac{1}{2}}$
neo-pentane	6.3	Trichloroethylene	9.3
Isobutylene	6.7	Tetrachlorethane	9.4
n-hexane	7.3	Tetralin	9.5
Diethyl ether	7.4	Carbitol	9.6
n-octane	7.6	Methyl chloride	9.7
Methyl cyclohexane	7.8	Methylene chloride	9.7
Ethyl isobutyrate	7.9	Ethylene dichloride	9.8
Di-isopropyl ketone	8.0	Cyclohexanone	9.9
Methyl amyl acetate	8.0	Cellosolve	9.9
Turpentine	8.1	Dioxane	9.9
Cyclohexane	8.2	Carbon disulphide	10.0
2,2-dichloropropane	8.2	Acetone	10.0
Sec.amyl acetate	8.3	*n*-octanol	10.3
Methyl isobutyl ketone	8.4	Butyronitrile	10.5
Dipentene	8.5	*n*-hexanol	10.7
Amyl acetate	8.5	sec-butanol	10.8
Methyl isopropyl ketone	8.5	Pyridine	10.9
Carbon tetrachloride	8.6	Nitroethane	11.1
Pine oil	8.6	*n*-butanol	11.4
Piperidine	8.7	Cyclohexanol	11.4
Xylene	8.8	Isopropanol	11.5
Dimethyl ether	8.8	*n*-propanol	11.9
Toluene	8.9	Dimethyl formamide	12.1
Butyl cellosolve	8.9	Hydrogen cyanide	12.1
1,2-dichloropropane	9.0	Ethanol	12.7
Mesityl oxide	9.0	Cresol	13.3
Isophorone	9.1	Formic acid	13.5
Ethylacetate	9.1	Methanol	14.5
Benzene	9.2	Phenol	14.5
Diacetone alcohol	9.2	Glycerol	16.5
Chloroform	9.3	Water	23.4

generally useful as a direct measure of the polar attractive bonding forces between polymer molecules.

Effect of Copolymerization

Copolymerization can be used to control the spacing between polar groups (ref 36). In random copolymers there is generally some repulsion between unlike comonomer units, which increases the intermolecular spacing, producing poorer packing, lower density, and softer properties, less than linearlly additive (ref 37); for example, styrene/methyl

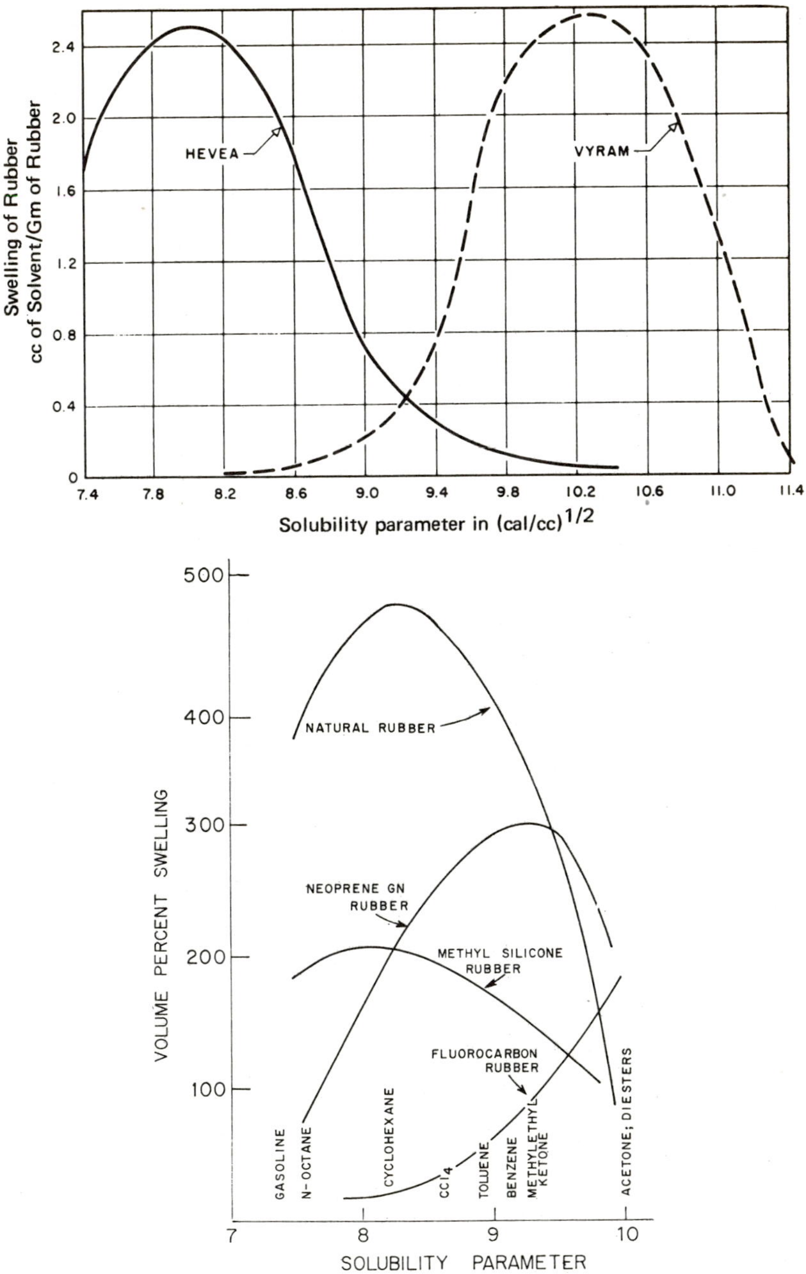

Figure 6-4. Determination of solubility parameters of polymers by swelling in solvents (ref 32,195).

acrylate copolymers give "ideal" glass transition temperature behavior, but ethyl and butyl acrylate copolymers give lower than linearly additive values (Figure 6-5). In extreme cases, repulsion may be so strong that properties actually go through a minimum, less than either of the

Table 6-7. Solubility Parameters of Polymers (ref 29)

Polymer	δ(cal/sec)$^{\frac{1}{2}}$
PTFE	6.2[c]
Poly(dimethyl siloxanc)	7.3
Polyethylene	7.9[o] 8.1[c]
Polypropylene	7.9
Polyisobutylene	8.05[o] 7.7[c]
Styrene-butadiene rubber (25% styrene)	8.09–8.6[o] 8.54[c]
Polyisoprene (natural rubber)	7.9–8.35[o] 8.15[c]
Polybutadiene	8.4–8.6[o] 8.38[c]
Polysulphide rubbers	9.0–9.4
Polystyrene	8.5–9.7[o] 9.12[c]
Polychloroprene	9.18–9.25[o] 9.38[c]
Nitrile rubbers (25% acrylonitrile)	9.38–9.5[o] 9.25[c]
Poly(vinyl acetate)	9.4[c]
Poly(methyl methacrylate)	9.0–9.5[o] 9.2[c]
Poly(vinyl chloride)	9.38–9.7[o] 9.55[c]
Bisphenol A polycarbonate	9.5[c.m]
Poly(vinylidene chloride)	9.8
Poly(ethylene terephthalate)	10.7
Cellulose (di)nitrate	10.56[o] 10.48[c]
Cellulose (di)acetate	11.35[c]
Epoxide resins	approx. 11.0
Acetal resin (homopolymer)	11.1[o]
Nylon 66	13.6
Polyacrylonitrile	15.4

[o]—observed, [c]—calculated from Small's data, [c.m.]—calculated but using obtained data with model organic carbonate

individual components: typical systems include styrene/methyl methacrylate, methyl methacrylate/acrylonitrile, and acrylonitrile/acrylamide (ref 39) (Figure 6-6). In rare cases there may actually be attraction between comonomer units, producing higher attractions and denser, harder properties than expected; examples are the higher-than-ideal glass transition temperatures observed in copolymers of vinylidene chloride/methyl acrylate, methyl acrylate/methyl methacrylate, and dimethyl siloxane/phenyl methyl siloxane (ref 40).

Table 6-8. Molar Attraction Constants at 25°C (ref 34)

Group	G Molar attraction constant G
$—CH_3$	214
$—CH_2—$(single bonded)	133
—CH<	28
>C<	−93
$CH_2=$	190
—CH= (double bonded)	111
>C=	19
CH≡C—	285
—C≡C—	222
Phenyl	735
Phenylene (*o*,*m*,*p*)	658
Naphthyl	1146
Ring (5-membered)	105–115
Ring (6-membered)	95–105
Conjugation	20–30
H	80–100
O (ethers)	70
CO (ketones)	275
COO (esters)	310
CN	410
Cl single	270
Cl twinned as in >CCl_2	260
Cl triple as in $—CCl_3$	250
Br single	340
I single	425
CF_2 in fluorocarbons only	150
CF_3 in fluorocarbons only	274
S sulphides	225
SH thiols	315
ONO_2 nitrates	~440
NO_2 (aliphatic)	~440
PO_4 (organic)	~500
†Si (in silicones)	~38

† Estimated by H. Burrell.

Polarity → Properties

Such secondary forces between adjacent polymer molecules restrict segmental rotation (ref 41,42). Since molecular cohesion and bond rotation energies are both in the range of 1–5 kcal/mol, polar forces between adjacent molecules are a significant factor reducing the inherent flexibility of the individual polymer molecule (ref 43). This effect of polarity, restricting segmental mobility and molecular flexibility, has pronounced effects upon polymer properties.

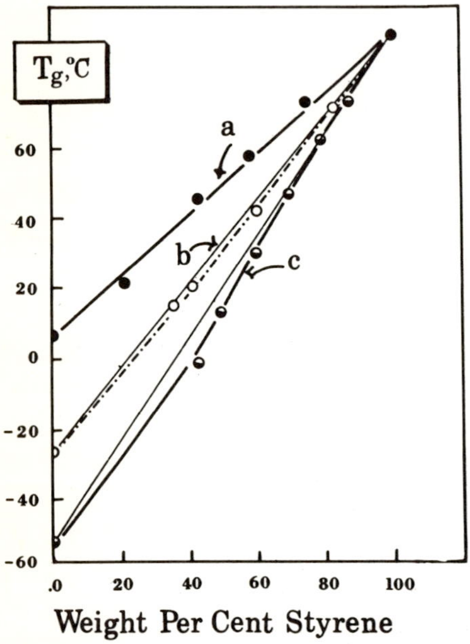

Figure 6-5. Effect of comonomer polarities upon glass transition temperatures (ref 38).

Dependence of glass-transition temperature, T_g, on composition for conventional copolymers of styrene and: a. methyl acrylate. b. ethyl acrylate, c. *n*-butyl acrylate,

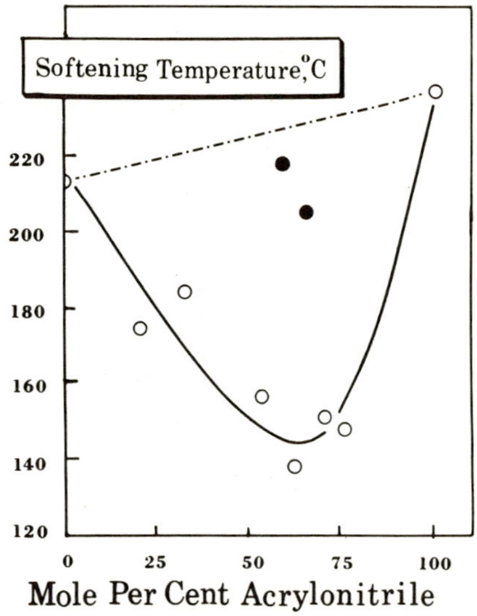

Figure 6-6. Extreme repulsion between comonomer polarities (ref 39). Softening points of copolymers of acrylamide and acrylonitrile: ○ conventional copolymers, ● block copolymers.

Polarity in polymers is generally accompanied by hydrogen bonding and crystallinity, forces which are much greater than simple dipole attraction, and thus tend to mask the simple effects of polarity in itself.

This makes it difficult to distinguish the simple effects of polarity alone upon polymer properties. In the discussion which follows, these effects may often be complicated or even obscured by the presence of hydrogen bonding or crystallinity or both. And many of them may have to be discussed more thoroughly under these later topics.

Processability

Polar attractions between adjacent polymer molecules increase the forces favoring crystallinity, and thus call attention to all the factors already considered relating processability and crystallinity (Chapter 5). In particular, strong polar attractive forces between polymer molecules in the crystal increase its melting point considerably. Up to a certain point this may be desirable for end-use properties. At very high degrees of polarity, however, crystalline polymers may be too high-melting for convenient processability, and may even melt above their thermal decomposition temperatures.

For solution processing, polarity requires the choice of solvents having solubility parameters close to those for the polymer itself. Generally solubility parameter of the solvent must be within 1 or at most 2 units of the polymer in order to dissolve it effectively.

In application of coatings and adhesives, similar polarities in the polymer and the substrate help to orient the polymer dipoles at the surface preferentially during processing, and thus improve the quality of the interfacial bond which is formed.

Mechanical Properties

Polar attraction between polymer molecules draws them close together, leaving less free volume (ref 44), and thus decreasing the ease of segment rotation and molecular mobility (ref 42). This decreased molecular mobility produces an increase in modulus (ref 45), stiffness (ref 46), yield point (ref 46), and tensile strength (ref 45). In copolymers, on the other hand, unlike units usually repel each other, producing greater intermolecular spacing, poorer packing, and somewhat lower density and hardness than would be expected (ref 37).

Adhesion and friction between two adjacent surfaces are directly related to mutual attraction between them, which in turn depends upon similarity in polarity (ref 47). Thus friction of a polymer against itself tends to be greater than against a dissimilar material; and friction of polymers against metals tends to be lower for nonpolar than for polar polymers. Unfortunately, however, these properties are complicated by other factors such as hardness, softening temperature, internal friction or mechanical damping, and shear strength. Consequently, actual coefficients of friction (Table 6-9) combine all these factors and often obscure the effects of polarity alone.

Table 6-9. Coefficients of Friction of Polymers (ref 48)

Polymer	Metal against Polymer	Polymer against Polymer
Polyvinyl chloride	0.4–0.9	0.45–0.55
Polystyrene	0.4–0.5	0.4–0.5
Polystyrene polyblend	0.38	
Polymethyl methacrylate	0.25	0.4
	0.4–0.5	0.4–0.6
Nylon 66	0.3 (0.36)	
Nylon 6	0.39	
Polyethylene (low density)	0.33–0.6	0.33–0.6
	0.6–0.8	0.1
Polyethylene (high density)	0.23	
Polyvinylidene chloride	0.68–1.8	0.8–2.0
Polyvinyl fluoride		0.1–0.3
Polytrifluorochloroethylene	0.56	
Polytetrafluoroethylene	0.04–0.10	0.04
	0.10–0.15	
Phenol-formaldehyde resin	0.61	
Rubber	0.3–2.5	

Thermal Properties

Polar attraction between adjacent polymer molecules generally requires greater energy to separate them and move them past each other, and thus raises their thermal transitions to higher temperatures.

Glass Transition Temperature Polarity generally increases the glass transition temperature (ref 42,44,46,49). This increase in glass transition temperature corresponds with a similar increase in solubility parameter, which is a more common measure of polar attraction between molecules (ref 50).

In fibers, polarity controls glass transition temperature to make it low enough for drawing and ironing, yet high enough to maintain orientation and provide crease retention (ref 45).

Copolymerization In random copolymers, repulsion between unlike comonomer units generally pushes them further apart, and this greater intermolecular spacing and poorer packing gives lower glass transition temperatures than would be expected (as previously discussed); on the other hand, in rare cases, attraction between the two comonomer units may give higher glass transition temperature than would be expected. In block and graft copolymers, the original glass transition temperatures of the parent homopolymers may remain, or partial miscibility may produce new glass transition temperatures closer to ideal (ref 51). Thus styrene and ethylene oxide are so dissimilar and immiscible that their block copolymers retain the glass transition temperatures of both hompolymers. Polystyrene grafts on rubber raise the glass transition temperature of the rubber 25 to 30°C, but also contain a polystyrene homopolymer phase which retains its original glass transition temperature. Styrene/butyl acrylate block copolymers show some mixing and repulsion in both phases, lowering both glass transition temperatures (Table 6-10).

Table 6-10. Glass Transition Temperatures of Styrene/Butyl Acrylate Copolymers (ref 52)

Polymer	Glass Transition Temperatures
Polystyrene	100°C
Poly(*n*-butyl acrylate)	16
Block copolymer	64 and −2

On the other hand, styrene/α-methylstyrene and acrylonitrile/acrylamide block copolymers contain such similar components that miscibility produces nearly ideal linear average behavior.

Brittle Temperature While mutual repulsion in copolymers lowers the expected glass transition and softening temperatures, it has the opposite effect on low-temperature brittleness. Thus the brittle points of dimethylbutadiene/acrylonitrile and of acrylonitrile/acrylamide copolymers are higher than ideal (ref 53), probably because mutual repulsion weakens the bonds between adjacent chains and consequently lowers the cohesive strength of the material under mechanical stress.

Melting Point Polar attractions between adjacent molecules in a polymer crystal increase the thermal energy necessary to overcome the stability of the crystalline structure and produce melting, and thus increase the melting point of the polymer (ref 46,54–57). Since similar cohesive energies exist in both the monomer and the polymer, there is a correlation in their crystalline melting points (ref 57). Since melting point depends upon total polarity, spacing the polar groups further apart decreases the melting point (ref 58) (Figure 6-7). For use in synthetic fibers, polar crystalline polymers should preferably melt below 300°C to permit good melt-spinning, yet above 200°C to withstand ironing (ref 45). In all of these observations, however, the interaction of other

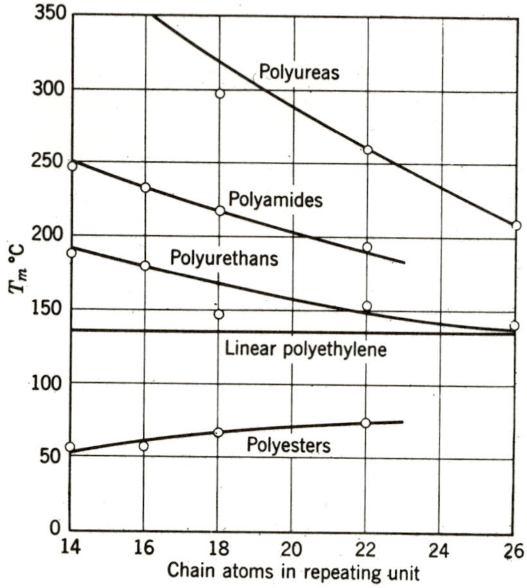

Figure 6-7. Spacing of polar groups affects polarity and melting point (ref 59). Trend of crystalline melting points in homologous series of aliphatic polymers.

structural features, such as molecular flexibility, may complicate and even obscure the effects of polarity alone.

Electrical Properties

Polarity is of course the primary factor in electrical behavior, but it is hardly the sole factor. In dielectric constant and loss, the polarity of the dipole determines its tendency to respond to an alternating electric field, but the mobility of the dipole determines its ability to make this response; paradoxically, strong dipoles create intermolecular attractions which simultaneously reduce mobility (ref 46). In electrical resistance, polar impurities are the most probable carriers of the electrical current, but they require considerable mobility in the polymeric matrix before they can produce any appreciable electrical loss. In breakdown, both of these mechanisms are involved, along with more complex structure-behavior relationships. Thus little data are available to isolate the simple effects of polarity on electrical properties.

Optical Properties

The response of polymer molecules to radiant energy, particularly in the infrared and visible wavelengths, is based on the temporary dis-

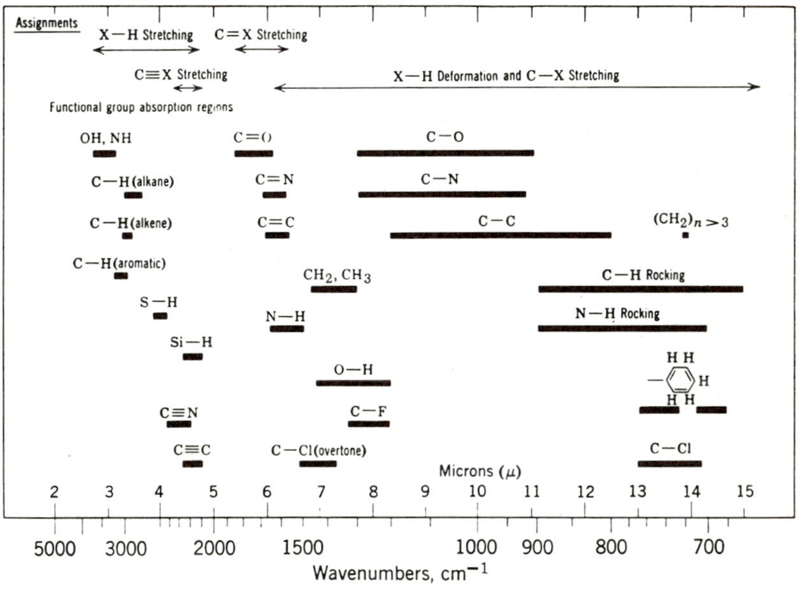

Figure 6-8. Infrared absorption bands useful in analysis of molecular structure (ref 60).

placement of valence electrons, and thus depends directly on the polarity of the bonds in the molecules.

INFRARED BEHAVIOR Infrared behavior is based on changes in dipole moment when a polar bond absorbs and releases energy (ref 60). This is very useful, both in spectroscopic analysis of molecular structure (Figure 6-8) and in preheating before thermoforming (ref 61,62).

Table 6-11. Dielectric Constants and Refractive Indexes of Polymers (ref 64)

	D	n	n^2
Polyethylene	2.28	1.51	2.28
Polystyrene	2.55	1.60	2.56
PTFE	2.05	1.40	1.96
Polyisobutylene	2.30	1.51	2.28

REFRACTIVE INDEX Retardation of the passage of light through a polymer is also controlled by the polarity of the bonds in the molecules, which thus determine the refractive index. Polarization of molecular structure, in an electromagnetic field, can be related to both dielectric constant D and refractive index n

$$\left(\frac{D-1}{D+2}\right)\frac{M}{\rho} = P = \left(\frac{n^2-1}{n^2+2}\right)\frac{M}{\rho}$$

where M is molecular weight and ρ is density (ref 63). Thus dielectric constant and refractive index are very simply related by the expression $D = n^2$ (Table 6-11). Just as electronegativities and dipole moments are additive (ref 18), so, too, molar refraction can be calculated by summing the refractions of individual groups in the polymer molecule (Table 6-12).

In all of these considerations, however, the effects of polarity are actually *intra*molecular, and do not depend upon the forces of polarity in *inter*molecular bonding.

Chemical Properties

Since polarity is primarily a form of chemical valence, polarity in polymers produces quite a number of effects upon their chemical behavior.

Table 6-12. Molar Refractions of Some Electron Groups* (ref 64)

Electron group	Refraction	Electron group	Refraction
C—H	1.705	C≡C	6.025
C—C	1.209	C—Cl	6.57
C=C	4.15	C—F	1.60

* Sodium D line.

SOLUBILITY AND SWELLING Solubility or swelling of a polymer in a solvent depends upon competitive intermolecular attraction between (*a*) solvent and polymer molecules vs (*b*) adjacent polymer molecules. Since two adjacent polymer molecules are clearly similar and exert maximum attraction for each other, no solvent molecule can compete as perfectly. If it is reasonably similar, however, it may still be able to intercept the intermolecular attraction between adjacent polymer molecules sufficiently to produce uniform solubility, or at least swelling in the case of lightly-cross-linked polymers.

These considerations are of great importance, first in solution-processing during manufacture of fibers, films, and adhesives; and later during the use of polymeric products under conditions which expose them to attack by potential solvents, either in industry or in the home.

Table 6-13. Solubility Parameters of Some Common Solvents* (ref 68)

Weakly hydrogen-bonding		Moderately hydrogen-bonding		Strongly hydrogen-bonding	
Solvent	δ_1	Solvent	δ_1	Solvent	δ_1
n-Hexane	7.3	Diethyl ether	7.4	2-Ethylhexanol	9.5
Cyclohexane	8.2	Diisobutyl ketone	7.8	2-Ethylbutanol	10.5
Carbon tetrachloride	8.6	Ethyl acetate	9.0	*n*-Butanol	11.4
Toluene	8.9	Methyl ethyl ketone	9.3	Isopropanol	11.5
Benzene	9.1_5	Tetrahydrofuran	9.5	*m*-Cresol	11.9
Chloroform	9.3	Acetone	10.0	Ethanol	12.7, 13.2
Ethylene dichloride	9.8	Dioxane	10.0	Acetic acid	12.9
Carbon disulfide	10.0	Dimethylformamide	12.0	Methanol	14.5
Acetonitrile	11.9	Ethylene carbonate	14.7	Water	23.4
Nitromethane	12.6				

* $(\text{cal cm}^{-3})^{\frac{1}{2}}$ at 25°C

Solubility Parameter Solubility and swelling of polymers in solvents are so intimately related to polarity that it is often defined in terms of the solubility parameter. In general, solubility parameters of the solvents should be within 1 or 2 units away from that of the polymer, in order to dissolve or swell it satisfactorily (ref 65–67) (Tables 6–13, 6-14). Thus in some cases high polarity may favor solubility (ref 26), while in other cases it may produce insolubility (ref 69,70). For use in fibers, polymers should be chosen with a solubility parameter which will resist dry cleaning solvents (ref 45).

Copolymers Amorphous copolymers have solubilities intermediate between those of the parent amorphous homopolymers, and may dissolve best in mixtures of solvents for each of the two types of monomer units.

Block and graft copolymers may dissolve in a solvent for either type of unit, simply collapsing the other type which is insoluble in such a solvent (ref 71,72); or they may require a solvent capable of dissolving both components independently (ref 73).

Table 6-14. Ranges of Solvent Solubility Parameters* which Will Dissolve Selected Polymers (ref 68)

<table>
<tr><th rowspan="2">Polymer</th><th colspan="3">Midpoint and Breadth of Range of Solubility Parameters to be Used with Solvents in which Hydrogen Bonding is:</th></tr>
<tr><th>Poor</th><th>Moderate</th><th>Strong</th></tr>
<tr><td>Poly(isobutyl methacrylate)</td><td>9.2 ± 1.8</td><td>8.7 ± 1.3</td><td>11.0 ± 0.5</td></tr>
<tr><td>Poly(butyl acrylate)</td><td>9.2 ± 1.8</td><td>8.7 ± 1.3</td><td>11.5</td></tr>
<tr><td>Poly(butyl methacrylate)</td><td>9.6 ± 1.4</td><td>8.7 ± 1.3</td><td>10.5 ± 1.0</td></tr>
<tr><td>Polystyrene</td><td>9.4 ± 1.2</td><td>9.0 ± 0.6</td><td>—</td></tr>
<tr><td>Poly(ethyl acrylate)</td><td>10.6 ± 2.0</td><td>10.4 ± 3.0</td><td>10.5</td></tr>
<tr><td>Poly(methyl methacrylate)</td><td>10.7 ± 2.1</td><td>10.7 ± 2.6</td><td>—</td></tr>
<tr><td>Poly(vinyl acetate)</td><td>10.6 ± 2.1</td><td>11.6 ± 3.2</td><td>14.5</td></tr>
<tr><td>Poly(methyl acrylate)</td><td>10.8 ± 1.9</td><td>12.0 ± 2.8</td><td>—</td></tr>
<tr><td>Nitrocellulose</td><td>10.9 ± 0.8</td><td>11.2 ± 3.4</td><td>—</td></tr>
<tr><td>Urea formaldehyde resin</td><td>—</td><td>—</td><td>10.1 ± 1.2</td></tr>
<tr><td>Natural rubber</td><td colspan="2">8.0</td><td>—</td></tr>
<tr><td>Polydimethylsiloxane</td><td colspan="2">7.6</td><td>—</td></tr>
<tr><td>Hydropolybutadiene</td><td colspan="2">8.1</td><td>—</td></tr>
<tr><td>Polyurethane</td><td colspan="2">10.0</td><td>—</td></tr>
<tr><td>Polyacrylonitrile</td><td>—</td><td>13.0 ± 2</td><td>—</td></tr>
</table>

* $(\text{cal cm}^{-3})^{\frac{1}{2}}$

Permeability Permeation of small molecules through a polymer generally depends upon chemical "similarity" (ref 74), particularly polarity. For example, permeability through polyethylene is lowest for highly polar materials and highest for materials of low polarity (Table 6-15). Permeability constant P is a function of the polymer F, the penetrant G, and the interaction between them H: $P = F \times G \times H$; specific interactions H arise between polar polymers and polar gases (ref 75). This is reasonably in accordance with solubility theory.

It has also been noted that increasing polarity and cohesive energy tend to produce a decrease in permeability (ref 76) (Table 6-16). This has been explained by pointing out that a polar polymer requires higher activation energy to permit diffusion (ref 77). Thus nitrile and chloro-

Table 6-15. Permeability of Classes of Compounds through Polyethylene (ref 74)

Alcohols	Lowest
Acids	
Nitro Derivatives	
Aldehydes & ketones	
Esters	
Ethers	
Hydrocarbons	
Halogenated hydrocarbons	Highest

prene rubbers are less permeable than natural and butadiene rubbers; and even the methyl group in rubbers contributes sufficient cohesion and activation energy to reduce permeability. Considering the low polarity of the methyl group, however, and the balanced polarity of polymers like polyisobutylene and polyvinylidene chloride, their exceptional impermeability appears to require further theoretical consideration to explain them fully.

STRESS CRACKING Solvent cracking occurs when a polymer is subjected to a moderate mechanical stress and immersed in a weak solvent, and ultimately breaks even under this moderate stress. The cause appears to be a lowering of the cohesive bond energies as the normal polymer-polymer bonds, which can sustain the mechanical stress, are replaced

Table 6-16. Permeability of Selected Polymers to Ethylene Oxide and Methyl Bromide (ref 76)

	P[cc (STP)/mm/cm^2/cm Hg $\times$ 10^8]	
Polymer	Ethylene oxide, 0°C, 330 mm Hg	Methyl bromide, 60°C, 395 mm Hg
Polyvinyl alcohol	0.0002	—
Polyethylene glycol terephthalate ("Mylar")	—	0.008
Polyvinylidene chloride (Saran)	—	0.08
Polychlorotrifluoroethylene ("Kel-F")	0.12	0.46
Cellulose acetate (15% DBP)	140	0.68
Polyamide (nylon 6)	—	0.84
Rubber hydrochloride ("Pliofilm")	73	3.4
Polyethylene (0.92 g/ml)	4.9	47.2
Ethyl cellulose (plasticized)	42	—
Polyisoprene (natural rubber)	—	153

by new polymer-solvent bonds which cannot contribute to the overall strength of the material (ref 78).

ADHESION Adhesion of a polymer to a substrate is strongest when the two have similar polarities. This is usually important in choosing the proper adhesive for each application (ref 79) (Table 6-17). Since most substrates have some polarity, polarity in polymers generally favors adsorption and adhesion (ref 81). In electrical encapsulation, polarity helps the polymer adhere to copper, steel, glass fiber, and other plastics

Table 6-17. Classification of Adhesives by Polarity (ref 80)

1. Low polarity
   a. Straight hydrocarbon. Asphalts, hydrocarbon rubbers, hydrocarbon resins
   b. Halogenated hydrocarbon. Chlorine-bearing rubbers, chlorine-bearing vinyl polymers
2. Medium polarity
   a. Nitrogen-, sulfur-, and silicon-containing. Nitrile rubbers, thioplasts, sulfur, silicones
   b. Oxygen-containing. Shellac, rosin, waxes, phenolic resins, furan resins, acrylics, vinyl esters and acetals
3. High polarity
   a. Oxygen-containing. Starches, dextrins, gums, cellulose esters and ethers, polyesters, polyvinyl alcohol, silicates, magnesium oxychloride, litharge-glycerol
   b. Amide and amine. Protein (hide, bone, and blood glues, fish glues, casein, soybean), polyamides, urea resins, melamine resins

used in constructing the electrical component (ref 82). In coating applications, polar groups in the polymer tend to orient toward a metal substrate and thus produce optimum adhesion (ref 83). While these generalities are useful, the many other factors in adhesion theory often complicate their interpretation and even obscure their significance.

Applications

In general, polarity should be low in elastomers, medium in plastics, and high in fibers (ref 54,84). In actual practice, interaction with other structural features modifies this requirement somewhat, and produces overlapping between these three classes of materials.

Thus polar attractions between polymer molecules are a fairly strong source of intermolecular bonding, and have significant effects upon properties and applications. In most polymers, however, they are accompanied and masked by the greater forces of hydrogen bonding and crystallinity, which produce even stronger intermolecular bonding between polymer molecules.

6-4 HYDROGEN BONDING

When a polymer contains basic electronegative atoms with semiavailable unshared pairs of valence electrons, it can donate these toward the hydrogen atoms of adjacent molecules to produce hydrogen bonding between the two molecules. Since most polymers contain such basic electronegative atoms, and nearly all organic polymers contain hydrogen atoms, hydrogen bonding is a common and important type of intermolecular bonding between polymer molecules. Furthermore, when a

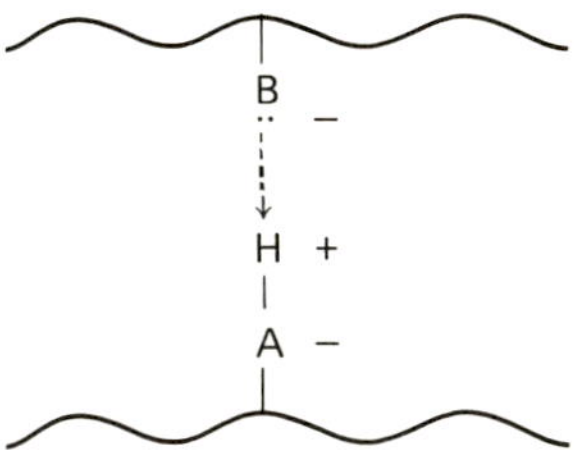

Figure 6-9.

liquid monomeric plasticizer is added to a polymer to soften its properties, the mechanism of plasticizer action generally involves the conversion of polymer-polymer hydrogen bonds into polymer-plasticizer hydrogen bonds. In general, hydrogen bonding between two polymer molecules reduces their mobility, while hydrogen bonding between a polymer and a plasticizer molecule increases the mobility of the polymer molecule.

Hydrogen Bonding between Polymer Molecules

The composition and structure of polymer molecules determine their ability to form hydrogen bonds between them, and these hydrogen bonds in turn affect the structure and properties of the polymer.

Structure → Hydrogen Bonding

Basic electronegative atoms with unshared pairs of valence electrons can contribute these electrons toward a neighboring acidic hydrogen atom to form a hydrogen bond between them (ref 13,85,86) (Figure 6-9).

The more electronegative the atom which donates these electrons, the stronger the hydrogen bond it will form (ref 87) (Table 6-18). The atom to which the hydrogen is attached by a covalent bond also affects hydrogen bonding: while hydrogen attached to carbon can form fairly strong hydrogen bonds, hydrogen attached to more electronegative atoms forms much stronger hydrogen bonds. In general hydrogen bonds are about 3 Å in length and 1.5–6 kcal/mol in strength (ref 3).

Table 6-18. Strengths of Hydrogen Bonds (ref 88)

Bond	Dissociation energy
C—H- - -N	3 kcal/mol
N—H- - -N	3–5
N—H- - -O	4
O—H- - -O	3–6
F—H- - -F	7

Thus most polymers contain some degree of hydrogen bonding. The total effect depends upon the strength of the individual hydrogen bonds and their concentration and number in the total polymer structure.

Interactions between Hydrogen Bonding and Structure

The composition and detailed structure of polymer molecules determine their ability to form intermolecular hydrogen bonds; and the formation of these hydrogen bonds, in turn, affects the larger features of polymer structure and behavior. Secondary valence forces involved in hydrogen bonding restrict segmental rotation of the polymer molecule, and thus decrease molecular flexibility (ref 89). When polymer molecules are oriented parallel to each other, particularly in the drawing of fibers, hydrogen bonding provides lateral forces to keep them in this parallel position (ref 90).

When polymers crystallize, hydrogen bonding between adjacent units in the crystal structure adds to the forces favoring formation of strong crystal bonds (ref 91). Whenever hydrogen bonding occurs in a crystalline polymer, the forces of crystallinity are so great that they are apt to mask and largely obscure the contribution which intermolecular hydrogen bonding makes toward polymer behavior and properties. Thus, while many of the effects of hydrogen bonding are credited to the forces of crystallinity alone, it should be remembered that they actually result from a combination of the forces of hydrogen bonding and crystallinity together.

Hydrogen Bonding → Properties

Hydrogen bonding provides a large number of fairly strong intermolecular bonds between adjacent polymer molecules. This greatly restricts their mobility and thus has important effects upon their properties. Most of the effects of polarity upon properties (Section 6-3) actually involve hydrogen bonding as well. Specific effects of hydrogen bonding which have been described most commonly include effects upon melting and solubility behavior during processing and later use, and upon mechanical strength and adhesion.

MELTING BEHAVIOR In general, formation of hydrogen bonds between adjacent chains in the polymer crystal restricts their mobility and raises their melting point markedly (ref 57,92). This is clearly seen in the high melting points of polyurethanes, polyamides, and especially polyureas (ref 93) (Figure 6-7). In all of these, increasing the length of the repeat unit between functional groups decreases the concentration of hydrogen bonding and consequently its effect upon melting point.

Replacement of hydrogen atoms by alkyl groups, on the nitrogens in polyamides, eliminates the hydrogens which contribute most to hydrogen bonding, thus reducing hydrogen bonding and producing much lower melting points. *N*-methyl nylons melt at least 100°C below unsubstituted nylons (ref 94).

SOLUBILITY Hydrogen bonding of polymer-polymer vs polymer-solvent pairs is a powerful force in solubility (ref 27,66,92). It exerts a strong modifying effect upon simple polarity as expressed by the solubility parameter (ref 67) (Tables 6-13, 6-14). The effect of hydrogen bonding on solubility is powerful enough to produce solution even when the solubility parameters of polymer and solvent differ considerably.

MECHANICAL STRENGTH Mechanical strength of a polymer depends on the size of the structural unit which is available to resist the stress applied to it, and thus increases with molecular weight (Section 3-2). Formation of secondary hydrogen bonds between polymer molecules increases the effective size of the structural unit which is available to resist mechanical stress, and thus produces high mechanical strength even at lower molecular weight (ref 95) (Figure 6-10).

Adsorption and Adhesion Hydrogen bonding between polymer molecules and adjacent materials adds a powerful force favoring adsorption and adhesion (ref 97).

These are some of the areas in which the effects of hydrogen bonding on properties have been particularly noted. Most examples of polarity in polymers also involve some degree of hydrogen bonding. Thus most

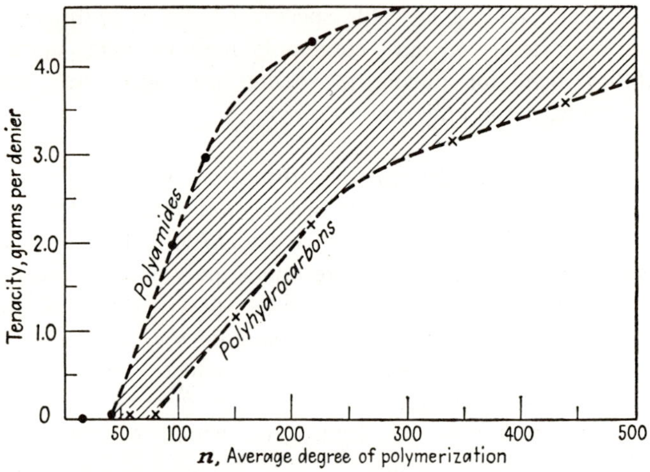

Figure 6-10. Interacting effects of hydrogen bonding and molecular weight upon tensile strength (ref 96).

of the effects of polarity on properties (Section 6-3) also involve hydrogen bonding to some extent. More intensive analytical studies are needed to distinguish between the two types of effects.

Plasticization

When a liquid monomeric plasticizer is added to a solid polymer to soften its properties, the mechanism of plasticization may very well involve the formation of hydrogen bonds between polymer and plasticizer molecules. The composition and structure of polymer and plasticizer molecules control their ability to form hydrogen bonds between them; and the hydrogen bonds thus formed modify the overall structure of the polymeric material, and therefore its behavior and properties.

Structure → Hydrogen Bonding to Plasticizer

Most successful combinations of polymers and plasticizers generally involve electronegative atoms in one, or most often in both, of the materials combined in this way. For example, in polyvinyl chloride the chlorine atom leaves the hydrogen atom on the same carbon relatively acidic and available for hydrogen bonding to the oxygen atom in an ester plasticizer (Figure 6-11). In cellulose esters the ester groups may have a similar effect, and the residual unesterified hydroxyl groups certainly provide very available hydrogens for hydrogen bonding.

To be compatible with each other, polymer and plasticizer must have very similar solubility parameters (ref 98) (Table 6-19). Dissimilar secondary plasticizers may be added only up to the point at which the average solubility parameter of the plasticizer mixture is within one

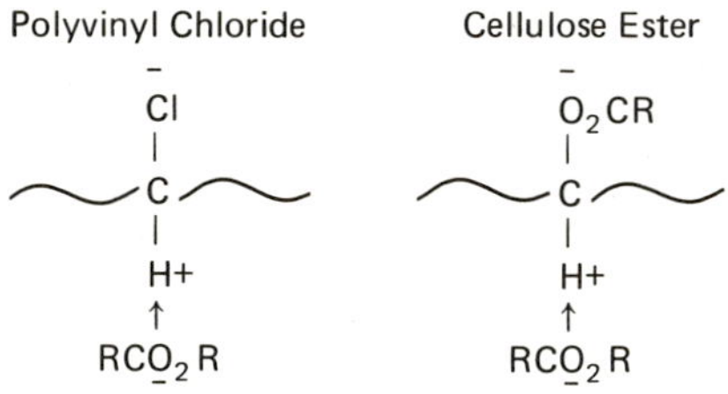

Figure 6-11.

unit of the solubility parameter of the polymer; beyond this point, polymer and plasticizer are no longer completely miscible. Plasticizers less similar to the polymer, in polarity and solubility parameter, may still be absorbed in small amounts and have important specialized effects upon properties. Polar polymers, particularly nylons and cellulosics, can absorb water from immersion or even simply from humidity in the atmosphere, with important effects upon their properties (ref 100).

Thus both hydrogen bonding and similar polarity are important in compatibility between polymers and plasticizers.

Effects of Plasticization upon Structure and Behavior

The plasticizing action of liquid monomeric plasticizers and absorbed moisture begins by their penetration between adjacent polymer molecules (ref 101,104). Such penetration requires similarity in polarity and solubility parameter, and in most cases it is greatly aided by hydro-

Table 6-19. Solubility Parameters for Some Common Plasticizers (ref 99)

Plasticizer	δ(cal/sec)$^{1/2}$
Paraffinic oils	7.5 approx.
Aromatic oils	8.0 approx.
Camphor	7.5
Diisoctyl adipate	8.7
Dioctyl sebacate	8.7
Diisodecyl phthalate	8.8
Dibutyl sebacate	8.9
Diethyl hexyl phthalate	8.9
Diisooctyl phthalate	8.9
Di-2-butoxyethyl phthalate	9.3
Dibutyl phthalate	9.4
Triphenyl phosphate	9.8
Tritolyl phosphate	9.8
Trixylyl phosphate	9.9
Dibenzyl ether	10.0
Triacetin	10.0
Dimethyl phthalate	10.5
Santicizer 8	11.0 approx.

gen bonding between the polymer and plasticizer molecules. It eliminates some of the intermolecular bonds between adjacent polymer molecules and replaces them by bonds between polymer and monomeric plasticizer molecules, thus reducing the forces of intermolecular attraction between the polymer molecules (ref 71,83,101–104) (Figure 6-12). This permits polymer molecules to slip more easily past each other (ref 104).

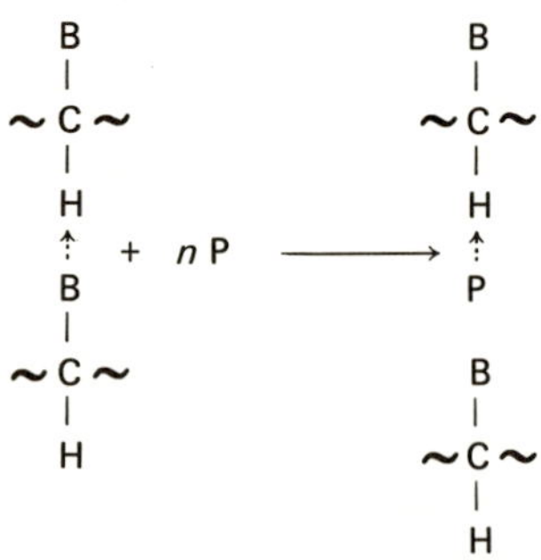

Figure 6-12.

The monomeric liquid plasticizer molecules have considerable mobility and large free volume, and they contribute this to the polymer in which they are absorbed (ref 105). This produces increasing mobility of segments of the polymer molecules, permitting them to flex more readily into different conformations (ref 102,103,106,107). This increased mobility in plasticized polymers increases the ease, probability, and frequency of both small and large deformations and thus affects most of their properties (ref 46). These effects are most frequently evident in mechanical, thermal, and electrical properties.

Plasticization → Properties

MECHANICAL PROPERTIES Incorporation of plasticizer into a polymer generally increases the mobility of the polymer molecules and has the expected softening effects upon mechanical properties. Thus it generally decreases modulus, hardness, and stiffness, and produces softer materials of increased flexibility, pliability, and distensibility (Section 3-2). It also reduces strength in general, such as yield point, tensile strength, compressive strength, and tear strength; and increases ductility and ultimate elongation. It increases toughness and resistance to impact, and reduces checking and chipping of coatings on rigid substrates. All of these changes result from the effect of plasticizer in reducing intermolecular bonding and thus increasing the mobility of polymer molecules.

THERMAL PROPERTIES Plasticization reduces bonding between polymer molecules and increases their mobility, thus decreasing the additional thermal energy required to produce motion of large segments. Thus plasticization generally lowers the glass transition temperature, broadens the transition region, shifts the modulus vs temperature curve to lower temperatures, and shifts the rubber-like properties to lower temperatures (Section 3-3). All of these effects may be due largely to the replacement of polymer-polymer hydrogen bonds by polymer-plasticizer hydrogen bonds.

ELECTRICAL PROPERTIES Plasticizers and absorbed moisture may increase molecular mobility by three Mechanisms: (1) They replace polymer-polymer hydrogen bonds by polymer-plasticizer hydrogen bonds, and thus permit greater mobility of the polymer molecule; (2) they introduce small mobile molecules of high free volume into the tangled long-chain polymer molecules, and thus lubricate the mobility of segments of the polymer molecule; (3) they act themselves as small molecules of high mobility. Any or all of these mechanisms may be involved in the effect of plasticizers upon electrical properties.

Electrical resistance and conduction probably depend upon transport of ionic impurities through the polymeric material (Section 3-4), but it is not yet clear to what extent this motion can be satisfied by mobility of the polymer molecules themselves (Mechanisms 1 & 2), or to what extent it depends upon the existence of a true liquid medium (Mechanism 3). Dielectric constant and loss depend upon mobility of small polar groups in the polymer molecule itself (Section 3-4), which could be entirely satisfied by Mechanism 1 and possibly also Mechanism 2; but of course the polar plasticizer molecules can also contribute their own mobility and ability to orient rapidly in an alternating electric field (Mechanism 3).

Thus hydrogen bonding between polymer and plasticizer could well be important in many of the effects of plasticizers upon electrical properties.

6-5 CRYSTALLINITY

Secondary valence forces—dispersion, induced dipoles, dipoles, and hydrogen bonds—are weak attractive forces between individual points in adjacent polymer molecules, much weaker than primary covalent bonds. In an amorphous, randomly-coiled polymer, these secondary attractive forces only occur whenever two chains happen to pass close to each other, and thus barely diminish the high mobility observed in elastomers. In a crystalline polymer, on the other hand, the regular tight parallel packing of adjacent chains produces a maximum number of points at which these secondary attractive forces can occur, and their total effect reduces molecular mobility very markedly. These cumulative effects of highly repetitive secondary attractions in crystalline structures are particularly evident in the increased rigidity and melting point, and decreased solubility, of crystalline polymers (Section 5-2).

For examples of specific polymers, a number of commercial materials illustrate these effects. Weak London dispersion forces provide the only attraction between aliphatic hydrocarbon polymers; but going from amorphous ethylene/propylene rubber to partly crystalline branched polyethylene to highly crystalline linear polyethylene, the increasing crystallinity causes an increasing proportion of these potential forces to become active, and thus produces marked increase in rigidity and strength. Polypropylene has only weak London dispersion forces between its chains: atactic polypropylene is amorphous, and very few of the points in the chains lie close enough to exert any attraction at all, so it is a soft gummy rubber; but isotactic polypropylene is so regular and crystalline that many points in the parallel helical chains lie in close proximity, and the cumulative effect of all these weak forces produces good rigidity and strength, high melting and heat distortion temperatures, and strong resistance to solvents. Commercial polystyrene is atactic and amorphous, and steric hindrance by the bulky phenyl side-chains provides rigidity which is readily overcome by thermal mobility at 100°C; but experimental isotactic polystyrene crystallizes into a regular structure in which overlapping resonance from repeating parallel phenyl groups accumulates to produce high-temperature rigidity up to a melting point of 240°C (Figure 6-29).

$$\sim\text{-O-}(CH_2)_4\text{-O-}\overset{\overset{\displaystyle O}{\|}}{C}\text{-NH-}(CH_2)_6\text{-NH-}\overset{\overset{\displaystyle O}{\|}}{C}\text{-}\sim$$

Figure 6-13.

Cellulose is a regular crystalline arrangement of parallel helices, and the three hydroxyl groups per monomer unit produce an accumulation of hydrogen bonding which is strong, high-melting, and insoluble; partial esterification to the acetate, used most often in commercial films and fibers, makes the structure more random and amorphous, reduces the frequency of effective hydrogen bonding, and gives a weaker, more fusible, soluble product; while complete esterification to the triacetate again produces a regular crystalline structure which, even without hydroxyl groups, can accumulate a high frequency of London dispersion forces and polarity to produce good strength and heat and solvent resistance once again (ref 108). Polyethylene terephthalate homopolymer, in commercial fibers and films, relies upon ~40% crystallinity to accumulate the repetitive London dispersion forces between parallel resonating *p*-phenylene rings, and the repetitive polarity of ester groups, to produce the high total attractive forces which give high modulus, strength, melting point, electrical quality, solvent resistance, and impermeability; copolymerization with isophthalic acid reduces regularity and crystallinity, and thus reduces the proportion of dispersion and polar forces which can be effective, and thus decreases all of these properties.

Finally, the first commercial polyurethanes (Figure 6-13) were so regular and crystalline that the accumulation of polarity and hydrogen bonding along the parallel chains made them primarily useful in fibers and bristles (ref 109); but the more recent and more common random amorphous structures (Figure 6-14) realize only a fraction of these potential attractions, and are most commonly elastomeric in nature. In all of these examples, the weak individual secondary attractions occur infrequently in the random amorphous structures, but accumulate rapidly in the regular repeating crystalline structures, making a tremendous difference in final properties. Thus the effects of crystallinity upon properties may be due primarily to this cumulative effect upon the weak secondary attractive forces between adjacent chains in most polymer molecules.

Actually, the forces and effects of crystallinity stand midway between the minimum effects of random secondary attractions and the

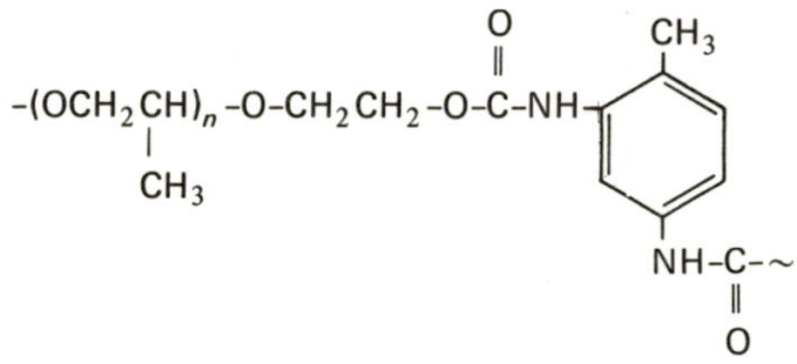

Figure 6-14.

maximum effects of primary covalent cross-linking. Because crystallinity represents the large sum of many small forces, it resembles cross-linking in many ways (ref 8,110), increasing tensile modulus (ref 111), tensile yield strength (ref 111), and ultimate tensile strength (ref 112); reducing creep (ref 113); giving longer slower relaxation times and flatter modulus vs frequency curves (ref 114); and overcoming the significance of the solubility parameter (ref 115). Because crystallinity is still only the sum of many weak secondary attractive forces, it is thermally dissociable unlike true primary covalent cross-links, and thus acts like meltable cross-links (ref 116,117). This combines the advantages of thermoplastic processing with many of the advantages of thermoset properties.

It may be worthwhile in closing to note some of the commercial polymers in which the benefits of crystallinity have proved most useful:

Polyethylene	Cis-polyisoprene
Polypropylene	Poly-*p*-xylylene
Cis-polybutadiene	Polyvinyl chloride

Polyvinylidene chloride	Polyformaldehyde
Polyvinyl fluoride	Poly-3,3-bis(chloromethyl)oxetane
Polytetrafluoroethylene	Cellulose
Polyvinyl alcohol	Polyethylene terephthalate
Polyacrylonitrile	Polyamide homopolymers
Polyethylene oxide	

6-6 IONIC BONDING

While most organic polymers are composed entirely of covalent bonds (ref 118), there are a few commercial polymers and a number of experimental ones which also contain some highly or completely ionic

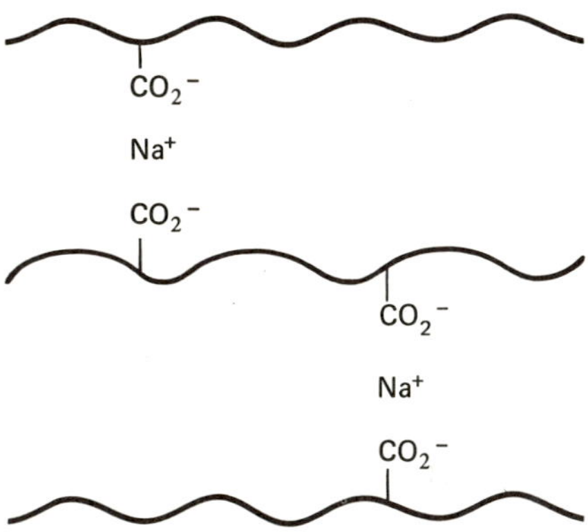

Figure 6-15.

bonds in the polymer molecule. Such ionic bonds, when they occur, are 2 Å–3 Å in length and 10–20 kcal/mol in strength (ref 3), and thus are capable of contributing considerably to the forces of intermolecular bonding.

Commercial ionomers, primarily ethylene copolymers with moderate amounts of sodium acrylate or methacrylate (Figure 6-15) behave as if they contained thermally-labile intermolecular cross-links, which melt during thermal processing but provide stiffness and toughness at room temperature (ref 119). Vinyl chloride terpolymers with vinyl acetate and maleic anhydride provide strong ionic bonding of organic coating to polar metallic substrates. The use of urea-formaldehyde and protein adhesives on polar substrates probably involves ion-ion and ion-dipole bonds (ref 81).

Such ionic bonds between polymer molecules are not only thermally fusible, but also chemically soluble in sufficiently polar solvents: weak electrolytes may dissolve in methanol, dioxane, or dimethyl-formamide; while hydrogen-bonded crystalline strong electrolytes may only dissolve in water (ref 120). Their growing importance in industry, as well as their obvious importance in biochemistry, indicate that we should expect to learn much more about them in the years to come.

6-7 CROSS-LINKING

The ultimate force in intermolecular bonding is of course covalent cross-linking between polymer molecules. Such cross-links are just as

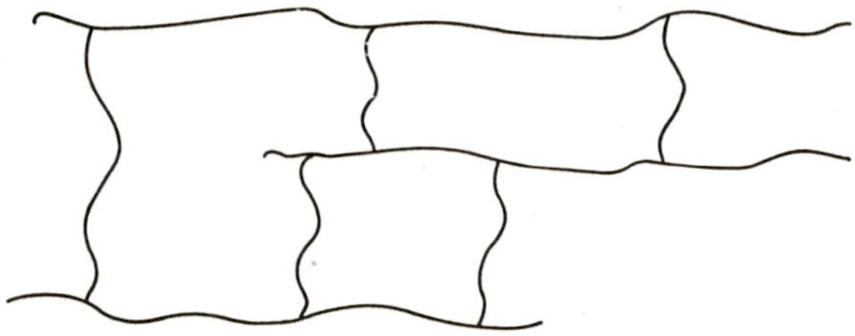

Figure 6-16. Schematic representation of a cross-linked, three-dimensional, network polymer (ref 121).

strong as the similar covalent bonds holding the molecule together, and thus convert a number of linear molecules into one large three-dimensional molecule (Figure 6-16). The primary covalent homopolar bonds are 1 Å–2 Å in length and 60–100 kcal/mol in strength (ref 2,3), and thus much more powerful than any of the lesser intermolecular bonding forces considered earlier.

Cross-linking has a number of effects upon properties which are quite general and independent of the level of cross-linking. Other properties depend very much upon the exact level of cross-linking which exists. Three distinct levels of cross-linking may be identified, and have distinctly different effects upon many properties: (1) Labile transient cross-links tend to open and close again under the influence of mechanical, thermal, and/or chemical stress; (2) light cross-linking joins the molecules together sufficiently to prevent them from flowing completely past each other, but not enough to restrict the motion of large segments within the polymer molecule; (3) heavy cross-linking joins the molecules together at so many points that even the motion of small segments is quite restricted. The effects of cross-linking on properties are best examined by considering first the effects of cross-linking in general and then the effect of each different level of cross-linking individually.

Cross-linking in General

The formation of strong primary homopolar covalent bonds between polymer molecules converts them into large three-dimensional cross-linked networks, and pulls the molecular chains closer together, leaving less free volume (ref 122), which restricts their mobility and thus generally produces harder properties.

Processability

The formation of cross-links between polymer molecules is the primary basis for dividing polymers into two general classes: thermoplastic which remain linear during processing, and thermosetting which become cross-linked during processing. Thermoplastic polymers are chemically stable and may be stored, processed, and used without change in their primary bond structure. Thermosetting polymers are chemically reactive, to permit the formation of cross-links during processing. Storage of such chemically-reactive polymers before processing often involves problems of shelf-life, caused by tendencies to cross-link prematurely during shelf-aging before processing. During processing the polymers should remain linear long enough to flow into their desired final form, but they should then cross-link rapidly and completely into the final chemically stable form, making them insoluble and infusible, and preventing further viscous flow (ref 121,123–125). These are conflicting demands upon the chemical reactivity of thermosetting polymers. They are met fairly well in commercial materials, but processors will continually desire better balance of premature stability vs high ultimate reactivity. Continuing research and development efforts by raw materials producers and formulators are making constant progress in improving this balance.

Mechanical Properties

In general the introduction of cross-links between polymer molecules reduces their mobility and thus increases their modulus (ref 126). In synthetic fibers, for example, this also gives higher hardness and better hand, recovery, and wrinkle resistance (ref 124).

Cross-linking generally affects large molecular deformations such as tensile strength (ref 46), but greater interactions between other effects may make the ultimate result unpredictable. Because it reduces elongation, cross-linking also often reduces cyclic stress fatigue life (ref 127).

Long-term modulus of polymers is directly proportional to the density of cross-links between their molecules (ref 128). Above the glass transition temperature, cross-linking reduces creep very dramatically (ref 129) (Figures 6-17, 6-18).

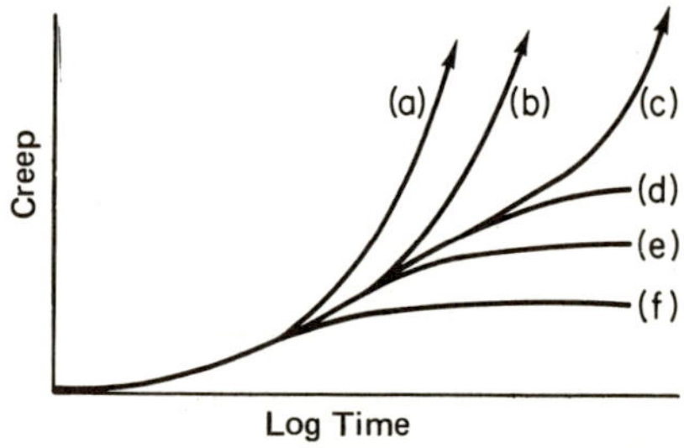

Figure 6-17. Effect of molecular weight and cross-linking upon creep (ref 131).

Master creep curves for various molecular architectures: (a) Low molecular weight linear polymer, (c) High molecular weight linear polymer, (d) Loose network polymer, and (f) Tight network polymer. Turner Alfrey and Edward F. Gurnee, *Organic Polymers*, © 1967. Reprinted by permission of Prentice-Hall, Inc., Englewood Cliffs, New Jersey.

These examples are typical of the way in which cross-linking reduces molecular mobility and thus "hardens" practical mechanical properties in general.

Thermal Properties

Glass Transition Temperature Cross-linking polymer chains pulls them closer together, decreasing the free volume between them, reducing segmental motion, and thus raising the glass transition temperature (ref 133,134). This effect has been quite generally observed (ref 135–137) (Figures 6-19 and 6-20), and has even been expressed mathematically in various ways. For example, $T_{g_x} = T_g + K_x\rho$ where T_{g_x} and T_g are the glass transition temperatures of cross-linked and linear polymer, K_x is a constant, and ρ is the number of cross-links (ref 138). Also, $T_g - T_g^\circ = (3.9 \times 10^4)/M_c$, where T_g and T_g° are the glass transition temperatures of cross-linked and linear polymer, and

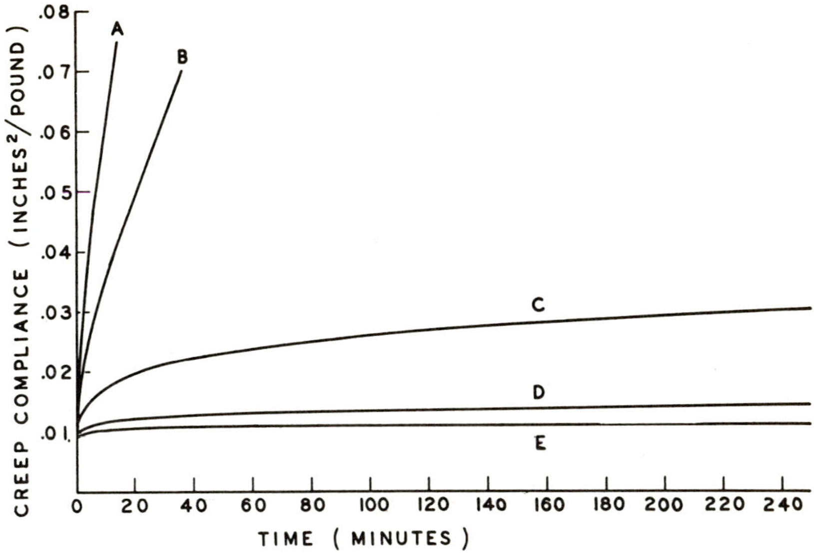

Figure 6-18. Effect of molecular weight and cross-linking upon creep of butyl rubbers (ref 130).

Creep of butyl rubbers of different molecular weights and degrees of cross-linking. A. Uncrosslinked, molecular weight = 315,000. B. Uncrosslinked, molecular weight = 480,000. C, D, E. Increasing degree of crosslinking.

M_c is the molecular weight of chain segments between cross-links (ref 139).

Aside from these general effects, some specific effects of cross-linking upon glass transition temperature may also occur. At very high degrees of cross-linking, the glass transition range may become too broad to

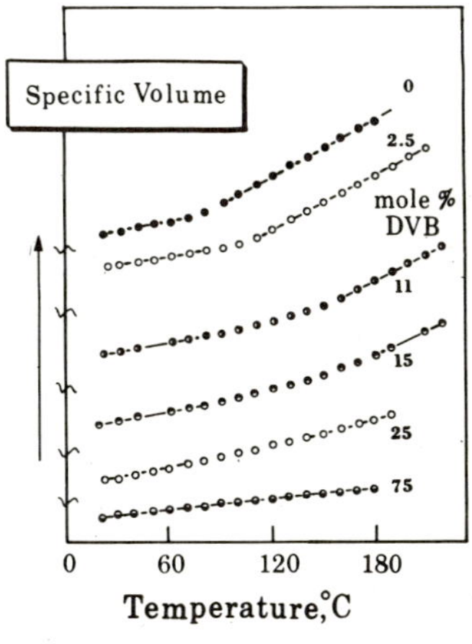

Figure 6-19. Divinyl benzene cross-linking of polystyrene: effects upon thermal expansion and glass transition (ref 132).

be expressed by such precise equations, or even to be observed at all (ref 138). Furthermore if the cross-links themselves introduce flexible groups, such copolymerization with flexible cross-links may actually lower the glass transition temperature (ref 138).

Modulus vs Temperature Curve The stable covalent bonds, by which cross-linking reduces segmental and molecular mobility, are thermally stable and insensitive up to quite high temperatures. Thus they have a profound effect upon the entire modulus vs temperature curve (Figure 6-21). They not only raise the glass transition temperature and the modulus above the glass transition temperature, but also affect

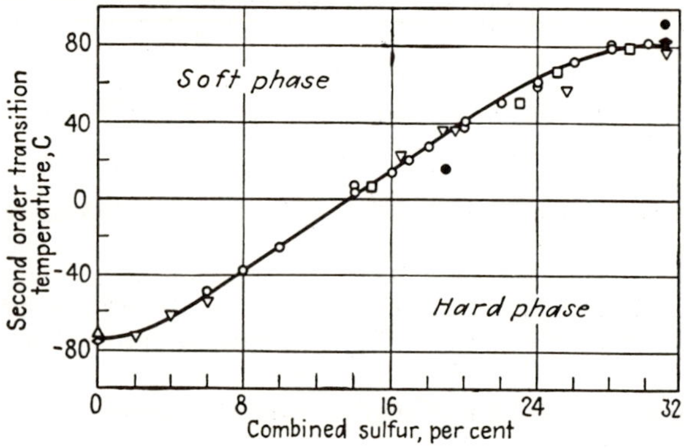

Figure 6-20. Sulfur cross-linking of hevea rubber: effect upon glass transition temperature (ref 137).

the large molecular deformations involved in viscous flow, making the polymer quite infusible.

Coefficient of Thermal Expansion It has already been observed that cross-linking draws the polymer molecules closer together and reduces the remaining free volume between them. This in turn reduces the coefficient of thermal expansion (ref 141) (Figure 6-19).

Chemical Properties

Permanent primary covalent cross-links between polymer molecules are a much stronger force than polarity and solubility parameter (ref 65), making polymers completely insoluble in common solvents (ref 142), and thus producing gelation when the cross-links form (ref 143). Cross-linking is an important factor in producing chemically resistant thermoset plastics, and elastomers which are resistant to fuel and oil. Cross-linking in fibers is important in producing moisture resistance and

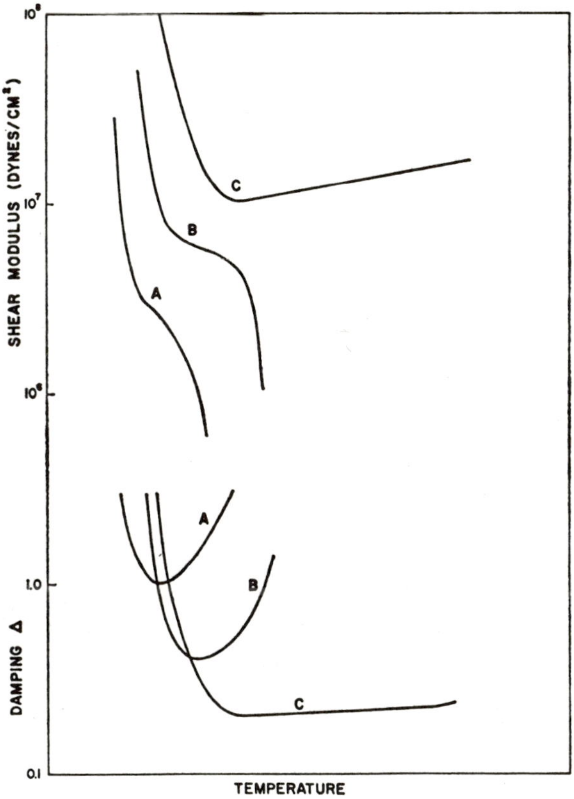

Figure 6-21. Effect of molecular weight and cross-linking upon modulus vs temperature curves (ref 140).

Schematic representation of the dynamic mechanical properties of amorphous polymers above the temperature of maximum damping. A. Low molecular weight polymer. B. High molecular weight polymer. C. Cross-linked polymer.

wrinkle resistance (ref 124). Cross-linking during the drying of oil paints provides solvent resistance, which permits the application of successive layers of paint. These are typical examples of the practical improvements in chemical resistance that result from insolubilization by cross-linking.

Labile Cross-Linking

Certain types of covalent cross-linking between polymer molecules are strong enough and permanent enough to have many of the conventional effects of cross-linking, and yet labile enough to permit occasional breaking and reforming of cross-links under stress. Thus the ~S—S~ cross-links in polysulfide elastomers tend to open under the influence of sustained mechanical stress, permitting gradual creep and stress

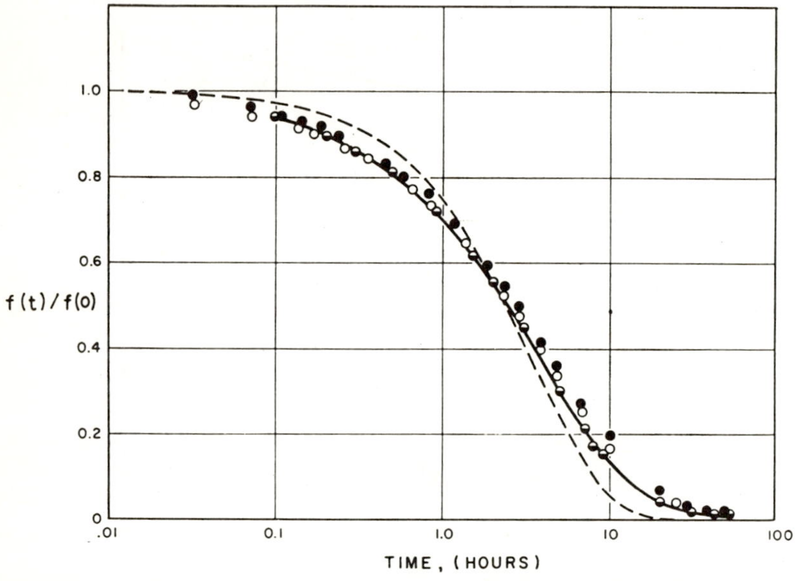

Figure 6-22. Stress-relaxation in polysulfide rubber (ref 146).
Relaxation of stress in polysulfide rubber H-11 at various elongations, and plot of exp $(-t/\tau)$ at 60°C.: ●, 10% extension; ○, 30% extension; ◒, 50% extension; ---- exp $(-t/\tau)$ with $\tau = 3$ hr.

relaxation, and then reform again to produce new stable dimensions (ref 144,145) (Figure 6-22). Such processes permit change in shape with no change in ultimate properties (ref 147).

Similarly it has been observed that cross-linked polyurethanes can still be remolded into new products, suggesting that the urethane bond is thermally labile under such conditions of temperature and pressure. In the reclaiming of rubber, treatment with high-pressure steam or hot mastication is sufficient to break the cross-links in vulcanized rubber and make it reusable, and in some cases even optimally desirable, in certain end products (ref 148). In such specialized cases, the temporary nature of labile cross-links is extremely important.

Light Cross-Linking

Low degrees of cross-linking, occurring perhaps once every 50–100 monomer units, are sufficient to bind the polymer molecules together

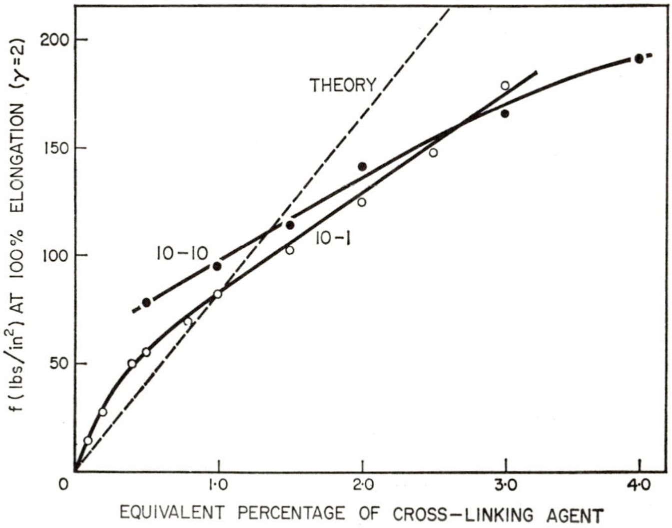

Figure 6-23. Concentration of cross-linking agent in natural rubber (ref 150).

Tension versus degree of cross-linking for natural rubber (% cross-linking $\propto 1/M_c$). Cross-linking agents are:

10 − 1 = decamethylene dis-methyl azodicarboxylate
10 − 10 = decamethylene dis-n-decyl azodicarboxylate.

and prevent them from flowing completely past each other, but not enough to reduce the flexibility of large segments within the molecules. These effects are particularly useful in elastomers, and occasionally also in fibers (ref 84).

Processability

Very low degrees of cross-linking will make polymer molecules insoluble and still leave them reasonably fusible and fluid in the molten state (ref 149). In most practical applications, however, somewhat higher

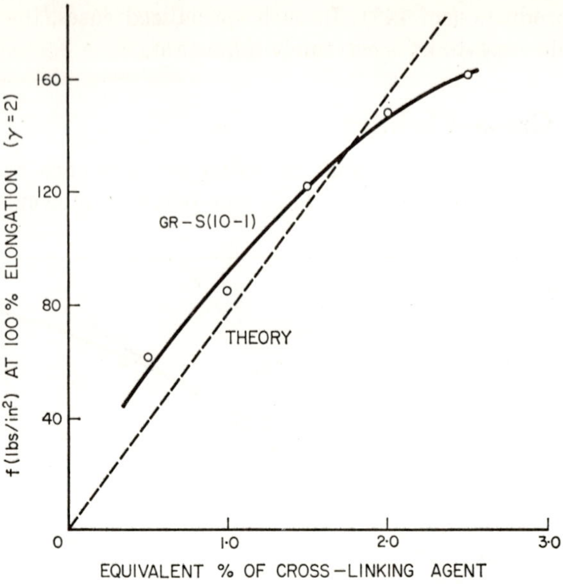

Figure 6-24. Concentration of cross-linking agent in GR-S rubber (ref 151).

Tension versus degree of cross-linking for GR-S rubber containing 72% butadiene cross-linked with decamethylene dis-methyl azodicarboxylate (% cross-linking $\propto 1/M_c$).

degrees of cross-linking are desired for maximum improvement of final properties, and these are obtained by vulcanization during the later stages of processing, after molten flow has been completed. The degree of cross-linking may be controlled by the concentration (Figures 6-20, 6-23, and 6-24) and reactivity (Table 6-20) of the cross-linking agent

Table 6-20. Accelerator Characteristics (ref 152)

Accelerator	Scorch Time: Measured at 244°F	Scorch Time: Calculated at 280°F	Vulcanization time at 280°F	Cure time at 280°F
Tetramethylthiuram disulfide	14	3.5	3.5	7.0
Tetraethylthiuram disulfide	22	5.5	4.5	10.0
MBT	10	2.5	7.5	10.0
MBTS	20	5.0	11.0	16.0
Ethylac	28	7.0	5.0	12.0

and of the reactive groups in the polymer (Figure 6-25); and it is also controlled by the temperature and time permitted for the cross-linking reaction (Figures 6-26 and 6-27), thus providing precise control over final properties (ref 156,157).

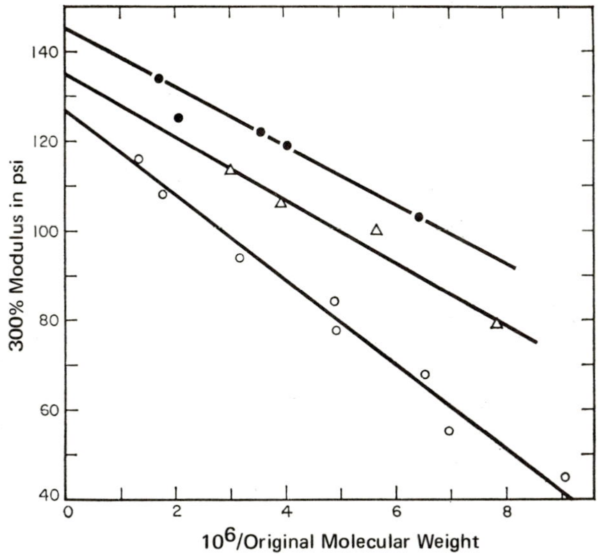

Figure 6-25. Concentration of reactive isoprene groups in butyl rubber (ref 153).

○ 0.5% isoprene units
△ 1.0 isoprene units
● 1.2 isoprene units

Mechanical Properties

Reversible Properties Light cross-linking has been most thoroughly studied in rubber, where it joins the molecules together and prevents them from flowing permenantly past each other, but still permits free motion of large segments within these molecules to retain their soft rubbery flexibility and elasticity (ref 119). Extent of cross-linking determines the restriction in such segmental mobility, and thus the modulus of the resulting vulcanized rubber (ref 158–160) (Figures 6-23 to 6-27). As a first approximation, modulus E is inverse to number

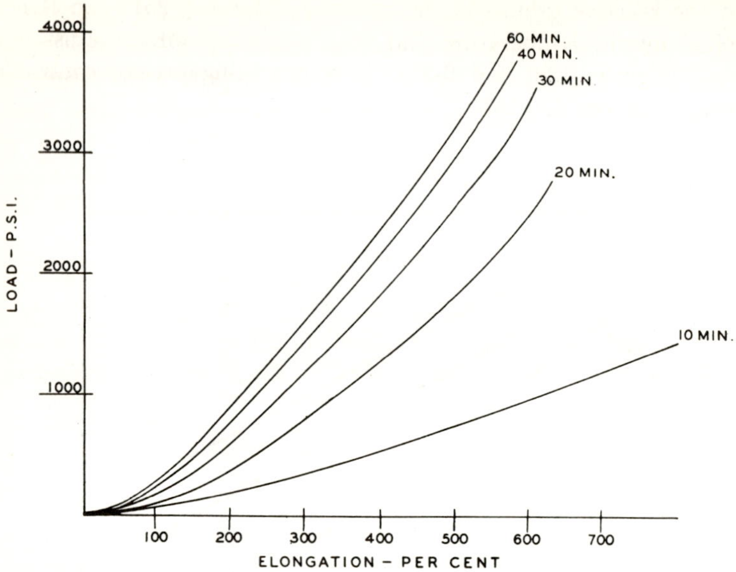

Figure 6-26. Effect of cure time upon stress-strain behavior of rubber (ref 154).

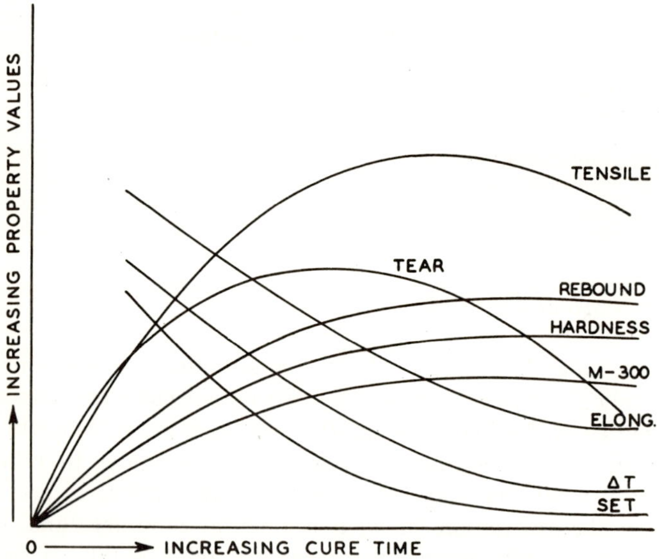

Figure 6-27. Effect of cure time upon properties of rubber (ref 155).

average molecular weight of polymer segments between cross-links M_c, thus $E = 3RTd/M_c$, where R is the gas constant, T the absolute temperature, and d the density (ref 161). More accurately the tensile stress σ, on the basis of the original cross-section, may be calculated from the more complete equation $\sigma = (RTd/M_c)(1 - 2M_c/\bar{M}_n)(\alpha - 1/\alpha^2)$, where is original number average molecular weight before cross-linking, and α is the ratio of stretched length to unstretched length (ref 162).

Ultimate Properties Such low degrees of cross-linking produce tremendous increase in strength, toughness, elasticity, and resistance to tearing and abrasion, along with gradual decrease in ultimate elongation (ref 155,163–165) (Table 6-21, Figure 6-27). These improvements are

Table 6-21. Effects of Vulcanization upon Properties of Rubber (ref 165)

Low viscosity[a]		High viscosity
High softness[b]		Low softness
High retentivity[b]		Low retentivity
Low modulus		High modulus
Low tensile strength	Vulcanization →	High tensile strength
High elongation		Low elongation
Low hardness[c]		High hardness
High compression set		Low compression set
Low rebound		High rebound
High solvent swell		Low solvent swell

[a] Mooney viscometer
[b] Goodrich Plastometer
[c] Durometer or equivalent

all important in the use of rubber in tires, mechanical goods, and many other applications.

In some cases, cross-linking has a conflicting effect in crystalline polymers, where the introduction of cross-links reduces regularity and crystallinity, and thus produces a softening of properties (ref 163,166,167). In such cases, increasing cross-linking and decreasing crystallinity may cancel each other out to some extent in their effects upon properties.

Long-term Properties In a linear polymer, long-term mechanical stress will cause a gradual disentanglement and slippage of molecules past each other. In a cross-linked elastomer, the stable covalent cross-links prevent this fairly well. In fundamental terms, vulcanization retards stress decay at constant strain (ref 161) (Figure 6-28), or creep extension at constant stress (ref 129) (Figures 6-17, 6-18). In practical terms, vulcanization reduces compression set (ref 169) (Table 6-21). Even in the best vulcanized rubber, however, there are still enough loose ends and unreacted linear molecules present to permit some residual creep (ref 170).

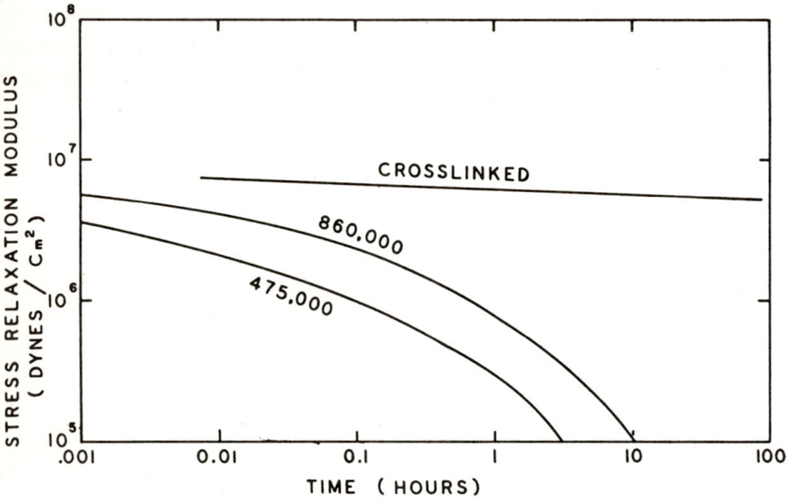

Figure 6-28. Effect of molecular weight and cross-linking upon stress relaxation of polyisobutylenes (ref 168).

Stress relaxation of a medium molecular weight (475,000) polyisobutylene, a high molecular weight (860,000) polyisobutylene, and a cross-linked polyisobutylene rubber at 30°C.

Thermal Properties

MODULUS VS TEMPERATURE Very light cross-linking may still permit fusion and molten flow of a polymer (ref 149). At any significant degree of cross-linking, however, the rubbery flow and liquid flow portions of the modulus vs temperature curve are completely eliminated (ref 171) (Figure 6-29). In fact, in cross-linked rubbers above the glass transition temperature, modulus may no longer be a function of temperature (ref 173).

HYSTERESIS Under alternating stress and release of stress, rubber molecules uncoil and recoil fairly reversibly; but there is still some internal friction leading to delayed response and conversion of mechanical energy into heat, as measured by the area in the stress-strain loop (ref 174) (Figure 6-30). In practical applications where this cycle repeats rapidly and constantly, as in thick bus and truck tires, the resulting temperature rise, reaching as high as 300°F, can cause rapid failure

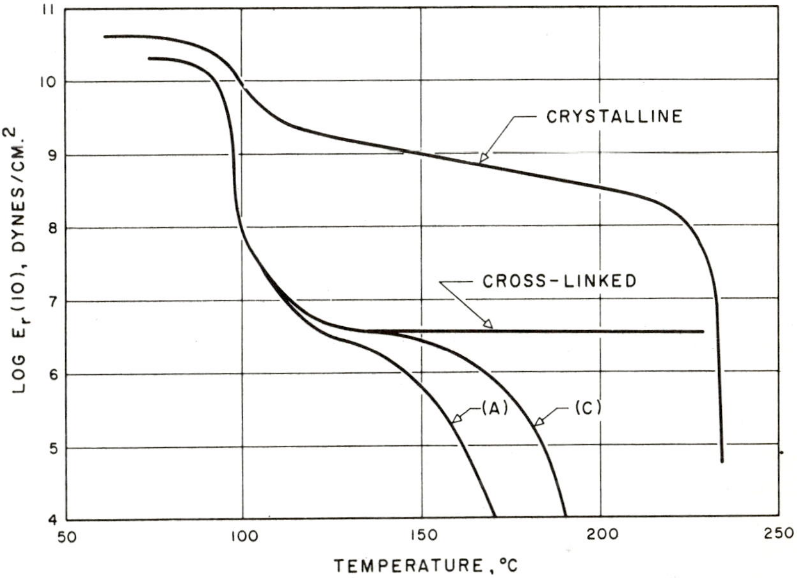

Figure 6-29. Effects of molecular weight, cross-linking, and crystallinity upon modulus vs temperature curves for polystyrene (ref 172).
$E_r(10)$ versus temperature for crystalline isotactic polystyrene, for polystyrene samples *A* and *C*, and for lightly cross-linked atactic polystyrene.

in use (ref 176). Vulcanization reduces this medium-range delayed elastic slippage between large chain segments, and thus reduces damping and hysteresis heat build-up (ref 11) (Figure 6-21).

Chemical Properties

Even the lightest degrees of cross-linking between polymer molecules are sufficient to keep them from dispersing and dissolving individually

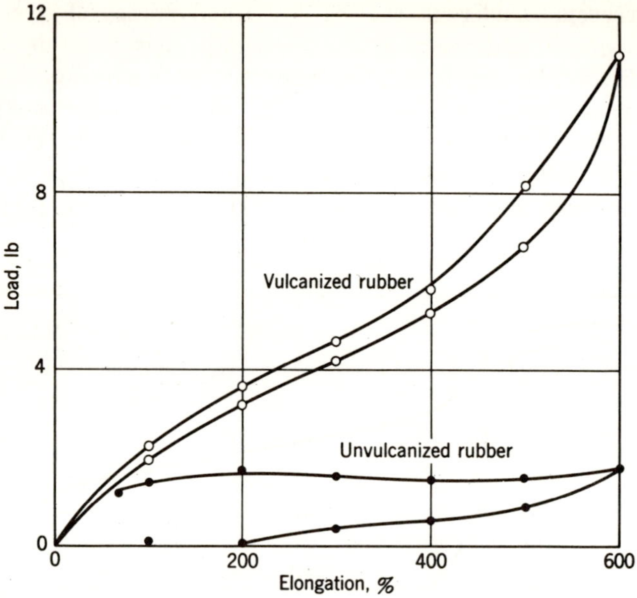

Figure 6-30. Hysteresis loops in stress-strain cycles on rubber (ref 175).

in solvents, and thus make the polymer quite insoluble (ref 149). Vulcanization of rubber makes it insoluble and reduces its swelling in common solvents, which is particularly important in resistance to fuels and oils (ref 175,177) (Tables 6-21 and 6-22). Cross-linking also reduces diffusion coefficient and therefore permeability, approximately in proportion to cross-link density or inversely to M_c the molecular weight between cross-links (ref 187).

Table 6-22. Swelling of Cross-linked Polyisobutylene Rubbers in Cyclohexane (ref 178)

Molecular weight of chain between cross-links	Sol fraction	Swelling ratio of gel
Uncrosslinked, M = 315,000	1.00	—
Uncrosslinked, M = 480,000	1.00	—
140,000	0.49	88.5
31,000	0.048	10.1
14,000	0.024	6.8

High Cross-linking

High degrees of cross-linking are achieved in the vulcanization of hard rubber, and in the cure of unsaturated polyesters, diallyl phthalate, alkyd resins, drying oils, urea-formaldehyde, melamine-formaldehyde, and phenolic resins. With cross-links forming on many, most, or even all of the monomer units in the polymer chains, these highly cross-linked polymers are completely three-dimensional networks, leaving very little mobility for polymer molecules, or even for any significant segment of the cross-linked polymer molecules, and the effects upon properties are extremely pronounced.

Mechanical Properties

High cross-linking restricts segment mobility so extremely that the polymer can exist only in the rigid glassy state (ref 119,179), and this has all of the expected effects upon mechanical properties. High degrees of cross-linking increase the modulus of a polymer in the rubbery plateau region (ref 180). High vulcanization of rubber produces high tensile strength and very low elongation (ref 181) ; high cross-linking in general reduces polymers to very low ultimate elongation (ref 182). Whereas impact strength of thermoplastics generally improves with increasing temperature, impact strength of heavily thermoset polymers may be completely independent of temperature (ref 183). Heavily cross-linked polymers show very little creep even over long-term tests up to many years (ref 178). All of these effects result from low segment mobility. Most generally useful in practice are the high modulus and creep resistance of these thermoset plastics.

Thermal Properties

Increasing concentration of cross-links between polymer chains reduces segment mobility and raises glass transition temperature tremendously, often above the decomposition temperature of the polymer (ref 102,119,178,182), so that the polymer cannot be rubbery at any temperature. Modulus remains almost constant up to decomposition, and the polymer actually decomposes before softening (ref 178,184) (Figure 6-31). Very high cross-linking can reduce creep even below the glass transition temperature (ref 170).

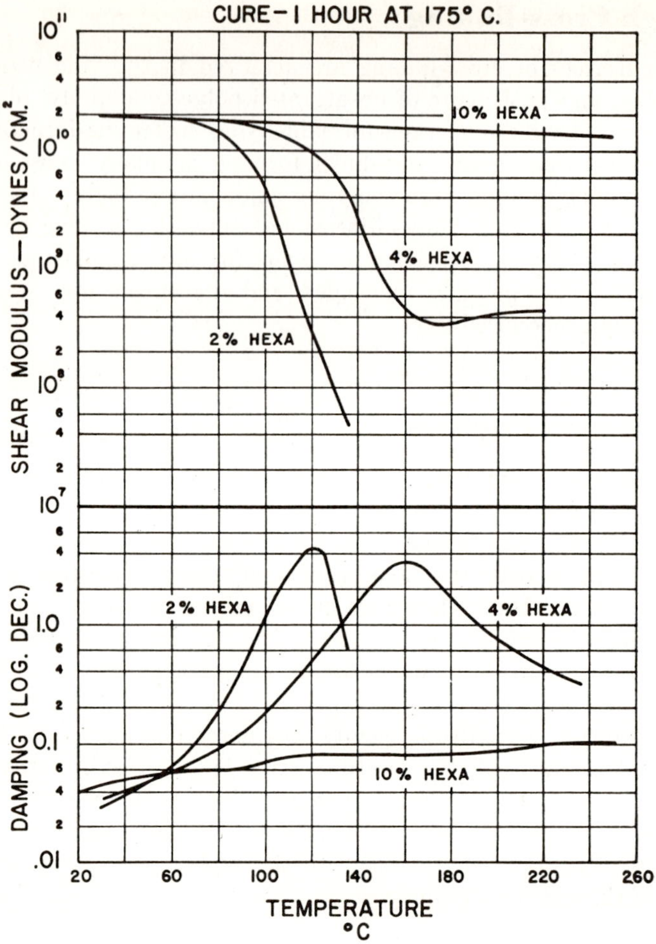

Figure 6-31. Cross-linking of phenolic resin by hexamethylene tetramine (ref 185).

Electrical Properties

When a partly-bound dipole tries to follow an alternating electric field by reorienting with each alternation of the field, viscous internal friction delays its vibrations and converts some of the electrical into thermal energy. This is measured as dissipation factor, Tan δ. In viscous uncured epoxy resins this dissipation factor is very high (Figure 6-32). As cure progresses, increasing cross-linking immobilizes the chain segments and dipoles so thoroughly that they no longer make any attempt at all to follow the alternation of the field. Consequently, dissipation factor decreases rapidly as cure approaches completion.

Chemical Properties

The low polarity and solubility parameter of most organic polymers makes them reasonably resistant to water and most inorganic reagents, but leaves them variably sensitive to swelling or even to solution by organic solvents. Cross-linking immediately prevents them from dissolving in organic solvents, and high degrees of cross-linking also reduce swelling to negligible proportions (ref 177). This makes thermoset plastics particularly desirable for their broad resistance to organic solvents.

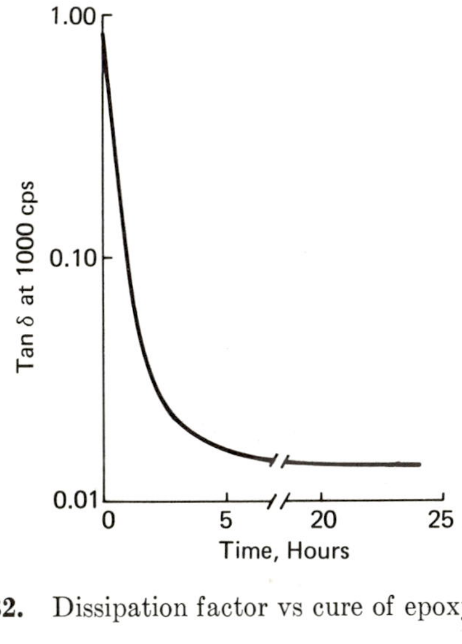

Figure 6-32. Dissipation factor vs cure of epoxy resin (ref 186).

High cross-linking reduces segment mobility to the point that sorption and diffusion of permeants in the polymeric matrix are both reduced, and thus permeability is reduced for both reasons (ref 187). Highly cross-linked electrolytes are insoluble but swollen in water, and find wide use as ion exchange resins (ref 188). In the application of thermosetting adhesives, creation of primary covalent bonds between adhesive and substrate may be important in the strength of the final bond (ref 81).

These are examples of the ways in which high degrees of cross-linking have useful effects upon the chemical properties of thermoset plastics.

6-8 SUMMARY

Finally, it is appropriate to compare the different types of intermolecular bonding, and to summarize their general effects upon polymer properties and applications.

Comparison and Summation of Intermolecular Attractive Forces

The intermolecular bonding forces between adjacent polymer molecules cover a broad range from very weak to very strong (Table 6-23,

Table 6-23. Average Bond Energies and Bond Distances for Different Types of Bonds (ref 3)

Type of Bond	Approximate Interaction Energy (kcal mole^{-1})	Approximate Range of Action (Å)
Homopolar	60	1–2
Hydrogen	1.5–6	3
Ionic	10–20	2–3
Van der Waals	1–2	3–5

Figure 6-33). London dispersion or Van der Waals forces are obscure in their mechanism and individually quite weak, but accumulate to a total attractive force of quite considerable size. Induced dipoles are of almost no significance in polymer chemistry. Dipoles exist in most polymers, but their major effects are usually overpowered and obscured by hydrogen bonding and crystallinity. Hydrogen bonds occur to a greater or lesser extent in most polymeric systems. Crystallinity introduces a high degree of parallelism and proximity between adjacent polymer chains, thus realizing the maximum frequency and effectiveness of dispersion, polar, and hydrogen bonding forces, and accumulating a total attractive force which is very high. Ionic bonds are rather rare in polymers, but have considerable strength wherever they do occur. Stable primary covalent cross-links between polymer molecules represent the ultimate

in intermolecular attraction, providing bonds between adjacent chains which are as strong as the bonds within the chains themselves.

The total force of intermolecular bonding ΣI between two adjacent polymer molecules is the product of three separate factors: $\Sigma I = S \times F \times M$, where S is the strength of the individual intermolecular attractions; F is the frequency with which they occur along the chain, or their concentration per unit length; and M is the total length which the two adjacent molecules share, which of course depends upon their size or molecular weight. Thus dispersion, polar, and hydrogen bond attractions S between adjacent molecules are individually quite

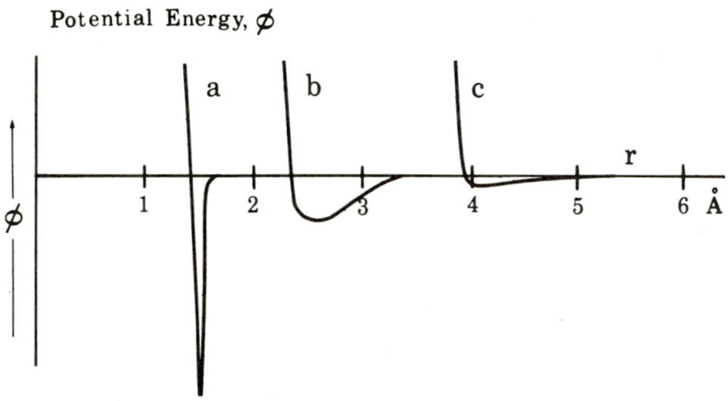

Figure 6-33. Potential energies and interatomic distances for different types of intermolecular bonding (ref 6).

Potential energy ϕ versus interparticle distance r for (a) Covalent bond, (b) Hydrogen bond, and (c) Van der Waals bond.

weak, but become of major importance with increasing molecular weight M, and take effect even more rapidly with the addition of crystallinity F.

General Effects upon Properties

The effects of each type of intermolecular bond upon polymer properties have already been considered. Many of these are general enough to apply to most or all types of intermolecular bonding and may be summarized as follows:

Processability

Processability is affected in several ways by intermolecular bonding. In general intermolecular bonding raises glass transition and melting temperatures, and thus requires the use of higher temperatures during processing. Intermolecular bonding also increases the difficulty of separating polymer molecules, and thus narrows the choice of solvents which may be used for solution processing of fibers, coatings, and adhesives.

Mechanical Properties

Mechanical properties are very dependent upon intermolecular bonding. Weak intermolecular forces produce soft flexible polymers, while strong intermolecular forces produce rigid glassy materials (ref 8,136). Internal friction in rubbery polymers is about 10^6 poises, whereas in amorphous solid polymers at or below their glass transition temperature it is 10^{12-13} poises (ref 9,189).

Rubbery stretching and elastic recovery depend upon loose entanglement and intermolecular attraction to limit viscoelastic flow. This intermolecular attraction may be achieved fairly well by simple entanglement of very high molecular weight molecules, better by secondary attractions, and best by primary cross-linking and temporary crystallization during stretching.

Ultimate strength properties depend upon the strength, times the number, of primary and secondary bonds which are serving to resist chain slippage between molecules (ref 190). Thus for stronger intermolecular bonds, properties begin to level off at lower molecular weight (Figure 6-10).

Thermal Properties

Melting vs Glass Transition Temperatures Increasing intermolecular bonding between polymer molecules reduces the freedom and flexibility of chain segments, requiring a higher temperature to overcome these bonds and restore the flexibility, and thus raising the glass transition temperature (ref 136,166,191,192). In crystalline polymers, intermolecular bonding also increases the attractive forces between adjacent polymer chains within the crystal, and increases the thermal energy required to separate these chains from each other and produce melting; thus increasing intermolecular attractive forces generally increase the melting temperature (ref 166,191,192). Since increasing intermolecular forces are apt to increase both the glass transition temperature and the melting temperature, the ratio between these in °K is apt to remain fairly constant, ranging from 0.5 to 0.67 (ref 49,191,192) (Figure 4-24).

Since many other structural factors are involved, however, it is not surprising that this ratio varies considerably in different types of polymers.

THERMAL MECHANICAL BEHAVIOR The normal relationship between modulus and temperature in linear amorphous polymers includes six characteristic regions (Figure 3-33): (1) At low temperatures the polymer is rigid and glassy, and has a high and fairly constant modulus. (2) With increasing temperature and thermal vibration, the molecules move far enough apart to produce sufficient free volume for occasional movement of larger segments, and the polymer passes through the glass transition region in which mobility and flexibility appear. (3) With further increase in temperature and segmental mobility, the polymer passes through a leathery transition region; and (4) with still further increase in temperature and mobility, it reaches a fairly constant rubbery plateau. (5) At still higher temperatures, as the whole molecules begin to slip past each other under high stress, the polymer exhibits rubbery liquid flow; and (6) at still higher temperatures, where whole molecules can slip easily past each other, the polymer exhibits true liquid flow.

All of these phases depend upon the ability of adjacent polymer molecules to move past each other, and thus are directly affected by the extent of intermolecular bonding (Figure 6-34): (a) Increasing polarity generally increases both the glass transition temperature and the melting temperature to a more or less proportional extent. (b) The appearance of a small degree of crystallinity extends the rubbery plateau up to the melting point of the crystalline region, while higher degrees of crystallinity introduce a stiffening effect which converts the entire rubbery plateau to a leathery state or even to one rigid in nature. (c) The appearance of a low degree of cross-linking prevents liquid flow completely, and extends the rubbery plateau up to the decomposition point; while higher degrees of cross-linking restrict segmental motion, eliminating the rubbery region and converting it to a leathery or even completely rigid glassy state, which remains constant up to the decomposition temperature. Thus both the type and the extent of intermolecular bonding have major effects upon the modulus vs temperature profile of polymers.

COEFFICIENT OF THERMAL EXPANSION Increasing the attractive forces between adjacent polymer molecules draws them closer and holds them more firmly together, and thus tends to decrease the coefficient of thermal expansion.

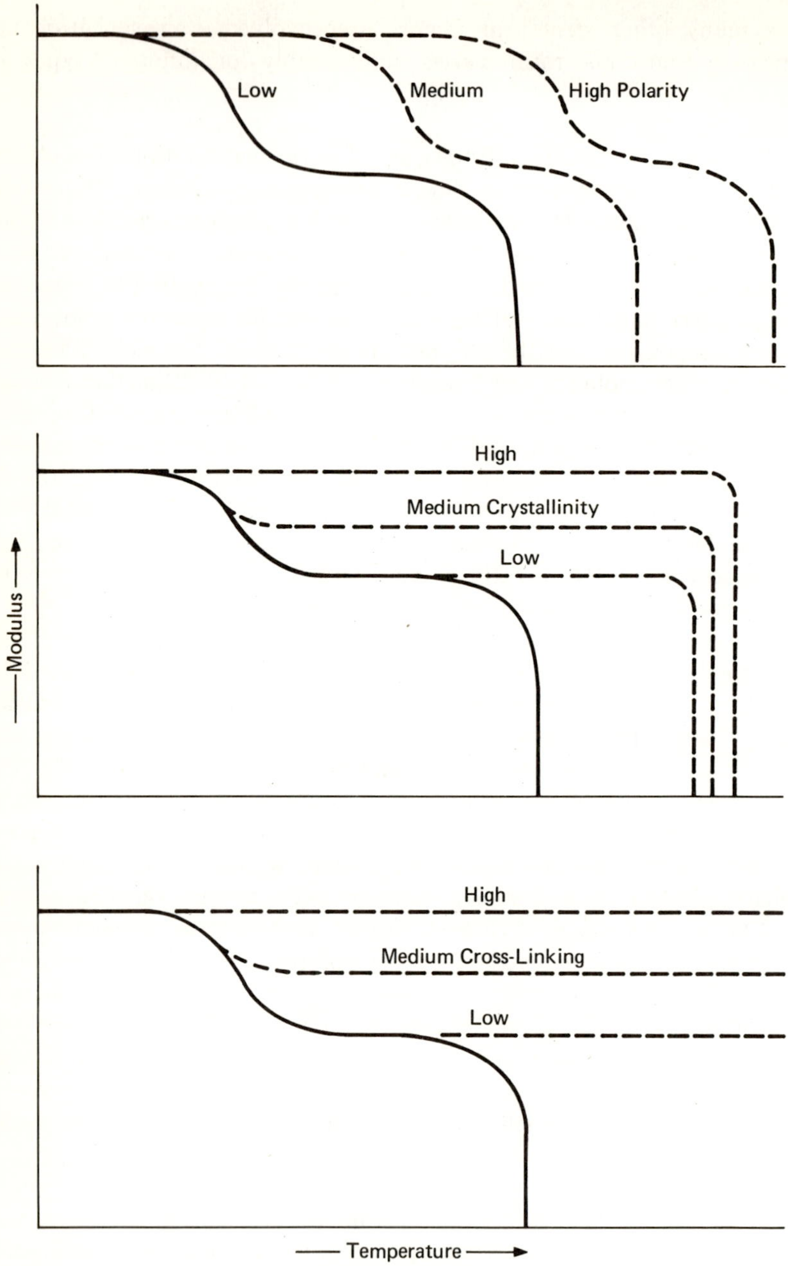

Figure 6-34. Effects of intermolecular bonding upon modulus versus temperature curves.

Chemical Properties

Increasing the degree of intermolecular bonding decreases the ability of the polymer to dissolve or even swell in solvents. It also decreases the mobility and spacing of polymer segments, and thus decreases the permeability of the polymer. Increasing the attractive forces between a polymer and its substrate tends to increase its adhesion, coefficient of friction, and abrasion.

These are typical of the general effects of intermolecular bonding upon polymer properties.

General Applications

Low intermolecular attraction (1–2 kcal/5 Å chain length) is most useful to provide softness and flexibility in rubbers, especially when combined with occasional cross-linking and preferably also temporary stretch-crystallization to provide strength and elasticity. Intermediate intermolecular attraction (2–5 kcal/5 Å chain length) is most useful in rigid plastics. High intermolecular attraction (5–10 kcal/5 Å chain length) is most useful in fibers, along with high crystallinity to maximize tensile modulus and strength (ref 8,193).

REFERENCES

1. Billmeyer, F. W., Jr., *Textbook of Polymer Science,* Interscience, 1962, pp 15–19.
2. Miller, M. L., *The Structure of Polymers,* Reinhold, 1966, p 140.
3. *Ibid.,* p 143.
4. Billmeyer, *op. cit.,* pp 19–21.
5. *Ibid.,* p 18.
6. Miller, *op. cit.,* p 142.
7. Brydson, J. A., *Plastics Materials,* Iliffe, 1966, pp 45–48.
8. Morton, Maurice, *Introduction to Rubber Technology,* Reinhold, 1959, p 54.
9. Golding, Brage, *Polymers and Resins,* Van Nostrand, 1959, p 547.
10. Nielsen, L. E., *Mechanical Properties of Polymers,* Reinhold, 1962, p 86.
11. *Ibid.,* pp 165–66.
12. Billmeyer, *op. cit.,* pp 17–18.
13. Golding, *op. cit.,* p 543.
14. Miller, *op. cit.,* pp 141–42.
15. Pauling, Linus, *The Nature of the Chemical Bond,* Cornell University Press, 1942, Chapter 2.

16. *Ibid.*, Chapter 5.
17. Brydson, *op. cit.*, p 62.
18. *Ibid.*, pp 95–96.
19. *Ibid.*, p 96.
20. Schmidt, A. X., & Marlies, C. A., *Principles of High-Polymer Theory & Practice*, McGraw-Hill, 1948, p 46.
21. Miller, *op. cit.*, pp 527–28.
22. Schmidt, *op. cit.*, pp 46–47.
23. Brydson, *op. cit.*, p 65.
24. Tobolsky, A. V., *Properties & Structure of Polymers*, Wiley, 1960, pp 65–66.
25. Billmeyer, *op. cit.*, p 21.
26. *Ibid.*, p 26.
27. Brydson, *op. cit.*, pp 65–67.
28. *Ibid.*, pp 69–71.
29. *Ibid.*, p 66.
30. Brandrup, J., & Immergut, E. H., *Polymer Handbook*, Interscience, 1966, pp 347–58.
31. Brydson, *op. cit.*, p 71.
32. Tobolsky, *op. cit.*, p 65.
33. Brandrup, *op. cit.*, pp 358–67.
34. Brydson, J., *op. cit.*, p 72.
35. *Ibid.*, pp 71–73.
36. Golding, *op. cit.*, p 544.
37. Miller, *op. cit.*, pp 476–77.
38. *Ibid.*, p 476.
39. *Ibid.*, p 477.
40. *Ibid.*, pp 477–78.
41. Brydson, *op. cit.*, pp 46–48.
42. Golding, *op. cit.*, pp. 555–56.
43. Billmeyer, *op. cit.*, pp 215–20.
44. Meares, Patrick, *Polymers: Structure & Bulk Properties*, Van Nostrand, 1965, p 258.
45. Billmeyer, *op. cit.*, p 231.
46. *Ibid.*, pp 208–9.
47. Nielsen, *op. cit.*, pp 223–24.
48. *Ibid.*, p 226.
49. Billmeyer, *op. cit.*, p 212.
50. Tobolsky, *op. cit.*, p 67.
51. Miller, *op. cit.*, pp 477–81.
52. *Ibid.*, p 479.
53. *Ibid.*, pp 480–81.
54. Billmeyer, *op. cit.*, p 20.
55. *Ibid.*, pp 212–15.
56. Miller, *op. cit.*, pp 527–29.

57. Nielsen, *op. cit.,* p 39.
58. Billmeyer, *op. cit.,* pp 215–17.
59. *Ibid.,* p 216.
60. *Ibid.,* p 91.
61. Bernhardt, E. C., *Processing of Thermoplasic Materials,* Reinhold, 1959, pp 471–73.
62. Modern Plastics Encyclopedia 1968, McGraw-Hill, 1967, p 777.
63. Brydson, *op. cit.,* pp 94–95.
64. *Ibid.,* p 95.
65. Baer, Eric, *Engineering Design for Plastics,* Reinhold, 1964, pp. 1041–42.
66. Billmeyer, *op. cit.,* pp 25–26.
67. Miller, *op. cit.,* pp 175–77.
68. *Ibid.,* p 176.
69. Billmeyer, *op. cit.,* p 227.
70. Golding, *op. cit.,* pp 19–20.
71. Billmeyer, *op. cit.,* p 225.
72. *Ibid.,* p 229.
73. Miller, *op. cit.,* pp 470–71.
74. Baer, *op. cit.,* p 660.
75. *Ibid.,* p 659.
76. *Ibid.,* p 662.
77. Billmeyer, *op. cit.,* p 228.
78. Baer, *op. cit.,* p 778.
79. Schmidt, *op. cit.,* p 636.
80. *Ibid.,* p 649.
81. Golding, *op. cit.,* p 696.
82. Volk, M. C., Lefforge, J. W., & Stetson, Russell, *Electrical Encapsulation,* Reinhold, 1962, p 27.
83. Payne, H. F., *Organic Coating Technology,* Vol. I, Wiley, 1954, p 18.
84. Billmeyer, *op. cit.,* pp 230–32.
85. *Ibid.,* pp 18–19.
86. Miller, *op. cit.,* pp 169–70.
87. Schmidt, *op. cit.,* pp 33–34.
88. Billmeyer, *op. cit.,* p 19.
89. Brydson, *op. cit.,* p 48.
90. Morton, W. E., & Hearle, J. W. S., *Physical Properties of Textile Fibers,* Butterworths, 1962, pp 21–23.
91. Meares, *op. cit.,* p 93.
92. McFarlane, S. B., *Technology of Synthetic Fibers,* Fairchild Publications, 1953, p 84a.
93. Alfrey, Turner, & Gurnee, E. F., *Organic Polymers,* Prentice-Hall, 1967, pp 38–39.
94. Billmeyer, *op. cit.,* p 219.
95. Schmidt, *op. cit.,* pp 363–64.
96. *Ibid.,* p 364.

97. Golding, *op. cit.*, p 697.
98. Brydson, *op. cit.*, pp 67–69.
99. *Ibid.*, p 68.
100. Nielsen, *op. cit.*, p 111.
101. Golding, *op. cit.*, pp 561–62.
102. Billmeyer, *op. cit.*, p 220.
103. Golding, *op. cit.*, p 559.
104. Penn, W. S., *PVC Technology*, McLaren, 1962, p 49.
105. Meares, *op. cit.*, p 266.
106. Brydson, *op. cit.*, pp 91–93.
107. Meares, *op. cit.*, p 233.
108. Modern Plastics Encyclopedia *op. cit.*, p 146.
109. Saunders, J. H., & Frisch, K. C., *Polyurethanes: Chemistry & Technology*, Part I, Interscience, 1962, p 4.
110. Billmeyer, *op. cit.*, p 206.
111. Brydson, *op. cit.*, p 58.
112. Morton, *op. cit.* (ref 8), pp 54–55.
113. Nielsen, *op. cit.*, pp 62–63.
114. Miller, *op. cit.*, pp 538–39.
115. Baer, *op. cit.*, p 1042.
116. Alfrey, *op. cit.*, p 85.
117. Tobolsky, *op. cit.*, p 198.
118. Billmeyer, *op. cit.*, pp 15–16.
119. Brydson, *op. cit.*, p 43.
120. Miller, *op. cit.*, p 624.
121. Billmeyer, *op. cit.*, p 8.
122. Meares, *op. cit.*, pp 257–66.
123. Billmeyer, *op. cit.*, p 210.
124. McFarlane, *op. cit.*, p 96.
125. Nielsen, *op. cit.*, p 166.
126. *Ibid.*, pp 165–67.
127. *Ibid.*, p 232.
128. Alfrey, *op. cit.*, p 72.
129. Nielsen, *op. cit.*, pp 60–61.
130. *Ibid.*, p 61.
131. Alfrey, *op. cit.*, p 71.
132. Miller, *op. cit.*, p 292.
133. Meares, *op. cit.*, p 265.
134. Miller, *op. cit.*, pp 292–93.
135. Brydson, *op. cit.*, p 49.
136. Golding, *op. cit.*, p 556.
137. Schmidt, *op. cit.*, p 189.
138. Miller, *op. cit.*, p 293.
139. Nielsen, *op. cit.*, pp 167–68.
140. *Ibid.*, p 165.

141. Miller, *op. cit.*, pp 296–97.
142. McFarlane, *op. cit.*, pp 87–97.
143. Billmeyer, *op. cit.*, pp 209–10.
144. Tobolsky, *op. cit.*, pp 206–10.
145. *Ibid.*, pp 255–59.
146. *Ibid.*, p 209.
147. Meares, *op. cit.*, p 362.
148. Morton, *op. cit.* (ref 8), Chapter 17.
149. Golding, *op. cit.*, p 19.
150. Meares, *op. cit.*, p 222.
151. *Ibid.*, p 223.
152. Morton, *op. cit.* (ref 8), p 121.
153. Meares, *op. cit.*, p 221.
154. Morton, *op. cit.* (ref 8), p 25.
155. *Ibid.*, p 28.
156. *Ibid.*, pp 25–28.
157. *Ibid.*, Chapter 5.
158. Meares, *op. cit.*, pp 220–23.
159. Nielsen, *op. cit.*, pp 83–86.
160. *Ibid.*, p 114.
161. *Ibid.*, pp 84–85.
162. *Ibid.*, pp 85–86.
163. Billmeyer, *op. cit.*, pp 225–26.
164. *Ibid.*, pp 526–27.
165. Morton, *op. cit.* (ref 8), p 110.
166. Alfrey, *op. cit.*, p 36.
167. *Ibid.*, p 52.
168. Nielsen, *op. cit.*, p 85.
169. Morton, *op. cit.* (ref 8), pp 27–28.
170. Nielsen, *op. cit.*, p 60.
171. Tobolsky, *op. cit.*, pp 75–78.
172. *Ibid.*, p 75.
173. Nielsen, *op. cit.*, p 109.
174. Billmeyer, *op. cit.*, pp 526–30.
175. *Ibid.*, p 527.
176. Morton, *op. cit.* (ref 8), p 27.
177. Nielsen, *op. cit.*, pp 60–62.
178. *Ibid.*, p 62.
179. Golding, *op. cit.*, p 545.
180. Tobolsky, *op. cit.*, p 77.
181. Billmeyer, *op. cit.*, pp 537–38.
182. Nielsen, *op. cit.*, p 110.
183. *Ibid.*, p 125.
184. *Ibid.*, pp 166–68.
185. *Ibid.*, p 167.

186. Baer, *op. cit.*, p 561.
187. *Ibid.*, pp 662–63.
188. Miller, *op. cit.*, pp 636–38.
189. Golding, *op. cit.*, p 553.
190. *Ibid.*, pp 572–73.
191. Brydson, *op. cit.*, p 55.
192. Miller, *op. cit.*, p 300.
193. Golding, *op. cit.*, pp 21–23.
194. Meals, R. N., & Lewis, F. M., *Silicones*, Reinhold, 1959, pp 57–59.
195. *Ibid.*, p 59.

Part Five

SUPERMOLECULAR STRUCTURE

Chapter Seven

Multiple Phases and Macrostructures

Beyond the smallest features of submolecular structure, beyond molecular size and flexibility, and beyond even the larger structural features of intermolecular order and bonding, there are a number of still larger structural features which may be best described as supermolecular. These in turn may range from relatively small structures, which are visible only through the optical microscope or even the electron microscope, up to those larger aspects of structure which are visible to the naked eye, and even to those very large structural features which are apparent on simple observation. All of these structural features, in increasing orders of magnitude, have important effects on the properties of polymers and their final products in different applications. They may be classified conveniently as follows: Chemically Homogeneous but Physically Heterogenous Microstructures; Temporary Heterogeneity for Processing; Chemical Microheterogeneity; Macrostructure—Macroscopic Dimensions of Individual Structures; Complex Multicomponent Total Macrostructures. Each of these five orders of magnitude includes a number of types of structural features which can now be distinguished and studied individually.

7-1 CHEMICALLY HOMOGENEOUS BUT PHYSICALLY HETEROGENEOUS MICROSTRUCTURES

Many polymers form a continuous solid phase which is chemically homogeneous and yet physically or morphologically heterogeneous. Two major examples are (1) Spherulites in crystalline polymers, and (2) Micelles in thermosetting polymers. These are two distinct types of structural phenomena and are best considered separately.

Spherulites in Crystalline Polymers

Nature of Spherulite Formation

When a crystallizable polymer is cooled below its melting point, the crystallites which form are so tiny that only the largest are visible even in the electron microscope. In most cases, crystallites are implied only by theoretical calculation from X-ray diffraction patterns. In many, perhaps most, crystalline polymers, however, crystallites do not exist simply as individual submicroscopic entities, but tend to aggregate into ordered structures of larger size called spherulites. These range from submicron and submicroscopic in size up to several millimeters or more (ref 1,2). These spherulites appear to be circular or spherical in form (Figure 7-1), and are probably composed of individual crystallites radiating outward from the center of the spherulite (ref 4,5). The interfaces between the spherulites may be interleaved with each other, producing overall bonding and strength (ref 6); or the crystallizing polymer may pack all of the low molecular weight and impure species into the interstices between the spherulites, leaving these as contaminated boundaries of lower strength and modulus (ref 7). Shrinkage during crystallization may further leave stresses and voids in these interstices, weakening them even more.

Factors in Spherulite Structure

The size and perfection of these spherulitic structures is very sensitive to the composition of the polymer and to the conditions under which it is crystallized. Random copolymerization generally decreases the size of spherulites (ref 8). Plasticizers, nucleating agents, contaminants, and ordered regions in melt or solution all affect crystallization and spherulite formation (ref 2). Nucleating agents are being used more and more frequently in commercial practice to control and decrease spherulitic size.

The most important factor in spherulite formation is temperature. Crystallization just below the melting point forms few nuclei, and each of the spherulites grows to large size. Crystallization just above the glass transition temperature forms many nuclei, and each of them can only grow to a spherulite of very small size (ref 9). Thus in general, slow cooling from the melt, and thermal annealing after crystallization, generally tend to form large spherulites; while rapid quenching from the melt generally tends to form small or even negligible spherulites (ref 10–12).

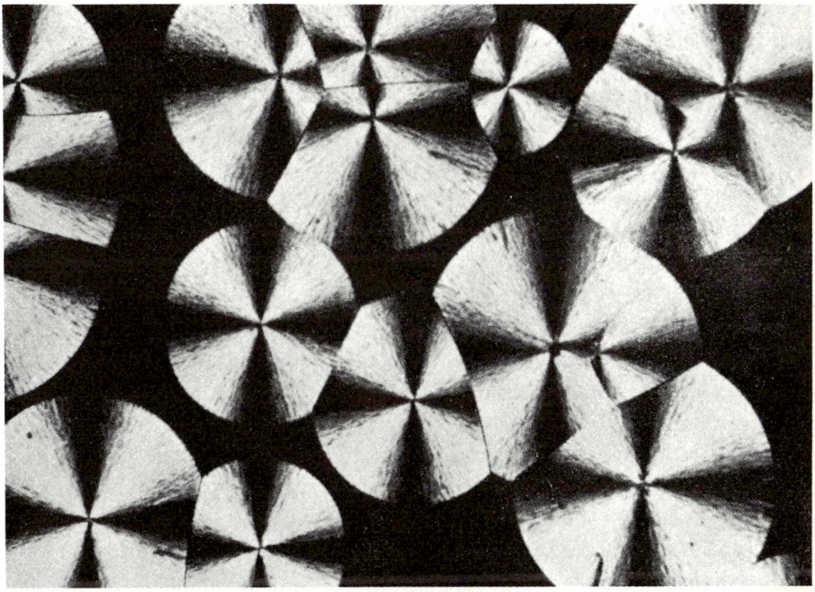

Figure 7-1. Spherulites.
Field of growing spherulites in isotactic polystyrene. Optical micrograph, crossed polarizers (ref 3). Turner Alfrey and Edward F. Gurnee, *Organic Polymers,* © 1967. Reprinted by permission of Prentice-Hall, Inc., Englewood Cliffs, New Jersey.

Even after crystallization and spherulite formation have been completed, spherulitic structure may be radically altered by mechanical stretching and orientation, such as is applied in the production of films and fibers. During stretching, spherulites first become elliptical in nature; at higher degrees of orientation they rupture (ref 13); and in highly drawn fibers they may be completely absent (ref 14).

Spherulite Structure → Properties

Spherulitic structure has profound effects upon the mechanical and optical properties of polymers. Decreasing the spherulite size in a rigid crystalline polymer will lower the modulus somewhat, but still give a relatively stiff product (ref 15). Most studies agree that smaller spherulites give softer ultimate mechanical properties, such as higher ultimate elongation and impact strength (ref 10,11,15–19). On the other hand some studies suggest that small spherulites give higher yield strength and brittle temperature, lower ultimate elongation, and greater tendency to break during cold-stretching (ref 10,17). Fracture sometimes goes between the spherulites, sometimes right through them (ref 10). The interstices between spherulites, containing low-molecular-weight impurities, stresses, and voids, are probably the weak interfaces where stress-cracking is most likely to begin (ref 7).

Spherulite size is also inverse to optical clarity (ref 20). Whenever the size of the spherulite approaches or exceeds the wavelength of visible light, the higher density and index of refraction in the crystalline spherulite, as compared to the amorphous interstices, causes scattering of the light, translucency, and even opacity in extreme cases (ref 21). Thus control of spherulite size has become extremely important in production of clear films, and growth of spherulites in solid moldings is responsible for the loss of clarity in most crystalline plastics.

Micelles in Thermosetting Plastics

Thermoset plastics are generally assumed to be fairly uniformly cross-linked and three-dimensional in nature. Electron microscope studies of fracture surfaces in thermoset polymers, however, indicate that they do not break in the smooth conchoidal manner normally expected from uniform glassy materials (Figure 7-2). Rather the fracture surface frequently indicates small hills and valleys, localized regions which have been named micelles. These rough micellar fracture surfaces suggest that fracture, instead of proceeding smoothly through a uniform composition, is actually seeking out the weak interfacial area between relatively strong micelles of thermoset material (ref 22,23).

Further consideration of the cross-linking reaction process indicates why this is likely to occur. When a reactive polyfunctional low-molecular-weight monomer, or intermediate B-stage polymer, is catalyzed and heated to produce cross-linking, this starting material is still small enough in molecular size to be soluble, fusible, and both chemically and physically homogeneous. As reaction begins, the statistical probability of forming each new bond is equally probable throughout the material. When molecular weight has grown to a certain critical size, however, the molecule is no longer small enough to remain uniformly dispersed in the liquid medium by Brownian motion, and begins to segregate out in the form of a rubbery gel. This will happen to each molecule at a somewhat different point in time during the total reaction. When a particular molecule exceeds the critical size for uniform solubility, and gels out in this way, it traps and takes with it most of the smaller molecular species within its geographic area.

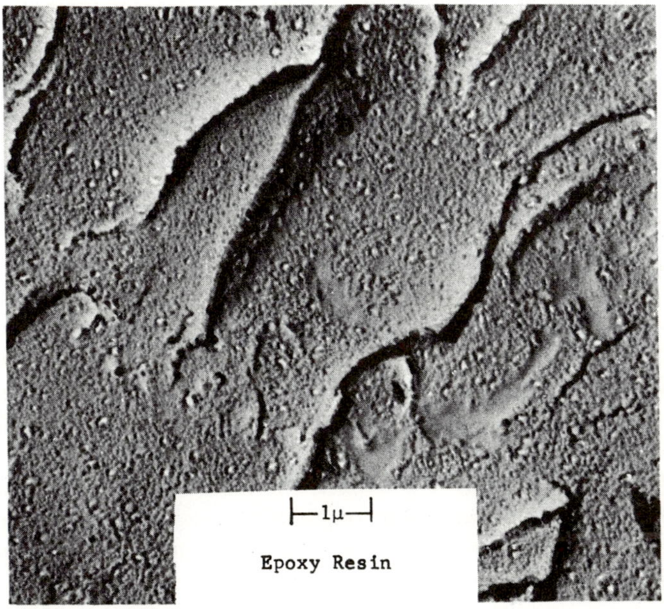

Figure 7-2. Micelles (ref 23). Reprinted from *Preprints* of the Division of Organic Coatings and Plastics, *23*(1), 395 (1963). Copyright 1963 by the American Chemical Society. Reprinted by permission of the copyright owner.

Further cross-linking reactions within this geographic area will tend to knit this gel particle more and more tightly within itself, as each new covalent bond which forms has a smaller dimension than the intermolecular spacing of the original separate reactive molecular species. As this gel particle continues to cross-link internally, it will shrink in size and squeeze out some of the remaining low-molecular-weight species. The total reaction mass will thus become a number of such gel particles each cross-linking internally and shrinking away from each other, squeezing lower-molecular-weight species into the interstices and leaving these to polymerize later. Thus the fully-reacted, final cured product will contain highly cross-linked particles separated by relatively weakly-filled and weakly-reacted interfacial areas.

In this way it is easy to see how a uniform initial chemical composition and reactivity can result in a physically heterogeneous cured product containing strong micelles separated by weak interfacial areas. This phenomenon may be of major importance in many highly-cured thermoset plastics such as allyl, epoxy, and phenolic plastics. It further suggests that the ultimate strength properties of such three-dimensional thermoset plastics might be greatly improved by long-range exploratory studies seeking ways to produce more homogeneous cross-linking.

7-2 TEMPORARY HETEROGENEITY FOR PROCESSING

Polymers are often prepared in a heterogeneous dispersed form which is of temporary but very significant value during processing, and are then converted during processing into a permanent continuous solid form. This temporary heterogeneity for processing is often both chemical and physical in nature. Three major commercial examples are (1) Emulsions, (2) Pastes, and (3) Powders. Each of these three forms of temporary heterogeneity is used in its own distinctive way during processing, but all are converted during processing into permanent continuous solid polymeric form in final end-use applications.

Emulsions

Commercial Materials

Many polymers are produced originally by polymerization in aqueous emulsion, and the resulting emulsions or latexes are directly usable in later processing. Typical examples are natural rubber, free-radical-polymerized synthetic rubbers, styrene/butadiene copolymers, vinyl acetate polymers, and acrylic polymers. Other polymers, such as solution-polymerized elastomers and alkyd coatings, are not produced originally in emulsion form, but can be converted into latex form later for improved specialized processing.

Structure

These polymer emulsions exist as stable dispersions of fine polymer particles in water. The polymer particles are generally submicron in size, averaging several tenths of a micron in most commercial systems. These are small enough to remain stably suspended in water by the thermal agitation and Brownian motion of the water molecules themselves. Since the polymers in these particles are usually soft and tacky enough to stick to each other, and agglomerate quite readily on contact, they are further separated from each other and stabilized by a monomolecular coating of surface active agent, most commonly the emulsifiers added to the original polymerization recipe, but frequently also further emulsion stabilizers added after polymerization, during compounding of the latex for use.

Processing

In practice these polymeric emulsions are generally converted to continuous solid finished form by removing the water by evaporation, bringing the polymer particles into contact and eventual fusion. Since in the later stages of evaporation the water must be removed from the surface of the solid polymer product, these products are most conveniently impregnations, thin films, coatings, adhesives, and open-celled foams, in all of which there is a high ratio of surface/volume.

Film Formation

In all cases, as the water evaporates, the polymer particles come closer together, until at some particular concentration they actually come into contact with each other. At this point the interstices between the barely-touching polymer particles are still filled with water, and the surfaces of the polymer particles are still coated and protected by surface active agent. As the water continues to evaporate, the soft polymer particles squeeze against each other to fill the remaining voids being left by the escaping water. If the polymer is below its glass transition temperature, its ability to squeeze beyond the spherical shape and fill these corner voids is distinctly limited, and the emulsion will generally dry to a discontinuous powder. For production of solid continuous polymeric products of commercial value, the polymer must either be inherently above its glass transition temperature at the temperature of evaporation of the latex, or it must be brought to this temperature by plasticization and/or heating. When the latex is evaporated above the glass transition temperature of the solid polymeric phase, continual removal of water from the intersticial corners between the spherical polymer particles will squeeze them against each other, forcing them more and more from spherical into cubical or more complex form. Under these forces, the protection of the surface active agent is no longer sufficient to maintain discrete boundaries between the soft mobile polymer particles, and the particles flow together under the force of the pressure and surface tension to form a single continuous solid phase (ref 24,25).

Applications

Such techniques are particularly useful for impregnation of porous materials; for production of continuous solid films; for application of continuous solid coatings as latex paints, paper coatings, leather coatings, and textile treatments; for the application of adhesives, particularly between porous adherends; and for the production of open-celled foams. In all of these cases, emulsions of polymers which are normally soft, flexible, or rubbery and therefore above their glass transition temperature, evaporate and fuse readily to form solid continuous polymers with high surface/volume ratio of great practical utility.

Pastes

Some combinations of solid polymers and nonvolatile liquid plasticizers are compatible at final equilibrium, but they display so little mobility and mixing at normal room temperature that they approach this equilibrium only very slowly. In such cases it is possible to disperse the finely-divided solid polymer in the nonvolatile liquid plasticizer to form a fairly fluid paste, which is relatively pourable and easy to convert into a variety of desired final shapes. This paste is sometimes called a dispersion, but this term is probably too broad and confusing for clear and definitive use. It may also be called a plastisol, which is probably the most common terminology. If a volatile liquid component is added temporarily to dilute the paste and increase its fluidity, it is then called an organosol. In either case the solid powdered polymer is dispersed in a liquid plasticizer temporarily for specific process technology, and later converted into the final solid form during processing.

Preparation

Paste technology is most commonly applied to plasticized vinyl chloride polymers, particularly the homopolymer (ref 26). Since most conventional vinyl chloride polymers interact fairly rapidly with liquid plasticizers even at room temperature, special care must be taken in the preparation of the solid powdered particles which will be used in such special paste technology. They are best prepared by emulsion polymerization at low temperatures, producing high-molecular-weight homopolymer in finely dispersed form. The emulsion is then spray-dried to produce very fine solid particles with a fused surface film. When these particles are stirred or ground into a liquid plasticizer, their extremely high molecular weight retards their mobility and ease of mixing with the plasticizer, and the fused surface film forms a fairly stable barrier to prevent any significant contact or mixing between the polymer and the plasticizer. Thus these particles remain fairly stably dispersed in the liquid plasticizer at room temperature. By proper choice of the polymeric powder and the liquid plasticizer, these pastes can remain stable for many months at room temperature.

Processing

During processing, the paste is spread or poured into its final form and then heated to a high enough temperature so that the mobility of polymer and plasticizer molecules becomes high enough to produce rapid fusion, generally at temperatures $\geq$150°C. At this temperature, not only do the polymer and plasticizer molecules have sufficient mobility to mix rapidly, but the tiny crystallites in the polyvinyl chloride also melt, eliminating their temporary "cross-linking" ability and permitting the individual molecules to move freely and disperse into the liquid plasticizer to form a true solution. Once the paste has been fused to a true solution in this way, it may be cooled back to room temperature, permitting the polyvinyl chloride molecules to form crystallites again, thus reestablishing the temporary pseudo-cross-links, forming a gel of liquid plasticizer molecules dispersed in a continuous solid polymeric phase, which produces the soft, flexible, and even rubbery quality of the finished plasticized vinyl product. Such paste technology is particularly useful for producing laminates of plasticized polyvinyl chloride with textiles, coatings of plasticized polyvinyl chloride on other materials, thin film products such as rubber gloves, and solid rubbery products such as dolls, playballs, and shoe soles.

Organosols

For some thin film and coating applications, it may be desirable to produce a less viscous, more fluid paste than can be obtained from a mixture of solid powdered polymer and liquid plasticizer alone. In such cases a volatile liquid diluent is added, and the paste is called an organosol. This volatile liquid diluent must be evaporated during heating and fusion, so these organosols are useful primarily in the production of thin films and coatings, where there is a high enough ratio of surface/volume to permit easy evaporation of the temporary diluent.

Rigisols

In other cases the ratio of liquid plasticizer to solid powdered polymer, which will form a fluid paste during processing, is so high that it produces a final fused plasticized polymer which is more soft and rubbery than is desirable for certain applications. In such applications, part or all of the normal stable plasticizer may be replaced by a reactive polymerizable plasticizer such as a glycol diacrylate or methacrylate. Such a reactive polymerizable plasticizer provides high fluidity during processing and then polymerizes during fusion to produce another rigid polymer and reduce or eliminate the amount of permanent liquid plasticizer in the finished composition. Thus it is possible to convert fluid pastes during fusion into fairly stiff or even rigid final products. These are sometimes called rigisols.

In all of these cases, a fairly stable dispersion of solid powdered polymer in nonvolatile liquid plasticizer provides fluidity for easy processing, and is then converted during processing by fusion into a homogeneous solid final form.

Powders

Most polymers are manufactured and sold in solid, granular, or powdered form, which is easy to pour, package, ship, unload, and feed into most conventional process equipment; and this process equipment fuses and converts them into continuous solid finished products. In a growing number of specialized processes, the original solid powder is ground to a very fine particle size, which can be handled in pseudofluid techniques and then fused into continuous solid form. These emerging techniques are frequently called powder processing or powder technologies.

Molding

Rotational molding is one of the most common examples of such powder process technology (ref 27). Large hollow plastic products are made from finely powdered polyethylene by tumbling it in a large heated rotating mold, in which the powdered polyethylene sticks to the rotating heated surfaces of the mold and fuses there to form a continuous solid wall.

Coating

A newer and more unusual technique which is growing in importance is fluid bed coating (ref 28). A product to be coated with plastic, generally a metal product, is preheated to a temperature far above the fusion point of the plastic which will be coated on to it. The finely-powdered polymer is held in a tank with a porous floor, and a stream of inert gas is blown continuously through the porous floor and up through the powdered polymer, suspending it in a fairly fluid state within the tank. The preheated metal product is dipped into this fluid bed of finely-powdered polymer, and the particles of polymer, which come in contact with the preheated metal surface, melt and stick to the surface. The temperature of the metal and the particle size of the polymeric powder determine the thickness to which such coatings will grow. In general the coating at this point is only partially fused, and the roughly-coated product must be further heated in an oven to produce complete fusion of the agglomerated polymer particles, raising them above their fluid temperature and permitting them to flow together by surface tension, to produce a solid continuous homogeneous coating.

Another use of such finely-powdered polymers for coating is flame-spraying. The powdered polymer is fired through an air gun, passing briefly through a flame which heats it above its fusion temperature, and impinging on the surface to be coated. The polymer is still in molten form when it strikes the surface, and fuses instantly to a continuous coating which solidifies rapidly on cooling.

Another technique for converting such finely-powdered polymers into continuous solid coatings is by electrostatic spraying (ref 28). Here the finely-powdered polymer is fired through an air gun which also applies a static charge to the polymer particles. When they strike the surface to be coated, the static charge causes them to adhere temporarily, but long enough for the desired process. They are converted from a loose powdered coating into a fused solid finished coating by oven-heating.

Still another use of such powder technology is the application of powdered polyethylene to textiles, followed by fusion in an oven, to produce a fairly continuous solid backing-film of polyethylene. This is used particularly in the production of auto carpeting.

Sintering

Some almost infusible polymers of very high melt viscosity, such as ultra-high-molecular-weight polyethylene (ref 29) and polytetrafluoroethylene (ref 30), may best be converted from powdery into continuous solid form by sintering at high temperature and pressure. If sintering is incomplete, leaving voids between the partially-fused particles, the final product will have low density, slightly low modulus, low tensile strength, low ultimate elongation, low impact strength, and very poor flex life (ref 31).

It has also been observed that polyvinyl chloride or polyacrylonitrile powders may be cold-pressed at high enough pressures to make them fuse together into completely transparent products; and yet the polymer molecules from adjacent particles have so little mobility, and are so poorly entangled, that the interfaces between these particles are still very weak and brittle. In such cases, increasing the pressure increases the glass transition temperature, and thus makes cold-pressing even less satisfactory (ref 32).

7-3 CHEMICAL MICROHETEROGENEITY

Many of the most useful polymeric products contain incompatible materials, finely dispersed throughout a continuous polymeric phase, in such a way that each of the incompatible phases contributes significantly to the desirable properties of the total product.

There is a growing belief that the combination of such chemically and physically incompatible phases functions in a very specific way to produce such optimum combinations of properties (ref 33). In almost all cases the most promising materials are combinations of a rigid, high-modulus, high-strength phase with a soft, flexible, low-modulus, extensible phase, in which one of these phases is continuous and the other is dispersed, and each is permitted to display a major proportion of its most desirable properties in the finished product. It is also important that the interface, at which these two incompatible phases meet, must be very well bonded in most cases, to produce the optimum combination of properties by making the two phases act as a unit rather than allowing them to respond separately to the stresses which the total composition is expected to withstand (ref 34). In most of such incompatible blends, the advantageous combination of properties is primarily mechanical and thermal in nature; but in certain special cases the most striking properties may be acoustical, electrical, and optical as well.

Several typical major classes of examples include: (1) Semi-compatible Polyblends, (2) Foams, and (3) Composites. In each of these, the nature and functions of the dispersed incompatible phases are quite different, and they are therefore best considered separately.

Semicompatible Polyblends

Structure

Miscibility between low-molecular-weight liquids requires fairly similar structure and/or polarity; but these requirements are quite easily met, and many combinations of liquids are completely miscible with each other. Among high-molecular-weight polymers, on the other hand, the requirements of similarity in structure and/or polarity are so stringent that very few combinations of polymers have any appreciable miscibility or compatibility, and compatible blends of two polymers are quite rare. When such compatible blends are found, the homogeneous mixtures which result have properties fairly similar to random copolymers of the same monomeric ingredients, and represent no significant practical advantage over simple random copolymers.

IMMISCIBILITY In most cases, combinations of two polymeric materials are far from being sufficiently similar in structure and/or polarity to have any appreciable miscibility or compatibility with each other. When two such incompatible polymers are thoroughly mixed with each other, the interfaces between the two incompatible species are generally very weak, and the overall mechanical properties of the blend are so poor as to be of little practical utility.

SEMICOMPATIBILITY Between these two extremes of compatibility and incompatibility, however, if two polymers have borderline compatibility with each other, one of them may be dispersed in the other and often produce an unusually advantageous combination of properties. In some cases the two polymers may just naturally happen to be sufficiently similar in structure and/or polarity to be somewhat compatible with each other. Typical commercial examples include the moderate compatibility of polyethylene with polypropylene, and of either of them with polyisobutylene and butyl rubber; the fair compatibility of butadiene elastomers with high styrene/butadiene copolymer resins; and the compatibility of polyvinyl chloride with certain nitrile rubbers. More often,

Figure 7-3. Interfacial blending of semicompatible polyblends.

however, the desired precise degree of semicompatibility is best achieved by graft copolymerization of one polymer onto the other.

TRANSITIONAL INTERFACE When two semicompatible polymers are mixed with each other, the molecules at the interface begin to migrate into and blend with each other to some extent, producing a transitional interface of improved strength and integrity (Figure 7-3). When a graft copolymer is used to improve the compatibility of two incompatible polymers, the graft copolymer probably forms the transitional interface between them (Figure 7-4). In either case the strength of the transitional interface is essential for optimum combination of the properties of each of the two incompatible phases.

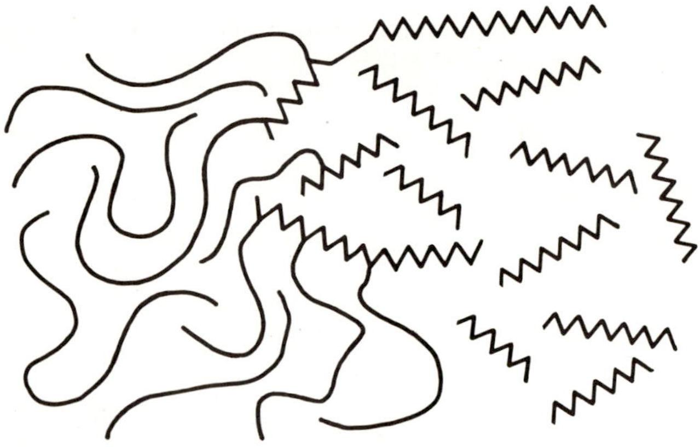

Figure 7-4. Graft-copolymer as transitional interface between semicompatible polymers.

Properties

High styrene/butadiene copolymers of fairly high hardness are sometimes dispersed into soft butadiene rubbers to improve the strength and toughness of these rubbers; in such compositions, the hard copolymer additive is frequently referred to as a reinforcing resin. In most commercial polyblends, however, a small amount of a relatively soft or even rubbery polymer is dispersed into a continuous matrix of a relatively hard, rigid polymer. Typical examples are impact styrene, ABS, rigid vinyls, impact acrylics, and impact phenolic plastics. In all of these, the rigid continuous major phase retains most of its original properties: high modulus, hardness, strength, hot strength, and chemical resistance. The dispersed soft or even rubbery phase, however, contributes certain critical properties, particularly processability and, most importantly, resistance to high-speed impact. Such balance of properties is not often obtainable in single pure polymeric materials, and is a major practical benefit from the use of such semicompatible polyblends.

PROCESSABILITY The relationship between semicompatible polyblends and processability has two critical aspects. In rigid polyvinyl chloride, semicompatible polymeric materials, which are soft or even rubbery in nature, are added to improve the processability of the relatively intractable pure homopolymer. These additives include nitrile rubber, acrylic, ABS, chlorinated polyethylenes, and ethylene/vinyl acetate copolymers. During high-temperature processing they provide increased fluidity, which produces much smoother, faster processing; but on recooling to the solid state for final use, they segregate as a minor dispersed phase within the rigid polyvinyl chloride matrix, and have no deleterious softening effect upon the rigid vinyl product.

Another and more general aspect of the relationship between structure and processability in semicompatible polyblends is the effect of processing on degree of dispersion. Since the two polymers are relatively incompatible, they can be dispersed only by providing considerable energy of mixing. During the manufacture of the semicompatible blend, sufficient mixing energy is supplied to bring the degree of dispersion close to its optimum point. Final properties are apt to depend very much upon the size of the dispersed particle, and the degree of blending at the interface where the transition occurs between the dispersed phase and the continuous phase. During continued further processing, such dispersion can be overdone; and an overprocessed polyblend may contain dispersed particles which are too small, or may even be too uniformly and compatibly dispersed into the continuous phase, thereby losing many of the advantageous properties which were originally intended. Thus the degree of processing applied to a semicompatible polyblend can be quite critical, and must be watched closely.

Mechanical Properties *Practical Properties* Mechanical properties are generally those which benefit most from the composition and structure of semicompatible polyblends. To the extent that the two polymers are incompatible, and the continuous rigid polymeric phase remains relatively pure, it retains most of its original hardness, modulus,

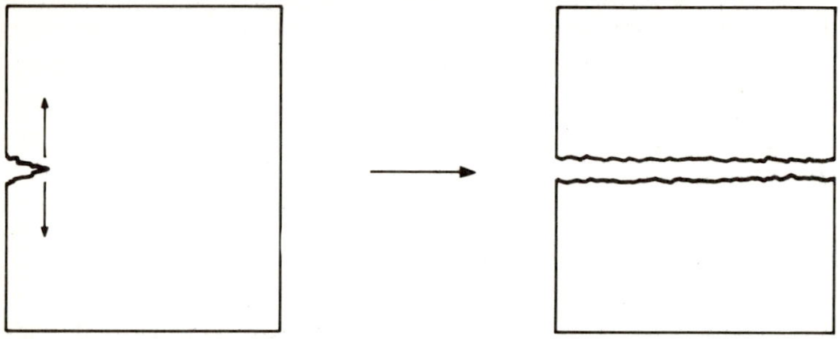

Figure 7-5. High-speed crack-propagation in brittle failure.

rigidity, and strength. To some extent, of course, the soft, rubbery additive is at least partially compatible in the continuous rigid phase, and does lower the numerical values of modulus and strength somewhat (ref 35,36); but at least the total material does remain relatively rigid and strong. More important, however, the presence of the dispersed rubbery particles produces a great increase in toughness, ductility, and particularly resistance to brittle failure under sharp, high-speed impact (ref 20,35,37–41).

Impact Strength Theory The tremendous improvement in impact strength is best understood by considering the nature of crack propagation during mechanical failure of a rigid plastic. Under high stress, failure begins at any flaw or inhomogeneity where the stress may concentrate, and the vectors all apply to a sharp corner or notch. In a rigid material which cannot respond rapidly to high-speed impact, the stress all concentrates at the point of such a crack or notch, and tears it apart in a single direction (Figure 7-5). As the crack propagates through the rigid matrix, all of the energy remains concentrated at the apex of the growing crack, and continues the propagation rapidly. In a pure homogeneous rigid material, such crack propagation can proceed rapidly entirely through the specimen, causing brittle failure.

In a semicompatible polyblend, containing tiny soft or even rubbery particles dispersed throughout the rigid plastic matrix, the growing crack soon comes to such a rubbery particle. If the interfacial bond and transition, from the continuous rigid phase to the dispersed rubbery phase, are reasonably gradual, strong, and integral, the mechanical energy from the growing crack will be transferred into the interior of the rubbery particle. Here molecular motion is much more rapid and large-scale, and the energy can be dispersed and dissipated rapidly and uniformly in all

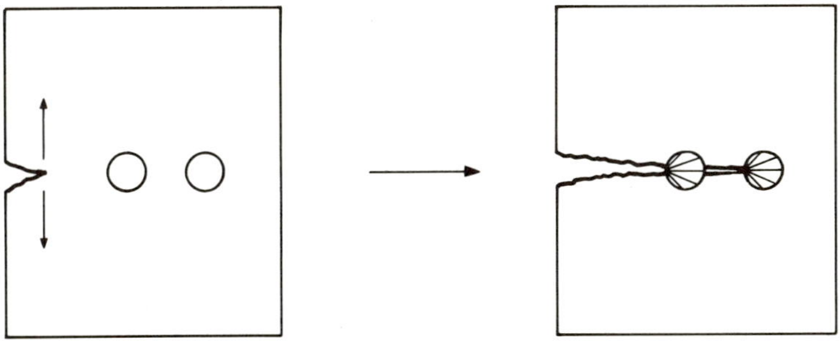

Figure 7-6. Energy dispersion by rubbery particles in a rigid matrix.

directions throughout the particle (Figure 7-6). To the extent that molecular motion is not quite fast enough to yield before the growing shock wave, some of the mechanical energy is converted into heat energy, raising the temperature, and softening and increasing the molecular motion still further. As the small dispersed shock waves travel radially out of the rubbery particle, there is no great stress-concentration at any one point. When these many small shock waves again reach the interfacial surface, where the rubbery particle meets the transitional area of change into the rigid phase, the stress at any one point is much smaller, generally insufficient to continue to propagate through the rigid phase. Even if the stress at any particular point at the interface should be sufficient to initiate and propagate another crack through the rigid phase, it would be considerably less than the original energy which initiated the breakage of the material; and by the time it reaches a second rubbery particle, it will be easily dissipated to a much lower harmless energy level. This mechanism makes it easy to see why such semicompatible polyblends, containing a soft dispersed rubbery phase in a rigid plastic matrix, can retain most of the hardness, modulus, rigidity, and strength of the pure rigid plastic, and yet show a tremendous benefit

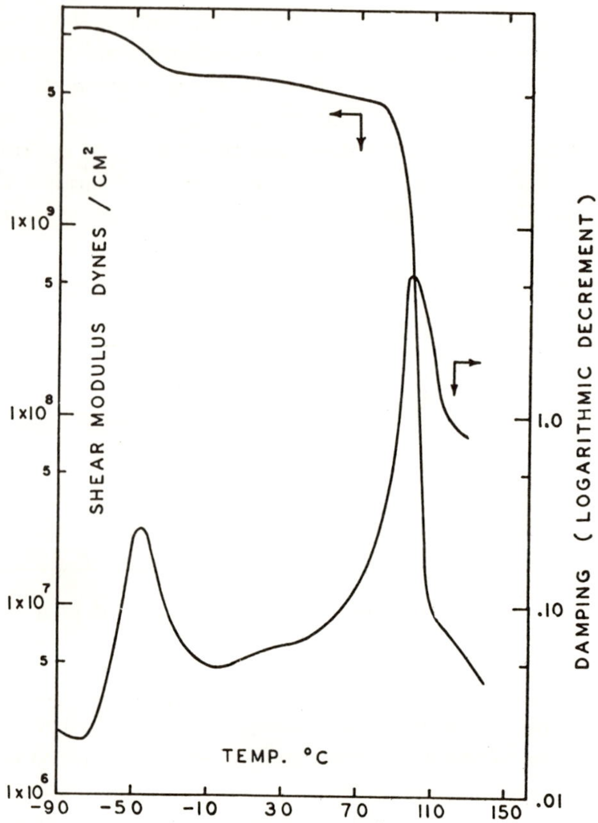

Figure 7-7. Modulus-temperature transitions in a semicompatible polyblend (ref 43).

Dynamic mechanical properties of an immiscible polyblend. Material is a mixture of polystyrene and a styrene-butadiene copolymer.

in ductility and impact strength due to the finely-dispersed soft rubbery phase.

THERMAL PROPERTIES Thermal properties also show the influence of the composition and structure of semicompatible polyblends. Whereas random copolymers and homogeneous polyblends show an intermediate transition between the pure components, semicompatible polyblends show two separate transitions in the modulus vs temperature curve (ref 36) (Figure 7-7). In such polyblends the continuous rigid phase still exhibits a glass transition temperature not far below the value for the pure material, while the dispersed rubbery phase exhibits a much lower secondary

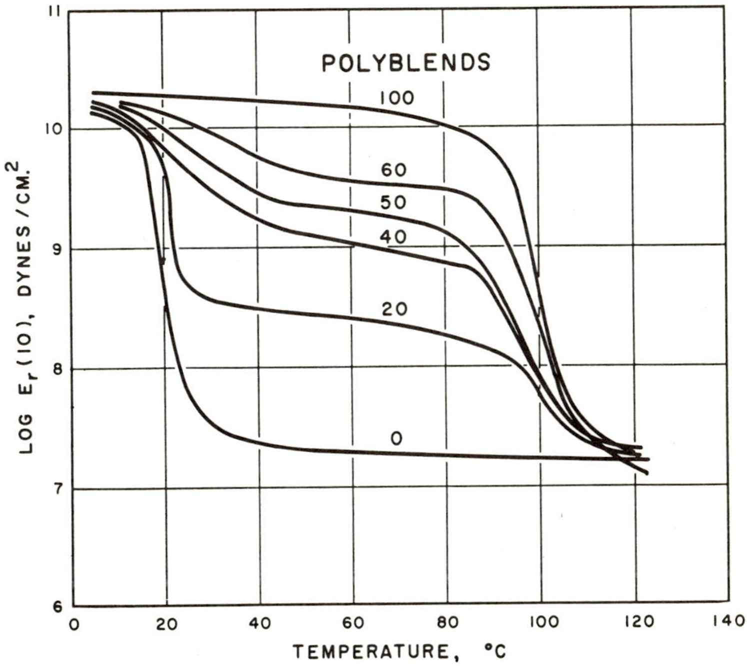

Figure 7-8. Modulus versus temperature in a series of semicompatible polyblends (ref 42).

$E_r(10)$ versus temperature for polyblends of polystyrene and a 30/70 butadiene-styrene copolymer. Numbers on the curves refer to the weight percent of polystyrene in the blend.

glass transition which is responsible for the high impact strength (ref 16) (Figure 7-8). The strength or size of the two transitions is proportional to the ratio of the two incompatible phases (ref 44). This effect of semicompatible polyblend composition and structure on the basic glass transition temperatures is the reason why such materials can retain most or all of the hot strength and heat deflection temperature of the original rigid continuous phase, and yet gain tremendously in impact strength even at low temperatures due to the dispersed rubbery phase.

Optical Properties The effects of semicompatible polyblending upon optical properties are generally undesirable. Since each incompatible phase has its own refractive index, light traveling through such a semicompatible polyblend will be refracted and dispersed many times as it enters and leaves the dispersed phase in its journey through the continuous phase. Thus most semicompatible polyblends are translucent to opaque. In many applications this may not be a serious problem, but wherever clarity is desired it creates a major problem for the plastics technologist. In rare cases it may be possible to design both the continuous phase and the dispersed phase so that they will have equal refractive indices, and thus give a transparent product even though it is composed of two incompatible materials. Even if the refractive indices are slightly different, it may be possible to achieve fair clarity in thin products. It must be recognized, however, that in general the optical properties of semicompatible polyblends represent one of their less desirable features, and thus constitute a continuing problem for the plastics technologist.

Fibers

One of the newer developments in semicompatible polyblends is the use of two incompatible polymers to form a single fiber. This new technique involves mixing two incompatible polymers in their molten form and extruding the molten blend through a spinneret, forming fibers in which the individual polymer components appear to separate partially into distinct fibrils within the overall fiber structure. Since this has been done primarily with condensation polymers, in which the functional group bonds still retain some chemical fluidity, it is quite possible that some interchange and copolymer formation occur between the two incompatible species, thus providing a semicompatible transitional interface which would strengthen the blend as it does in more conventional impact plastics. Further study is needed to determine the structure, properties, and optimum applications of such polyblended fibers (ref 45).

Foams

Nature

Foamed plastics derive their unique properties from (1) their unusual thin-walled cellular construction, and from (2) the fact that they are primarily solid polymers by weight and primarily gaseous by volume. Some properties are determined primarily by the type of cellular construction, others by the composition of the polymer, and still others by the composition of the gaseous phase.

Probably the greatest factor in foam structure is the density, or the ratio of solid to gaseous phase; the other major variable in foam structure is the relative proportion of open and closed cells. Most of the common polymers can be foamed, and transmit many of their bulk properties to the final foam. The gas phase is originally the blowing agent—air, nitrogen, carbon dioxide, steam, pentane, or fluorocarbon gas—but during aging these gases are very apt to migrate out of the foam, while atmospheric air migrates in to replace it. Each of these structural features has important effects upon certain properties of the foamed plastic.

Production

Processes for producing plastic foams involve three general steps: (1) Preparation of the polymeric material in a viscous liquid state, (2) introduction of fine bubbles of gas to produce expansion, and (3) solidification of the foamed plastic to stabilize the foamed structure.

Viscous Liquid Polymer Thermoplastic polymers are generally brought to the viscous liquid state by heating. Thermosetting polymers are prepared by partial prepolymerization to a viscous liquid intermediate state. In some cases the solid polymer is dispersed in a liquid medium to produce the viscous liquid state, as in the foaming of rubber latex and of vinyl plastisols.

Gas Bubbles The gas used for foaming may be introduced mechanically or chemically. Air can be introduced mechanically into vinyl plastisols by whipping or beating. Pentane is absorbed into polystyrene beads and then evaporated thermally to produce pentane gas and cause foaming. Fluorocarbons are refrigerated and compressed to the liquid state, mixed with polyurethane intermediates, and evaporated into the gaseous state by the exothermic polymerization reaction to produce foaming. Chemical production of gases to cause foaming is most often accomplished by thermal decomposition of organic azo compounds, which liberate nitrogen gas upon heating (Table 7-1). Carbon dioxide is often produced by decomposition of carbonates, or by the reaction of isocyanates with water during the formation of polyurethanes. In all cases it is generally desirable to introduce the gas in fine bubbles to produce a fine cell structure.

Table 7-1. Reactive Foaming Agents (ref 46). Reprinted by permission of Modern Plastics Encyclopedia, McGraw-Hill, Inc.

Chemical description	Structure
A: Azo-compounds	
1) Azobisformamide (Azodicarbonamide)	$H_2N—CO—N{=}N—CO—NH_2$
2) Azobisisobutyronitrile	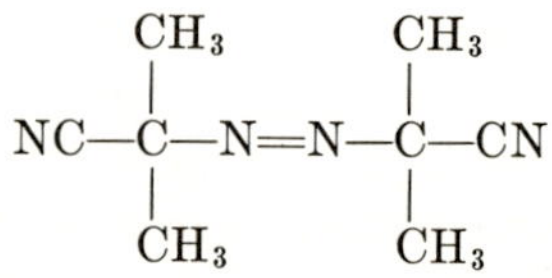
3) Diazoaminobenzene	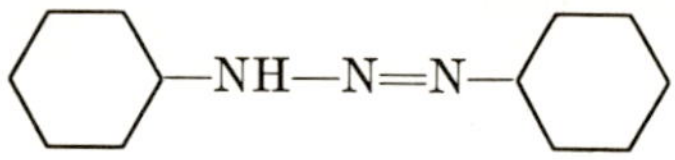
B: N-Nitroso-compounds	
4) *N*,*N*′-Dimethyl-*N*,*N*′-dinitroso-terephthalamide	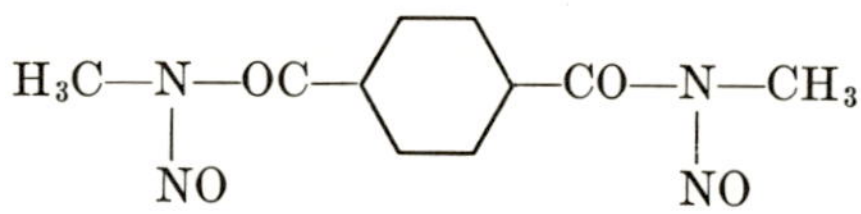
5) *N*,*N*′-Dinitrosopentamethy-lenetetramine	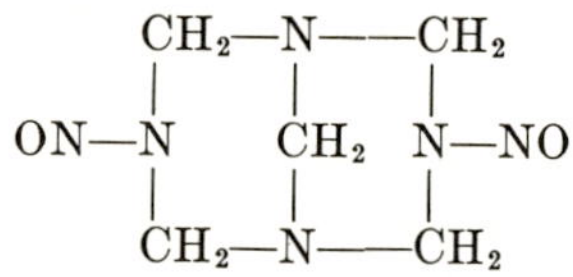
C: Sulfonyl hydrazides	
6) Benzenesulfonyl-hydrazide	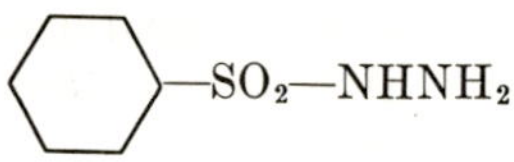
7) Toluene-(4)-sulfonyl hydrazide	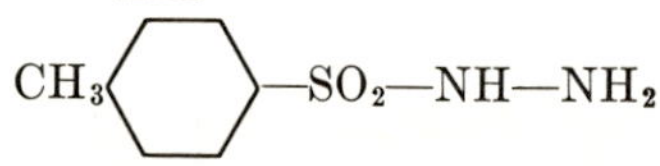
8) Benzene-1,3-disulfonyl hydrazide	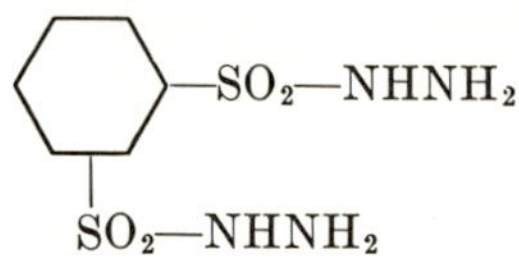
9) Diphenylsulfon-3,3′-di-sulfonyl hydrazide	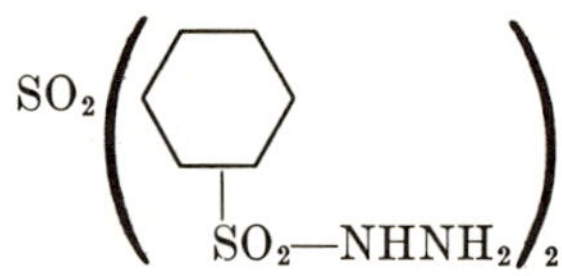
10) 4,4′Oxybis (benzenesulfonyl hydrazide)	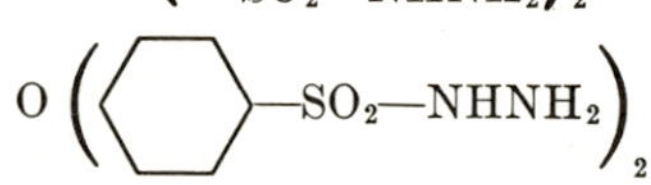

Solidification After foaming of the viscous liquid polymer, it must be solidified to stabilize the foam structure which has been produced.

Thermoplastics are solidified by cooling, but this represents a distinct problem in removing heat from the center of a highly insulating structure. Endothermic evaporation of gases during foaming may be used as an aid in such cooling.

Thermosetting polymers can be cross-linked and cured during foaming, and thus brought to the final solid state even at higher temperatures. This requires careful control of the rate of reaction and cross-linking along with the rate of foaming, but it does simplify the problem of stabilizing the final structure.

For the production of open-celled foams, one other step must be inserted between foaming and final solidification—a crushing step to break the cell walls before they are converted from the viscous liquid to the final solid state. This step must be performed at precisely the right moment, after the foam has been established, but before the cell walls have solidified to a strength which could not break by simple crushing.

Fabrication and Finishing Processability of foams involves also the fabrication and finishing of foam products. Machining and shaping of foams is generally very easy to accomplish. Bonding by adhesives, painting, and decorating, on the other hand, may become very delicate operations because of the sensitivity of the flimsy foam structure toward heat and organic solvents; so specialized materials and techniques are generally required.

Structure

Foam structure has seven individual components: (1) Polymer Composition; (2) Gas Phase Composition; (3) Foam Density; (4) Percent of Open and Closed Cells; (5) Cell Size; (6) Cell Geometry; and (7) Orientation of Cell Walls. These structural components can be varied independently in most cases. Some are more important in one type of property, some in another. The interaction of these structural factors can be seen from a review of their combined effects upon foam properties.

Properties and Applications

Density The most obvious and striking characteristic of plastic foams, particularly at low density, is their light weight. This is used directly when low density and closed cell structure are combined to provide buoyancy for flotation applications (ref 47,48).

Rigidity and Strength Cellular construction distributes mechanical stresses in such a way that foamed structures are generally much stiffer than an equal weight of solid material (ref 49). On the other hand, modulus and mechanical strength of foamed plastics are generally

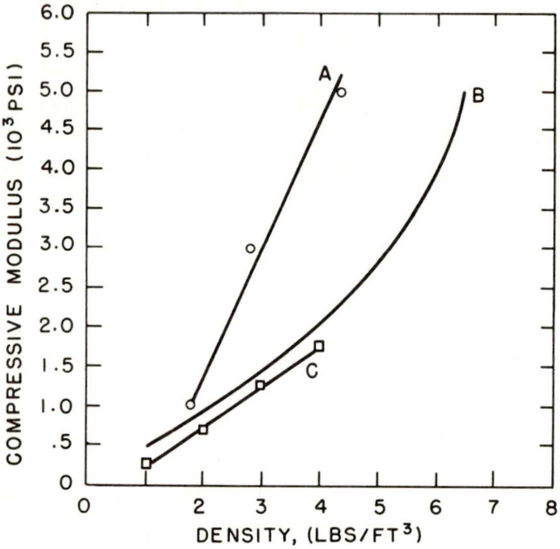

Figure 7-9. Modulus versus density for polystyrene foams. Compressive modulus: A. extruded; B. molded; C. molded. (ref 50)

almost proportional to foam density (Figures 7-9 to 7-12), but are considerably lower than would be predicted from density alone; for example, approximately: $E = E_o(d/5d_o)$ and $TS = (TS)_o(d/5d_o)$, where E and TS are the modulus and tensile strength of foams, E_o and $(TS)_o$ of the solid polymers, and d and d_o are the densities of foam and solid polymer (ref 53). In addition to density, the modulus of the polymer itself is also important in determining whether the foam will be rigid or flexible. In closed-cell structures, the gas exerts a pressure on the walls of the cells, and contributes to rigidity and strength of the foam structure (ref 54); conversely, open-celled foam structures are unusually flexible, even when made from high-modulus polymers. Cell size also affects mechanical properties (Figures 7-13 and 7-14); but these effects

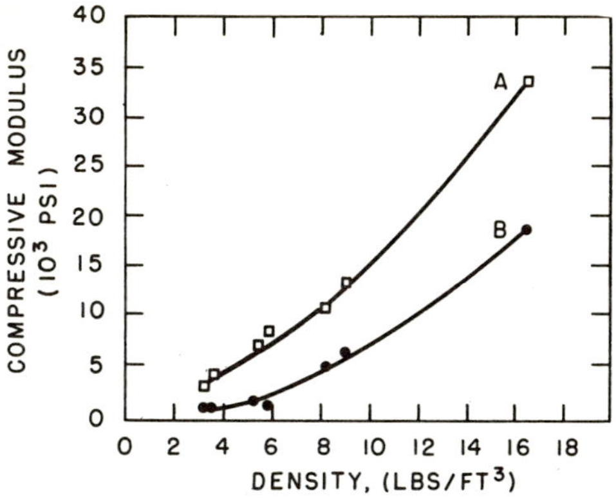

Figure 7-10. Modulus versus density for polyurethane foams. Compressive modulus: A. polyether (‖); B. polyether (⊥) (ref 50).

are complex and not well understood. Orientation of the polymer structure in the cell walls during blowing undoubtedly contributes to the overall mechanical properties of the final foam. Furthermore, asymmetric orientation of the total foam structure, caused for example by uniaxial rise during blowing, produces anisotropic structure and properties in the final foam (ref 54).

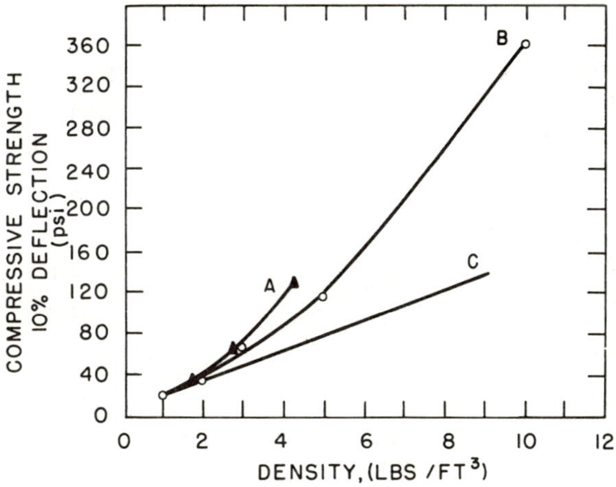

Figure 7-11. Strength versus density of polystyrene foams. Compressive strength at 10% deflection versus density for polystyrene foams: A. extruded; B. molded; C. molded (ref 51).

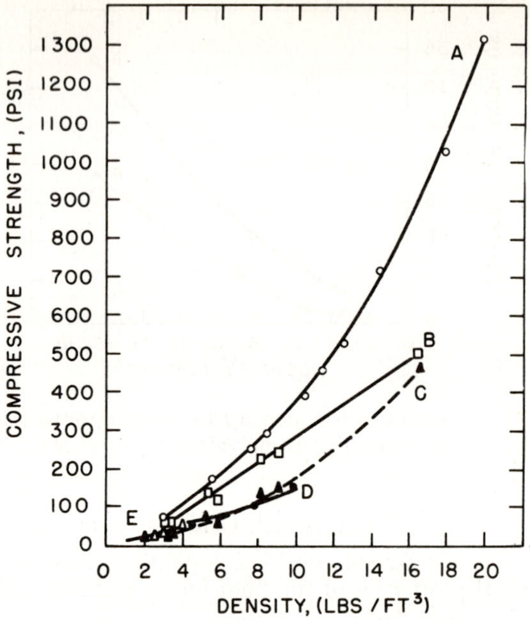

Figure 7-12. Strength versus density of polyurethane foams. Compressive strength at yield versus density for urethane foams: A. ○ polyether measured ⊥ at 10% deflection; B. □ polvether (‖); C. ▲ polyether (⊥); D. ● polyether (⊥); E. △ polyester (ref 52).

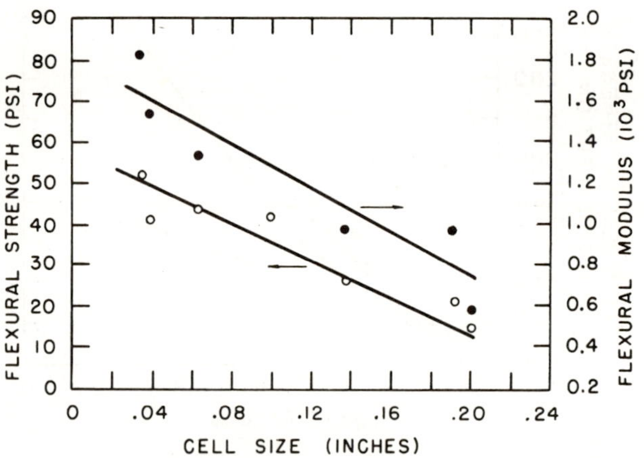

Figure 7-13. Flexural properties versus cell size of polystyrene foams. Extruded foam, 1.58 ± 0.1 lb/ft³ density (ref 55).

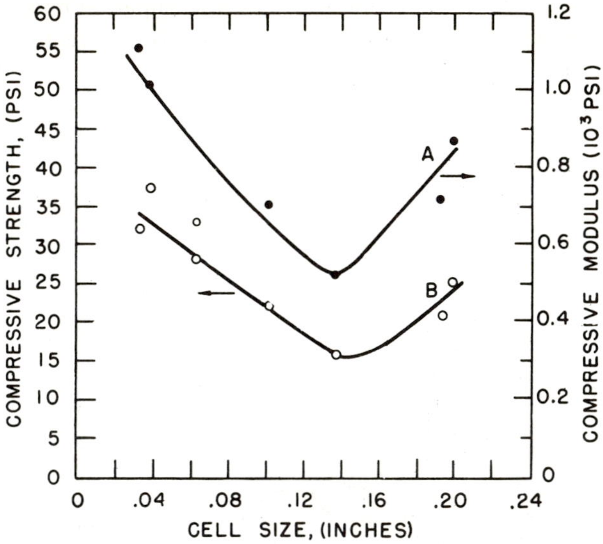

Figure 7-14. Compressive properties versus cell size for polystyrene foam. Extruded foam, of 1.58 ± 0.1 lb/ft³ density (ref 56).

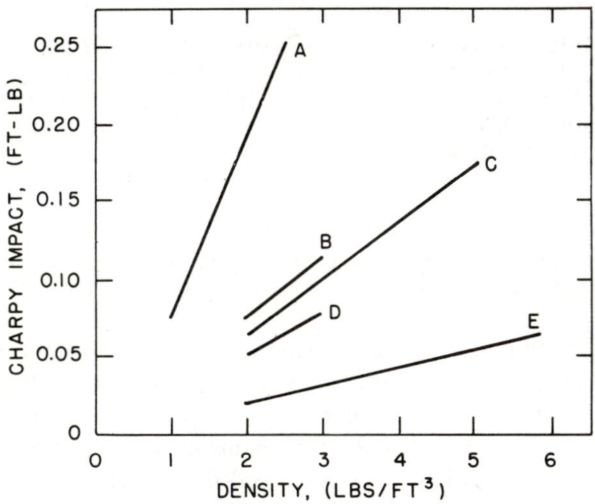

Figure 7-15. Charpy impact strength versus density for foams of: A. molded polystyrene; B. polyether urethane; C. polyester urethane; D. urea-formaldehyde; E. phenol-formaldehyde (ref 57).

Impact Cushioning While conventional impact strength tests may be directly related to foam density (Figure 7-15), the major significance and utility of plastic foams is their ability to absorb shock and provide cushioning, particularly in the packaging of delicate products (ref 58). Both rigid and flexible foams are used for this purpose. When a flexible foam is used, the rubbery resilience is an obvious aid in cushioning mechanical shock. When a rigid foam is used, it is usually deformed

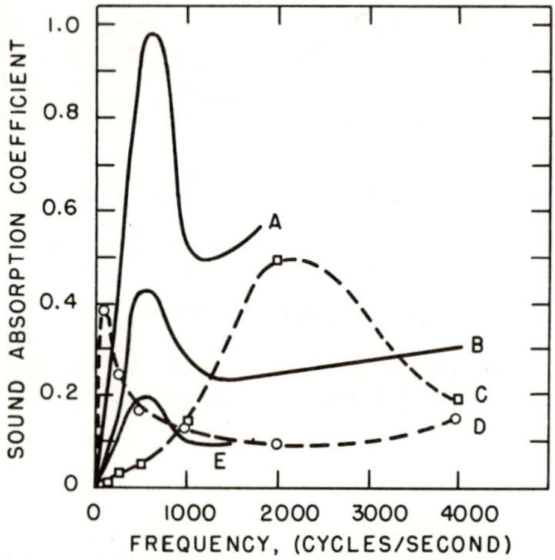

Figure 7-16. Acoustic absorption spectra of foams (ref 65).
Sound absorption coefficient versus frequency for foams of: A. urethane, 5 lb/ft^3; B. phenolformaldehyde, 8 lb/ft^3; C. extruded polystyrene, 2 lb/ft^3, #4 mounting; D. extruded polystyrene, 2 lb/ft^3, #7 mounting; E. molded polystyrene, 2 lb/ft^3.

permanently by the shock; and its protection of the delicate articles packaged in it is generally explained by the energy absorbed in crushing the cell walls of the foam itself.

Acoustic Behavior Plastic foams do not reflect sound waves, and are thus useful for reducing reverberation in noisy rooms. Plastic foams may also absorb some sound, but only at specific frequencies (Figure 7-16). On the other hand, low density foams provide no barrier to sound transmission; for this a high density is needed (ref 65).

Thermal Insulation The most outstanding thermal property of plastic foams is their highly efficient performance in thermal insulation

(Table 7-2). Paradoxically, their ability as thermal insulators depends primarily upon the gas trapped in the foam, rather than upon the plastic in the cell walls. Heat flow depends upon the thermal conductivity of the material and upon the distance the heat must flow through that material and is expressed as $Q \sim k/x$, where Q is quantity of heat flow, k is thermal conductivity, and x is the distance the heat must flow (ref 60). In a plastic foam the trapped gas provides both the lowest thermal conductivity and the greatest distance through which the heat must flow, and thus exerts by far the major effect on thermal insulation. Closed-cell structures are highly preferable because they greatly reduce convection of gas in the cells (ref 59). It is obvious that foams blown with carbon dioxide and especially with fluorocarbon gases will provide the best thermal insulation; however, during long-term use there is a tendency for air to diffuse through the cell walls into the foam, and for carbon dioxide or fluorocarbon gas to diffuse through the cell walls out of the foam. Diffusion of air and particularly of moisture vapor raises the thermal conductivity of the gas phase in the foam (Figure 7-17). For long-term performance at maximum efficiency, the foams must be extremely well sealed to prevent such diffusion (ref 62).

Table 7-2. Thermal Conductivities of Various Substances Expressed in (Btu in.)/°F ft^2 hr (ref 59). Turner Alfrey and Edward F. Gurnee, *Organic Polymers*, © 1967. Reprinted by permission of Prentice-Hall, Inc., Englewood Cliffs, New Jersey.

Marble	14–20	Typical rigid plastic foam	
Glass	5.5	(containing air)	0.22
Typical solid plastic	1.0		
Air	0.18	Typical rigid plastic foam	
CO_2	0.12	(containing CCl_3F)	0.15
CCl_3F	0.06		

Other Thermal Properties Effects of temperature on mechanical properties of plastic foams are parallel to their effects upon the solid polymer of which the foam is composed (Figures 7-18, 7-19); in general, modulus and strength are inverse to temperature. Specific heat of plastic foams depends primarily upon the plastic, because the weight of gas is negligible (Table 7-3). Coefficient of thermal expansion of a plastic foam also depends primarily upon the plastic itself, unless there is collapse of the cell walls (Table 7-4).

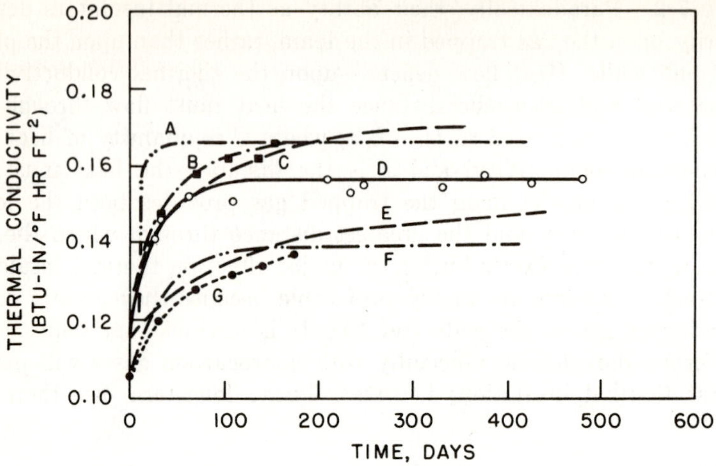

Figure 7-17. Effect of aging upon thermal conductivity of polyurethane foams (ref 61).

A. polyether, 2 lb/ft³, aged at 122°F, 6″ × 6″ × 1″; B. polyether, 2.05 lb/ft³, aged at 73°F, 8″ × 8″ × 1″, (‖) direction; C. polyether, aged at 110°F, 0 or 100% humidity; D. polyether, 2 lb/ft³, aged at 140°F, 8″ × 8″ × 1″; E. polyether, aged at 73°F, 0 or 100% humidity; F. polyether, 2 lb/ft³, aged at 77°F, 6″ × 6″ × 1″; G. polyether, 2.05 lb/ft³, aged at 73°F, 8″ × 8″ × 1″, (⊥) direction.

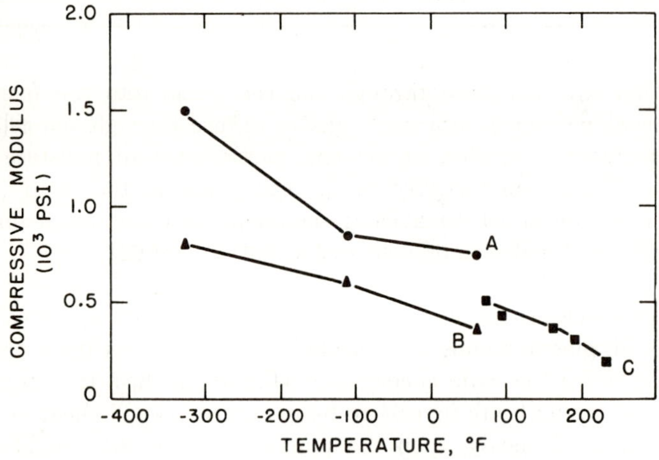

Figure 7-18. Modulus versus temperature of polyurethane foams (ref 63).

Compressive modulus: A. 1.8 lb/ft³ polyether; B. 1.5 lb/ft³ polyether; C. polyether.

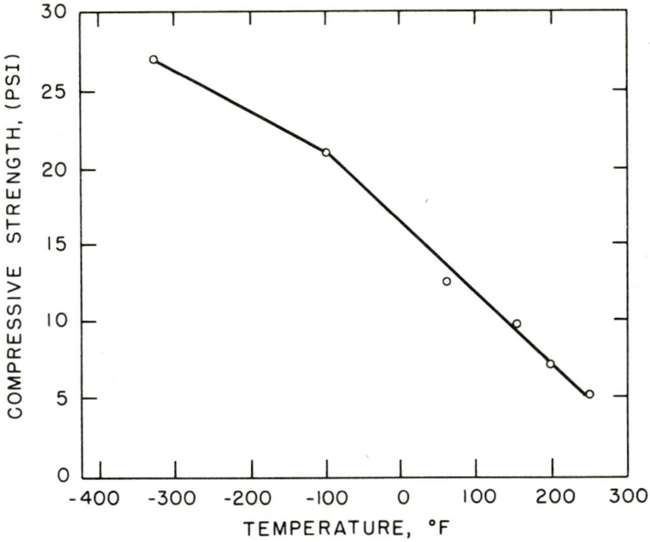

Figure 7-19. Strength versus temperature of polyurethane foam. Compressive strength at yield versus temperature for 1.5 lb/ft^3 polyether urethane (ref 56).

Electrical Properties Electrical insulation, like thermal insulation, depends primarily on the large volume of gas rather than the small volume of solid plastic in the plastic foam. Foam filled with air approaches the low dielectric constant and dissipation factor of air itself, and the function of the thin plastic walls is simply to keep the air

Table 7-3. Specific Heats of Plastic Foams (ref 64)

Foam composition	Temp., °F	Specific heat, Btu lb^{-1} °F^{-1}
Cellulose acetate—air	72	0.36
Phenol-formaldehyde—air	72	0.38
Polyethylene—air	72	0.5
Polystyrene—air	72	0.29
	40	0.27
	—	0.27
Silicone—air	72	0.31
Urethane—air	72	0.29
Urea-formaldehyde—air	72	0.40

Table 7-4. Coefficient of Linear Thermal Expansion of Plastic Foams at Low Temperatures (ref 64)

Foam type	Density, lb/ft^3	Temperature range, °F	Coefficient of linear thermal expansion, °F^{-1} × 10^{-5}
Molded polystyrene	>1.25	−148 to 32	5.0
Urethane	5.0	−148 to 32	3.9
Urethane (polyether)	—	−320 to 75	3.0
Urethane (polyester)	—	−320 to 75	2.2
Epoxy	—	−148 to 32	2.5

in place. Closed-cell foams of low polarity are particularly desirable to reduce water vapor transmission as well (Table 7-5).

Optical Properties Gas bubbles in a plastic foam have much lower density and therefore lower refractive index than the solid plastic cell walls. Therefore light passing through the foam is refracted and scattered every time it enters or leaves another cell (ref 67). This makes plastic foams much less transparent than solid plastics, generally translucent to opaque (ref 68). Light transmission is generally inverse to thickness as well (Figure 7-20). Correspondingly plastic foams require

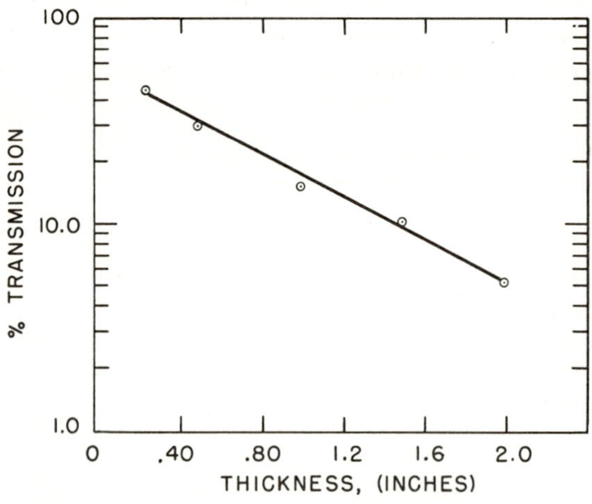

Figure 7-20. Light transmission through extruded polystyrene foam, 1.8 lb/ft^3, .04 to .08-in. cell diameter (ref 68).

Table 7-5. Electrical Properties of Plastic Foams (ref 66)

Plastic phase	Density, lb/ft³	Dielectric constant				Dissipation factor (× 10^{-4})					
		10^3cps	10^4cps	10^5cps	10^9cps	10^3cps	10^4cps	10^5cps	10^6cps	10^8cps	10^9cps
Cellulose acetate	6–7	1.10—1.12				20 to 30					
Epoxy	5.0	2.0				50					
Polyethylene	32	1.48	1.49	1.49	1.49	3.3	3.3	3.3	3.3	3.3	3.3
	2.0				1.05						2.0
	18				1.575						33
Polystyrene	1.5*	1.05				<5.0	<5.0				
	1.5*	1.017				1.0					
	9.8*				1.161						1.7
	4.3†				1.077**						6.0**
	1.9†			<1.05	<1.05			<4.0			
	2.8†			1.07	1.07			<4.0			
	4.3†			1.07	1.07			<4.0			
Silicone	3.5			1.09				28			
Polyether-urethane	9.3				1.193**						38.0**
	9.8				1.206**						35.0**
	1.5–3.0		1.05					3–13			

* Molded polystyrene foam.
† Extruded polystyrene foam.
** Measured at 24 × 10^9 cps.

more colorant than solid plastics, because the foam is opaque and the colored phase is more dilute (ref 69).

CHEMICAL PROPERTIES Chemical properties of plastic foams depend primarily upon the polymers themselves. In fact, use of special blowing gases such as fluorocarbons is limited to polymers which can resist their solvent action, such as polyurethane foams. In the plastic foams, expansion stresses created during foaming, and the high surface area of the cell walls, combine to make solvent attack somewhat more serious than on the solid plastic (ref 70).

Water absorption and transmission are inverse to density, and are greatly retarded in closed-cell structures (ref 71).

PRICE Plastics economics generally recognizes that cost per unit weight is less important than cost per unit volume. This consideration becomes critically important in plastic foams. If a higher-cost material—and even a higher-cost foaming process—can yield useful properties at lower density, the overall economics may still be highly favorable (Figure 7-21). Foaming to very low density is an obvious way to reduce costs in many cases, but can easily exceed the processor's ability and produce poor structures and poor properties. Experience has shown that this does not permanently represent an absolute lower limit of useful density. Rather, as foam processing techniques improve, this lower limit decreases steadily; and lower and lower foam densities of good structure may be expected in the future.

Composites

Most polymeric materials are not used in pure homogeneous form, but are mixed with incompatible solid powders or fibers before fabrication into end products. These are most generally called fillers. When they improve mechanical properties they are generally called reinforcements.

Although the original use of fillers and reinforcements began and developed in a practical and empirical way, there is now a growing theoretical interest in the broader concept of composites in general. In the most general terms, these are finely-dispersed mixtures of two incompatible phases, one of high modulus, the other of low modulus (ref 73). In most commercial polymeric products, a high-modulus filler or reinforcement is dispersed in a low-modulus polymeric matrix.

The effects of the dispersed phase upon the mechanical properties

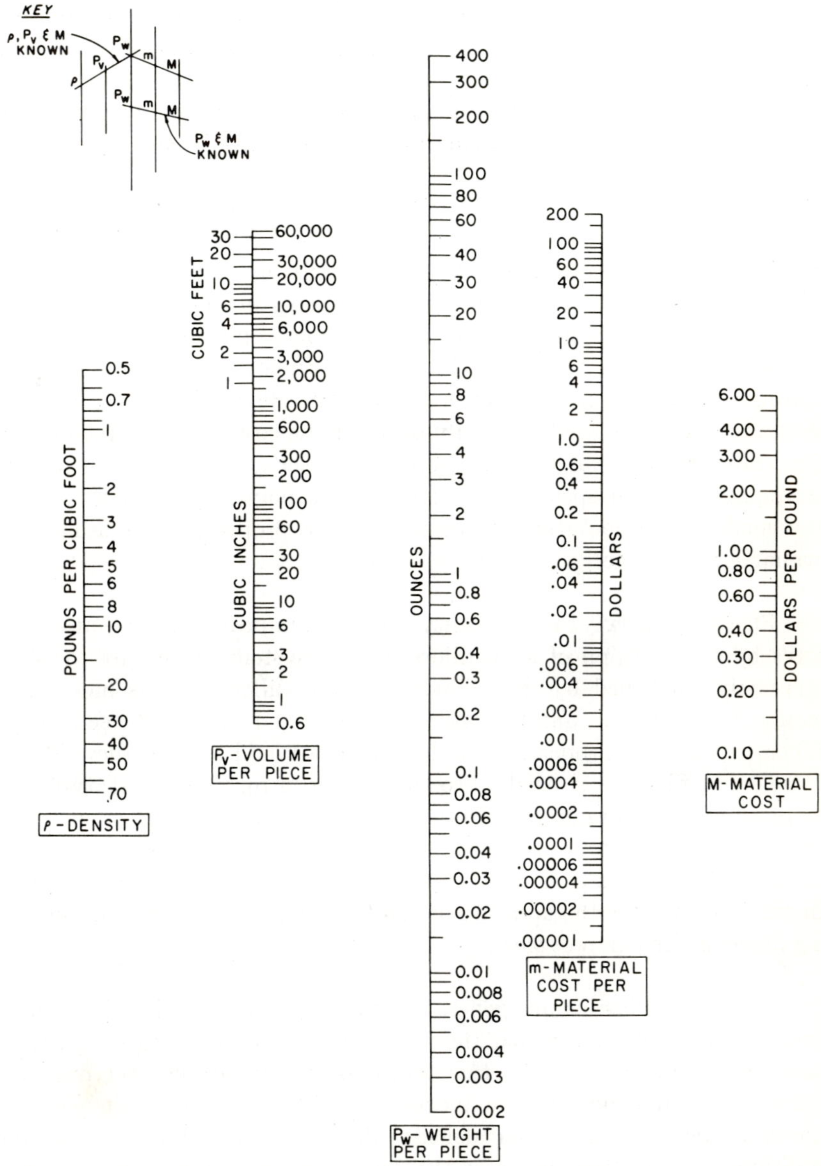

Figure 7-21. Economic nomograph for plastic foams (ref 72).

of the composite depend upon the size and shape of the dispersed particles, their surface area, and ultimately upon the mechanical or chemical bonding between the dispersed phase and the polymeric matrix at the interface at which they meet. In most cases the properties of composites represent an advantageous combination of the best properties of each component.

For convenience, the high-modulus dispersed phase may be classified as (1) Extenders, (2) Reinforcing Pigments, and (3) Fibers.

Extenders

When an inorganic powder is mixed with an organic polymer, the two are generally incompatible, and the inorganic powder remains as a finely-dispersed solid phase in the polymeric matrix. In most cases it does not improve the strength of the polymer, and is referred to as a simple filler or extender. It is a popular fallacy that such low-cost inorganic fillers are used primarily to lower the cost of the total composition. This is not usually true. The compounder generally selects a filler to improve certain specific properties of his polymer, and cost is only a secondary consideration. Effects on other properties are usually of primary importance.

PROCESSABILITY Processability may be improved in a variety of ways by the addition of solid fillers. Liquid systems which are too fluid are easily thickened by the addition of solid fillers; this is most commonly practiced in unsaturated polyesters and in paint technology. Where thixotropic properties are desired in a liquid system, the addition of selected fillers can easily produce and control this, most typically in plastisols and organosols. Polymers which are too nervy and rough during processing may be smoothed tremendously by the addition of proper fillers; this is particularly true in extrusion of rubber for example. In many such ways the addition of solid fillers is a great help in improving processability in polymers.

DENSITY Plastics are generally much lighter than the older metal and ceramic materials which they replace; and in some cases, particularly in consumer products, this lightness is undesirable and produces a feeling of flimsiness in the plastic product. Solid inorganic powders generally have much higher density than organic polymers; and the addition of such fillers increases the density of the filled polymeric composition, producing a desirable increase in overall density of the final product.

MECHANICAL PROPERTIES *Modulus* In a pure polymeric material, each segment of the polymer molecule has a certain relative freedom to rotate and migrate, and is restricted only by the equally free rotation and migration of adjacent similar molecular segments. In a filled polymer, on the other hand, some polymer molecules lie directly adjacent to high-modulus particles of filler, which have relatively no mobility at all. A polymer molecule lying near such a rigid surface is restricted to one-half of its normal ability to rotate and migrate. Solid high-modulus inorganic fillers thus produce a significant increase in the modulus of the filled polymer (ref 35) (Figure 7-22); thus $G = G_1v_1 + AG_2v_2$, where G is the shear modulus of the filled polymer, G_1 and G_2 are the moduli of pure polymer and filler respectively, v_1 and v_2 are the corresponding volume fractions, and A is called the adhesion factor and varies between 0 and 1 (ref 75). Along with this fundamental increase in modulus, fillers therefore also produce practical improvements in hardness and in resistance to indentation of the plastic product. At very high loading, the filler particles may actually rub against each other, producing broad transition regions and high damping of mechanical motion (ref 76).

Strength Most inorganic powders, whether crystalline or glassy, have sharp jagged edges which, combined with their high modulus, act as stress-concentrators in the polymeric matrix. When mechanical stress is applied to the polymer, it concentrates at such sharp points and causes failure and breakage. Thus many simple fillers or extenders actually lower the ultimate strength and elongation properties of the polymeric material (ref 35). Such loss of ultimate properties is generally undesirable; and use of solid fillers must be restricted to minimize such losses, or at least keep them within reasonable bounds. Since such plastics are not generally used up to the point of ultimate failure, this is generally quite readily accomplished.

Complex Mechanical Properties More complex mechanical properties, such as lubricity and abrasion resistance, may be profoundly affected by the choice of the proper solid fillers. Addition of graphite, molybdenum disulfide, and more recently polytetrafluoroethylene powders, are highly effective in increasing the lubricity of the polymer,

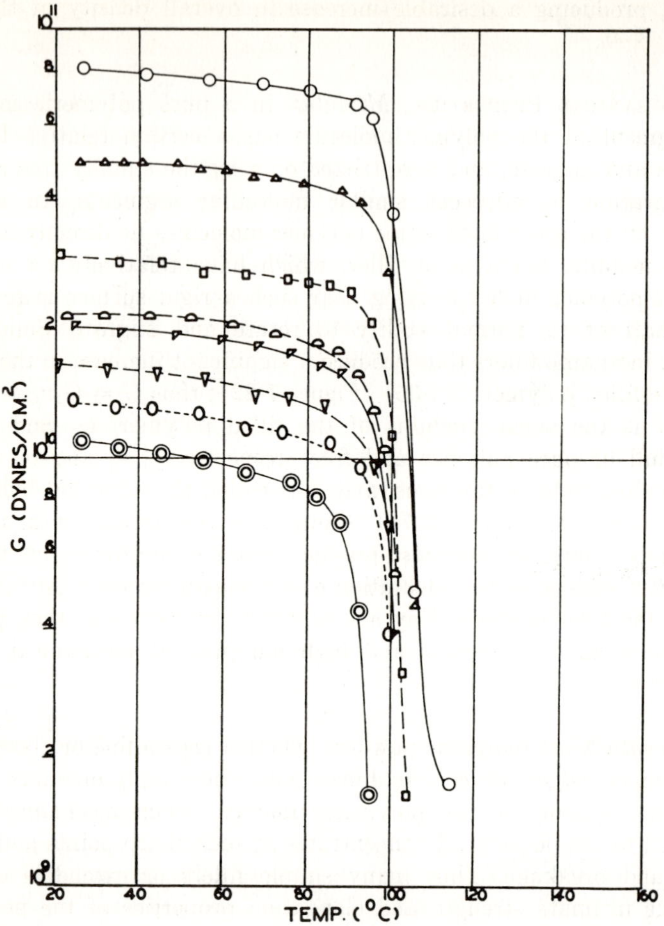

Figure 7-22. Effect of fillers upon modulus of polystyrene (ref 74). Shear modulus versus temperature: ◎ Polystyrene control. ▽ 20% mica. △ 40% mica. 0 20% calcium carbonate. ⌓ 40% asbestos. ○ 60% mica. ▽ 20% asbestos. □ 60% asbestos.

and producing the highly self-lubricating surface which is very useful in mechanical applications such as gears, bearings, and cams.

On the other hand, abrasion resistance of polymers is often greatly improved by the addition of considerable quantities of hard high-modulus solid fillers. In extreme cases the polymeric matrix may simply serve as a binder for the mounting of abrasive particles, as in the use of epoxy and phenolic resins as binders for friction surfaces and abrasive materials.

THERMAL PROPERTIES Since the presence of solid fillers restricts the mobility of adjacent polymer molecules, it is not surprising that such solid fillers frequently increase the heat deflection temperature of the plastic composition somewhat. Highly conductive fillers, such as graphite and powdered metals, produce considerable improvement in thermal conductivity of the usually highly insulating polymeric material. Dense fillers may also produce considerable increase in specific heat of the total composition, which can be useful in heat-sink applications.

The most important effect of fillers on thermal properties, however, is their ability to reduce the coefficient of thermal expansion. As compared with metals and ceramics, organic polymers generally expand and contract about ten times more rapidly with increases and decreases in temperature. Particularly when plastics are used in combination with such metal or ceramic materials, this difference in coefficient of thermal expansion produces serious stresses and frequent failures. Addition of powdered inorganic fillers to polymers produces a proportional decrease in the overall coefficient of thermal expansion of the total composition, and thus brings it much closer to the desirable low values observed for metals and ceramics (ref 77,78).

ELECTRICAL PROPERTIES Carbon black, graphite, and powdered metal fillers have occasionally been used to produce electrical conductivity through plastics (ref 79,80). Polymers are such excellent insulators, however, that it is generally believed essential for the conductive filler particles to make actual contact with each other throughout the material in order to conduct electricity effectively from one side to the other. Such compositions are primarily used to prevent the development of static charges on nonconductive polymeric products such as rubber tires and moving belts.

OPTICAL PROPERTIES Inorganic fillers generally have much higher densities, and therefore higher refractive indexes, than organic polymers (Table 7-6). Light passing through a filled polymer is refracted, dispersed, and scattered, making it translucent to opaque (ref 21). Thus fillers can not be used in a transparent product; but they are often used purposely to produce opacity, particularly in the formulation of paints and in protection against ultraviolet degradation in weathering. Conversely, use of a filler with a refractive index close to that of the polymeric matrix produces very slight scattering and very little opacification, and may be useful in applications such as the use of silica as a flatting agent to reduce gloss in coatings.

One major application of the effect of fillers on optical properties is in the production of opaque colors. Some fillers are used in large amount for their effect on other properties, and just incidentally produce a good uniform color in addition, for example calcium carbonate whites and carbon blacks. The full range of colors used in most plastics and coatings however, is obtained by very small concentrations of very-fine-particle-size high-surface-area highly-efficient colored pigments. These must be chosen for dispersability in the polymer, stability during processing and use, and absence of any deleterious interactions with the polymer, particularly chemical degradation reactions. Their effects on other properties are generally negligible, because they are used in such low concentrations.

Table 7-6. Refractive Indexes of Inorganic Fillers (ref 81)

Filler	Index
Titanium dioxide	2.5
Zirconium oxide	2.4
Zinc sulfide	2.4
Antimony oxide	2.2
Zinc oxide	2.01
Basic carbonate white lead	2.00
Basic sulfate white lead	1.9
Barytes	1.64
Calcium sulfate	1.6
Calcium carbonate (chalk)	1.6
Talc	1.57
China clay	1.56
Silica (quartz)	1.55
Diatomaceous earth	1.43

PRICE While simple fillers and extenders are chosen primarily for their beneficial effect on properties such as these, a secondary consideration in all cases and a primary consideration in some cases is the low cost of most simple inorganic fillers, frequently lowering the overall cost of the total final plastic composition. Many common fillers are so low in price that, even with their higher density, they still reduce the overall cost of the plastic product. When a filler can be chosen both to improve the properties and to reduce the cost, the overall effect is highly desirable. When an occasional compounder chooses a low-cost filler and uses a high concentration primarily to reduce cost, the results can be disastrous.

In general, it has been observed that rubbery types of polymers have the highest tolerance for high loading with solid fillers before they show any major deterioration of properties. This is probably because the mobility of the rubbery molecules permits them to conform more closely to the surface of the filler particles, and to absorb and disperse the mechanical stresses at the interface which might otherwise cause failure.

It has also been observed that the deteriorating effects of solid fillers may be mitigated, delayed, or minimized by treatment of the filler surface to produce a more compatible interface between inorganic filler and organic polymer. This is often readily accomplished by treatment with reactive organosilanes (ref 82). The reactive organosilane probably produces a more gradual transitional interface, across which mechanical stresses can be transferred more readily from one phase to the other, preventing their separation and failure.

Reinforcing Pigments

In certain special systems, the addition of a solid inorganic powder to an organic polymer actually improves the overall mechanical properties. The filler is then called a reinforcing filler or reinforcing pigment. The most significant effects are obtained by adding carbon black to rubber. Carbon blacks ranging from 15 mμ to 275 mμ in size are added to diene rubbers in high concentrations to improve their tensile strength and resistance to tearing and abrasion (ref 83–85). The effect applies to noncrystallizing elastomers in general, particularly silicones (ref 86), and to other fillers such as silica (ref 85–88).

The mechanism of reinforcement may be visualized in various ways. Fine particle size, high surface area, and good dispersion of the filler in the polymer should all contribute to a greater degree of interfacial interaction, and fine-particle-size fillers generally give the greatest increase in modulus and strength (ref 84). Some of the more effective carbon blacks and silicas appear to form chain-like agglomerates rather than individual particles, and these chain-like structures may contribute considerably to the overall strength of the composition. Secondary valence forces are probably formed between the π electrons in the carbon black and the π electrons in the double bonds of diene rubbers. The rubber may even form primary covalent bonds with the carbon black (ref 89). Vulcanization may serve to form bonds between the rubber and the carbon black (ref 90). Any or all of these mechanisms may be important in the reinforcing effects which are observed.

Carbon black fillers act like vulcanized cross-links in their improvement of rubbery properties (ref 35). Ultimate failure in elastomers is a tearing process, and the presence of fillers serves to block the growth of such a tear, giving high ultimate properties (ref 91).

Reinforcement by pigments has been most successful in elastomers, much less successful in conventional plastics materials. The reasons for this difference may depend upon some of the mechanisms involved, such as those discussed above. The development of effective reinforcement for plastics would certainly be highly desirable.

Fibers

Dispersion of fibers in a polymeric matrix can produce tremendous improvements in mechanical and thermal properties.

MECHANISM When mechanical stress initiates separation at a flaw in a low-modulus polymeric matrix, stress concentration at the flaw causes the separation to move rapidly through the sample and cause complete failure. Incorporation of fibrous reinforcement changes the phenomenon completely. When the stress moving through the low-modulus polymeric matrix reaches the high-modulus fiber, it is transmitted throughout the fiber instantaneously, and thus dispersed over its entire length. Furthermore, the high length to diameter ratio of the fiber gives it a high surface to volume ratio which produces further dispersion of the mechanical stress over this entire surface area. Thus when the stress reenters the polymeric matrix at all points over this interfacial area, it is so dispersed and so diluted as to be relatively harmless and ineffective (ref 16). This mechanism (1) disperses mechanical stress throughout the sample to prevent the initiation of failure, and (2) stops the propagation of a crack once failure has begun.

FABRICS The most dramatic improvements in properties are observed when glass fibers are used to reinforce epoxy and polyester resins (ref 92). Glass-cloth-reinforced plastics are generally superior to metals in strength to weight ratio, but remain softer than steel in their modulus properties (Table 7-7). Filament-wound plastics benefit particularly from their highly anisotrophic structure, and may equal or exceed steel in strength to volume ratio. Perhaps most dramatic of all may be the new three-dimensional fabric-reinforced plastics (ref 94), but these are extremely difficult and costly to produce.

SHORT FIBERS Proceeding in the other direction toward easier processing techniques, the use of chopped glass mat provides easier processing but lower properties than long fibers (Table 7-7). Much older and longer-established are the uses of organic fibers such as α-cellulose in urea and melamine resins, wood flour in phenolic resins, and asbestos in vinyls and other polymers. In all of these cases the fiber structure

Table 7-7. Properties of Metals vs Glass-Fiber-Reinforced Plastics (ref 93)

				Glass fiber-reinforced resin		
Property	High-strength steel	Aluminum alloy	Titanium	Mat laminate	Cloth laminate	Filament-wound
Density, lb/in.^{3}	.285	.041	.163	.055	.062	.075–.080
Ultimate strength $\times$ 10^3, psi	250	80	150	30	90	250–350
Impact strength, ft-lb in.				20	60–80	60–80
Modulus of elasticity $\times$ 10^5, psi	300	110	160	8–20	18–50	50–70
Thermal conductivity	105	512		13	15	
Strength/density $\times$ 10^4, in.	88	88	92	54	145	310–430

provides a mechanism for dispersing concentrated mechanical stresses, and thus reducing the possibility of failure in the plastic matrix.

BONDING BETWEEN FIBER AND RESIN In the use of reactive fibers such as cellulose and nylon, there is the additional possibility that the thermosetting polymer may actually form primary covalent bonds to the fiber and provide additional strength in this manner. In the use of glass fibers, such reactivity is not naturally available, and much of the early use of glass-fiber reinforcement resulted in failure at the interface between the inorganic glass fiber and the organic polymer. This has generally been remedied by treating the surface of the glass fiber with reactive organosilanes or methacrylato chromates. This treatment applies an organic surface covalently bonded to the inorganic fiber beneath, and much more homogeneously compatible with the organic polymer above the interface. In many cases the organosilane is chosen to provide additional reactivity with the organic polymer, and thus permit the possibility of direct covalent bonding from the glass fiber through the reactive interface to the thermosetting polymer on the other side. Proper use of such reactive finishes greatly improves the strength of the interfacial bond and the properties of the overall reinforced plastic (ref 95,96).

REINFORCED THERMOPLASTICS More recently, short glass fibers have been incorporated into thermoplastic polymers to improve their properties as well. Although these fibers cause problems in processability, and are often too short to produce any major improvement in impact strength, they can produce significant increase in modulus, strength, and creep resistance (ref 97,98).

Possibly their most significant beneficial effect is their ability to reduce the coefficient of thermal expansion. This is probably due to two different mechanisms, acting to reinforce each other: (1) The glass fiber itself has a much lower coefficient of expansion than the polymeric matrix, and thus reduces the average coefficient of expansion in the total composition. (2) Furthermore the refusal of the glass fibers to expand and contract sharply with temperature may actually provide a mechanical restraint on the dimensional changes of the low-modulus polymeric matrix. In both of these ways, reinforcement by even very short glass fibers serves to bring the coefficient of expansion of polymers much closer to the desirable low values in metals and ceramics.

OPTIMUM FIBER LENGTH The optimum length of these fibers has still not been determined. While longer fibers will probably give greatest strength and particularly impact resistance, shorter fibers will generally be easier to process. Some research studies even suggest that the shorter fibers may give the best properties. In extreme cases very short microscopic fibers or "whiskers," such as single crystals of metallic and metal oxide fibers, are proving highly desirable, particularly in the development of new materials of extremely high modulus and heat resistance (ref 99); and despite their extremely short length, they may still be used to provide tremendous improvement in plastic properties. The ultimate significance of fiber length is therefore far from finally determined.

7-4 MACROSTRUCTURE: MACROSCOPIC DIMENSIONS OF INDIVIDUAL STRUCTURES

Scientists generally study the properties of polymeric materials as if they were independent of the size and shape of the total product.

Thus processability is often described in terms of temperature, pressure, time, and viscosity. Modulus and strength measurements are divided by cross-sectional area and reported as if they were independent of the size of the product. Low-temperature embrittlement, heat deflection, and thermal degradation are likewise reported in absolute numbers. Surface and volume resistivity are distinguished from each other, but each is reported in absolute units. Optical clarity is usually discussed without reference to the thickness of the sample through which light is being transmitted. Chemical resistance is generally discussed without reference to the amount of surface exposed. Even price is all too often judged without reference to the quantity or form in which the polymeric material will be delivered. It is true that the most precise tests always specify the exact dimensions of the test specimen, but the significance of these dimensions is generally not considered when the test results are extrapolated for use in practical applications.

Actually most properties and behavior of polymeric products depend greatly upon the size and shape of the product, and thus of the sample on which these measurements are made. The quantitative relationships between properties, size, and dimensions are very poorly understood; but the qualitative effects should at least be noted with care, and may thus give an insight into the fundamental mechanisms and relationships involved.

The subject is most conveniently approached by considering the number of major dimensions in the polymeric product—One-Dimensional Fiber, Two-Dimensional Film, and Three-Dimensional Solid Product—and the ways in which the size and shape of these macrostructures affect their fundamental properties and practical behavior.

One-Dimensional Fiber

When the bulk of a polymeric material is reduced to the form of a fiber, either by biosynthesis or by mechanical extrusion, its macrostructure becomes essentially one-dimensional in nature, with profound effects upon most of its properties.

Flexibility

Most obvious of course is flexibility. Nearly all fibers are made from polymers which have a high modulus and are completely rigid in conventional solid three-dimensional form. Yet when reduced to the thinness of a single fiber, this same modulus is now applied to such a small cross-sectional area that only a negligible amount of force is required to bend the fiber, and it is therefore considered very flexible. On the other hand, a tensile stress along the one macrodimension of the fiber—its axial length—still encounters the full effect of its inherent modulus, and thus makes the fiber strong and tough and able to resist deformation and retain its shape.

Strength

The ultimate tensile strength of a polymer in fibrous form is generally much higher than it is for the same polymer in bulk three-dimensional form. This may be due to a number of mechanisms. Man-made fibers are highly stretched during production, creating a high degree of uniaxial orientation; and it is possible that even natural fibers are subjected to equally effective orientation effects during their production. This orientation produces parallel uniaxial alignment of molecules and crystallites within the fiber, and thus increases the mechanical strength in the axial direction. Tensile failure probably begins at microscopic or submicroscopic flaws in the fiber structure; the number of such flaws in a thin fiber will be much lower than the number of flaws in a three-dimensional bulk sample of polymer, and thus the frequency of failure will be much less (ref 100). Furthermore those flaws which remain will be axially oriented during the drawing of the fiber, and thus become much less sensitive to tensile stress than the randomly-arranged flaws in a three-dimensional bulk sample of polymer, where many of the flaws would suffer from the full transverse effects of the tensile stress (ref 101) (Section 5-3).

It is also possible that the surface of a polymeric material may be stronger than the interior (ref 100). This skin effect is still rather mysterious and obscure, but it may depend upon surface energy, case hardening, atmospheric oxidative cross-linking, or other mechanisms. In any case, the ratio of surface to volume in a one-dimensional fiber is much higher than in a three-dimensional polymeric product. And thus improvement of tensile strength due to the skin effect should be greatest in a one-dimensional fiber.

As a result of any or all of these mechanisms, the ultimate tensile strength of polymers in fibrous form is generally much higher than it is for the same polymers in bulk three-dimensional form.

Abrasion is a major cause of wear and failure in fiber products. While much of the problem may be related to fabric structure, fundamentally it arises from the inverse effect of orientation upon transverse strength (Section 5-3).

Fiber Structure

In actual practice, polymers are hardly ever used in the form of simple individual fibers (Table 7-8). Rather, these fibers are gathered into much more highly organized structures in all practical applications. This grouping of individual fibers into larger, more organized macrostructures goes through a succession of stages of increasing complexity; and in each of these stages, size, shape, and more complex features of design become increasingly important in ultimate properties and performance. The diameter of the individual fiber determines its stiffness and glitter (ref 103). The cross-section of the fiber may have many important effects on its ultimate behavior, and quite a variety of cross-sections have been observed in nature and produced commercially (Figure 7-23). The longitudinal shape of the individual fiber is generally modified from rod-like to wavy by crimping in order to improve softness, flexibility, and

Table 7-8. Supermolecular Structural Features of Individual Fibers (ref 102)

Dimensions	Structural features
10 cm	Fiber lengths
1 cm	
0.1 cm	Spacing of crimps in wool
100μ	Spacing of convolutions in cotton
10μ	Fiber diameters
1μ (10^{-4} cm)	Major structural features
	Molecular length
0.1μ	Laminae
100 Å	Microfibrils
	Crystalline and noncrystalline regions
10 Å	
1 Å (10^{-8} cm)	Molecular width
	Atoms: C,N,O / H
0.1 Å	

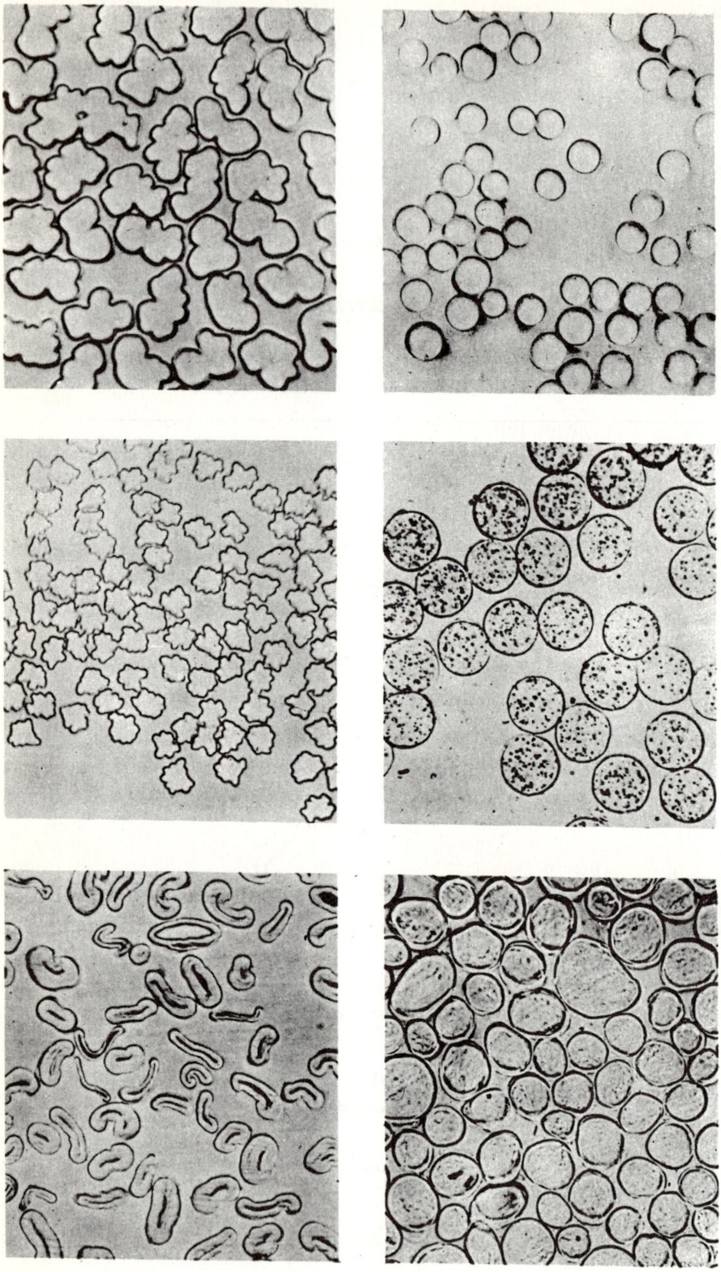

Figure 7-23. Cross-sections of common fibers (ref 104). Photomicrographs, $\times 250$, showing cross-sectional shapes of fibers. Top, left to right: cotton, viscose, acetate. Bottom, left to right: wool, 66 nylon (pigmented), poly(ethylene terephthalate).

resilience (ref 105). The overall length of the individual fiber has important effects upon properties: while long staple is generally easier to process, short staple provides more ends which in turn provide a feeling of warmth (ref 106).

Fabric Structure

Individual fibers may be gathered directly into nonwoven products such as paper, felt, and low-cost fabrics. In most fabrics, however, the individual fibers are too flimsy and unsubstantial to be handled as such, and are first gathered into strands, threads, or yarns. These in turn are converted into highly-structured fabrics by a wide variety of weaving patterns, which again have very significant effects upon performance. Thus, flexibility, drape, and permeability are all far greater than in films of the same composition, and these properties are obviously directly dependent upon fabric structure. Low packing density of the fibers in the yarn provides for easy tailoring (ref 105). Construction of yarn and fabric also affect abrasion resistance (ref 107), luster (ref 108), and comfort (ref 109). Going even one step further in complexity of construction, there is a growing tendency among fabric manufacturers to laminate two or more fabrics to each other to produce further improvement and refinement of properties and performance. These qualities cover a very broad range from the most fundamental and quantitative measures of polymer and fiber properties up to the most qualitative and psychological factors in ultimate performance (Table 7-9). Such properties and performance characteristics can actually be itemized and characterized in much greater detail (ref 111).

Two-Dimensional Film

When a bulk solid polymer is produced in the form of a thin film, this is essentially two-dimensional in macrostructure, and the resulting properties may be significantly different from those of the same polymer in conventional three-dimensional form. Equipment and techniques for producing and processing film are quite different from those used on solid bulk plastics.

Polymers which are rigid in bulk three-dimensional form are generally quite flexible in the form of thin film; and polymers which are soft and flexible in three-dimensional form may actually be soft, weak, flabby, and tacky in thin film form. While the total strength of a film depends directly upon its thickness, the unit strength of a film may be considerably higher than that of the bulk polymer (Table 7-10). This is particularly true when the films are stretched and oriented during production. It may also be due to the smaller number of microscopic and submicroscopic flaws which are present in a thinner sample of material, and/or to skin effects such as case-hardening, atmospheric oxidative cross-linking, and surface energy.

Table 7-9. Properties and Performance of Textile Fabrics (ref 110). Reprinted by permission of Fairchild Book Division, Fairchild Publications, New York, N.Y.

Abrasion resistance	Impact strength
Bleaching	Ironability
Bulking power	Knittability
Crepeing	Launderability
Crease retention	Luster
Covering power	Moisture regain
Comfort factor	Pilling resistance
Cross sectional shape	Refractive index
Dermatitic properties	Resilience
Dimensional stability	Resistance to insects
Dry cleaning behavior	Resistance to mold, rot, and bacteria
Drying rate	Shrinkage
Durability	Sizing behavior
Elastic recovery	Static characteristics
Electrical properties	Stiffness
Fatigue resistance	Stress/strain characteristics
Finishing behavior	Sunlight resistance
Fibrillation	Toxicological properties
Fulling	Uniformity
Flammability	Warmth
Flex resistance	Weaving behavior
Heat resistance	Wicking behavior
Heat settability	Wrinkle resistance
Hysteresis	Wrinkle recovery

Fairly good clarity may be achieved in thin films, even though it is quite unattainable in thick sections of the same polymers. This is due not only to the thickness of material which the light must penetrate, but also to the ease of quenching and stretching during production, which produces much finer crystal size.

Chemical resistance of films may be lower than that of solid plastics, simply because they expose a much higher ratio of surface to volume. Permeability of liquids and gases through films is generally considered

Table 7-10. Effect of Film Thickness upon Properties (ref 124). Reprinted by permission of Modern Plastics Encyclopedia, McGraw-Hill, Inc.

	Cellulose acetate, light gages (1-mil)	Cellulose acetate, medium gages (10-mil)	Cellulose triacetate light gages (1-mil)	Cellulose triacetate, medium gages (10-mil)	Cellulose acetate, butyrate, light gages (1-mil)	Cellulose acetate, butyrate, medium gages (10-mil)
Thickness range, in	0.00088–0.002	0.005–0010	0.00080–0.002	0.003–0.020	0.0011–0.002	0.005–0.050
Maximum width, in.	40	40	40	40	40	40
Area factor[a]	22,000	2,160	21,200	2,140	23,300	2,310
Tensile strength,[b] p.s.i.	13,000–15,000	8,000–10,000	12,000–15,000	9,000–11,000	7,000–9,000	5,000–6,000
Elongation, percent	25–35	25–30	25–35	25–40	50–60	60–100
Bearing strength,[c] 1-mil thickness, wide points	40–60	250–300	50–70	300–350	40–70	170–220
Tearing strength,[d] gm.	5–10	150–200	5–10	100–300	5–10	250–350
Folding endurance[e]	500–600	5–10	1,000–2,000	5–25	800–1,200	20–30
Water vapor permeability[f]	1,200	170	790	140	790	190
Dielectric strength, volts/mil	4,400	1,400	5,000	1,400	5,500	1,600

[a]—Square in. per lb. of film. Figures may vary with resin plasticizer ratio. [b]ASTM D882, procedure B. [c]—ASTM D774. [d]—Elmendorf test Figures represent pull required to continue tear of film after starting. Force required to start is too high to measure on same equipment and therefore is not normally determined. [e]—Using the M.I.T. Folding Endurance Tester. [f]—ASTM D697; values are in grams/24 hr./m.^{2} at 25°C.

to be inverse to thickness. Actually the relationship is much more complex because of skin effects and because of the separate effects of sorption and desorption at the surfaces vs diffusion through the interior of the film.

In all of these and many more ways, the two-dimensional macrostructure of films produces very different properties from the bulk form of the same polymer.

Three-Dimensional Solid Products

Even among three-dimensional solid products, properties are often highly dependent upon the total size and shape of the material.

Size and Shape

Processes such as molding and casting become increasingly difficult in larger sizes. Melt fracture during extrusion is directly related to size and shape of extrudate (ref 112). Mechanical properties such as modulus and strength are generally reported in specific values per unit area, and total properties of the entire product are therefore assumed to be directly proportional to this area; actually this relationship is hardly well proved, and requires considerable further study. Thermal properties are generally reported independently of the size and shape of the sample, even though in critical studies it may be recognized that values such as brittle temperature and deflection temperature do depend upon these variables. While most electrical properties are reported as if independent of sample size and shape, electrical breakdown is clearly related to thickness of the sample (Figures 7-24, 7-25). Chemical properties, although generally reported as if they were characteristic of the bulk material, actually depend upon attack at the surface and therefore depend to a large extent upon the relative amount of surface to volume exposed. Even a simple property like price is inversely related to quantity of material traded.

Surface

Skin effects, already noted as predominant in fibers and films, remain very important even in three-dimensional solid products. Beneficial effects of the surface skin on mechanical strength and chemical resistance are sometimes noted. These may be due to surface energy, orientation of the surface during processing such as injection molding, surface quenching when thermoplastics are injected into cold molds, and high cure of thermosetting materials at the surface in hot molds. These mechanisms may explain the beneficial effect of the surface on mechanical strength, friction, resistance to abrasion, and chemical resistance.

Conversely, many poor properties of three-dimensional solid products may often be ascribed largely to the skin effects, such as air-inhibition of polymerization and cure; surface oxidation and corrosion

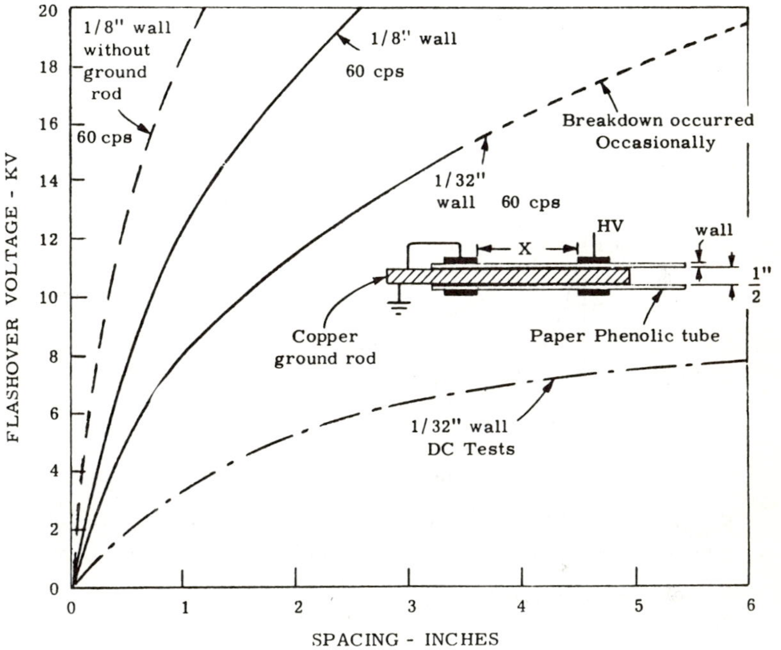

Figure 7-24. Effect of geometry upon flashover voltage of phenolic/paper tubes (ref 113).

in general; and the presence of irregularities, flaws, and scratches caused during handling and use. Strength properties are generally one-tenth to one-hundredth of those expected from theoretical calculations, and this discrepancy is generally explained by the presence of flaws, physical discontinuities, weak points, microheterogeneities, cracks, notches, embedded particles, or voids. Cracks as small as 1 μ in length and molecular in width are often postulated as the point at which failure begins. Mechanical stress on the total sample produces stored elastic energy in the sample, which is released by growth of these cracks, producing new surface and new surface energy. Probability of such flaws is directly related to the size of the specimen, which explains why larger samples fail at lower stress than smaller ones (ref 115,116).

Optical clarity also depends upon a smooth surface (ref 20).

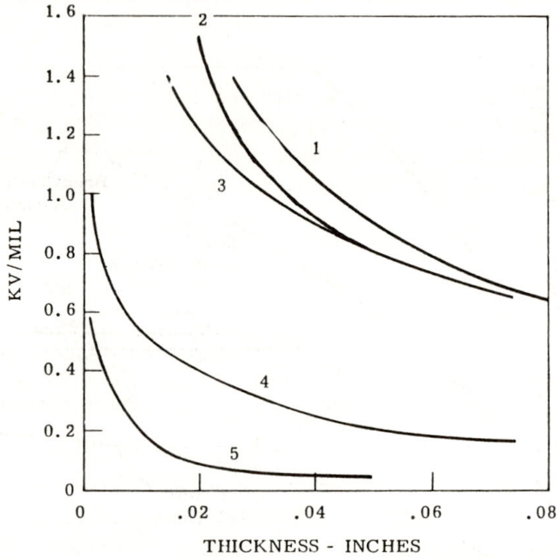

Figure 7-25. Effect of thickness upon dielectric breakdown (ref 114). Short-time breakdown strength versus thickness—comparison of materials.

1. Annealed polystyrene
2. Polyfluoroethylene-propylene (DuPont FEP)
3. Polyethylene
4. Paper-phenolic laminate
5. Wet paper-phenolic laminate

Overall Design

A major factor in the properties of three-dimensional solid products is the overall design of the product.

While rigidity can obviously be increased by increasing thickness, this also increases weight and cost. Alternatively, rigidity may be increased by the design of ribs, arches, or corrugations, without increasing the weight and cost of the final product.

Similar considerations apply to ultimate strength. Certain ultimate properties, particularly impact strength, are particularly sensitive to radius of curvature; a sharp notch concentrates mechanical stress, particularly under rapid impact, and accentuates the probability of brittle failure. Some plastics, particularly ductile materials, may be more notch sensitive than others; and this is an important factor to be considered in product design (ref 117).

Acoustic reverberation and dispersion depend largely upon the design of the surface. Design factors are equally important in determining the thermal and electrical properties of the total product.

In these and many more ways, the size and shape of total three-dimensional solid products are very important in determining their properties and performance, and deserve much greater and more detailed consideration than are generally given to them at present.

7-5 COMPLEX MULTICOMPONENT TOTAL MACROSTRUCTURES

The most specialized and most beneficial uses of polymeric products are often obtained by highly sophisticated and complex designs, often by combining polymeric products with other material. Several typical examples include (1) Membranes, (2) Functional Films, and (3) Laminates. These may be examined as typical of such highly sophisticated applications.

Membranes

Thin films or layers of polymeric material may be designed to produce precisely controlled differential permeability to different materials. This control may be achieved by chemical and/or physical design of the membrane.

Chemical Semipermeability

The most sophisticated, complex, and least understood are the biological membranes in both plant and animal cells, which control the biochemistry of life processes.

Synthetic membranes are finding increasing use in industrial processes. The most highly developed are probably the cellulose acetate membranes used for desalination of water by reverse osmosis. Here a membrane about 1000 Å thick provides the best balance between rapid and economical passage of water on the one hand vs high and efficient rejection of salt on the other hand. But since a 1000 Å membrane is too thin and weak to be self-supporting, the most successful commercial membranes provide a 4-mil porous cellulose acetate backing to give strength to the solid thin skin (Figure 7-26) (ref 118).

A wide variety of other membranes are being developed for potential commercial applications. In most of these, the membrane is essentially

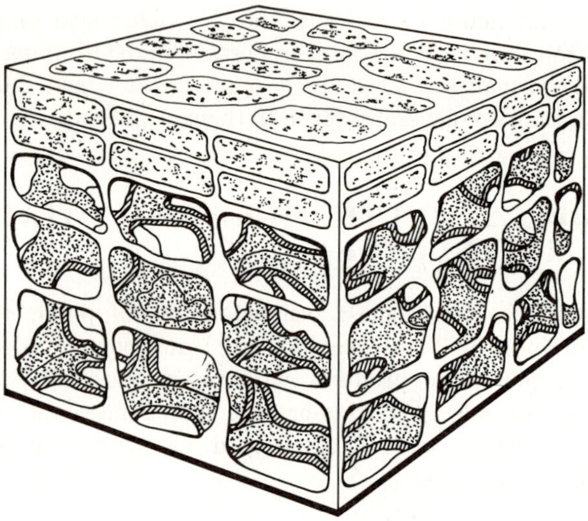

Figure 7-26. Schematic model of cellulose acetate asymmetric membrane used for desalination of water by reverse osmosis (ref 125).

solid and not porous in structure, and the control of differential permeability is achieved primarily by chemical composition. Some of the theories of permeation even here are based upon physical as well as chemical concepts.

Physical Porosity

Many membranes are designed with coarser physical porosity as well. In highly porous media such as paper and textiles, permeation is primarily by convection (ref 119). Filter paper is a typical example of such a porous semipermeable membrane operating primarily by physical mechanisms.

Battery separator plates, whose main function is mechanical separation of electrodes, combined with easy permeability by aqueous acid and salt, may be made by partial impregnation of phenolic resin into paper, enough to provide a chemically-resistant coating on the paper fibers, but not enough to destroy the physical porosity of the paper. Or they may be produced by incorporating tiny starch particles or salt crystals in a rigid vinyl sheet before forming, and then leached out of the sheet by boiling water after forming.

A dry blend of polyvinyl chloride resin particles, which have absorbed liquid plasticizer, may be heated and compressed just enough to fuse the particles to each other without completely closing the porous openings between them (ref 120).

These are typical of coarser membranes in which physical porosity is used primarily for separation of larger particle sizes.

Functional Films

While many polymers are produced in the form of unsupported films for packaging and other applications, two important fields of applications deposit these films on a substrate to combine certain functional properties of the film with the mechanical properties of the substrate: (1) Coatings, and (2) Adhesives.

Coatings

Many polymers are deposited as thin film bonded to a solid substrate, to produce specific effects. Decorative coatings, such as paints and inks, employ the polymer as an adhesive binder for the colored pigments which are primarily used to produce the decorative effect. Protective coatings use the polymer primarily to prevent corrosive attack upon the substrate.

Other coatings have a variety of specialized uses. One highly sophisticated application is microencapsulation, in which tiny particles of a material are covered by a protective coating of polymer, which keeps them isolated until use. The contents are then freed by breaking the protective coating during use (ref 121).

Adhesives

A wide variety of polymers are applied as a thin layer between two solid substrates, and employed primarily to bond and hold these two substrates close to each other. To be a good adhesive, a polymer must first have sufficient fluidity, polarity, and/or reactivity to flow and wet the surface of both substrates; and, when the bond has been completely formed, the polymer must have in addition sufficient cohesive strength to hold itself as well as the substrates together. Different theories of adhesion may emphasize mechanical effects, polarity, reactivity at the interface, or actual mixing between materials across the interface (ref 122).

Laminates

The best combination of properties for many specialized products is often obtained by laminating a film or sheet of polymeric material to films or sheets of other materials.

Polymer/Polymer

Lamination of two polymeric films to each other can often achieve a more desirable balance of properties than could ever have been obtained by modifying the structure of a single polymer film. One layer may provide mechanical properties while another provides controlled permeability, heat sealability, printability, fogging resistance, chemical resistance, or other important qualities. Occasionally laminates may involve three or more layers of different materials.

Polymer/Metal Foil

Lamination of polymeric films to metal foils can combine the electrical conductivity and impermeability of the metal foil with the heat-sealability, solvent-sealability, abrasion resistance, low conductivity, decorability, and corrosion resistance of the polymer. Laminations of polymer sheets with metal sheets are important in electrical applications and in rigid, high-strength, corrosion-resistant structures.

Polymer/Fiber

Polymers may also be laminated to paper or fabric, thus combining the strength and decorative effects of the paper or fabric with the impermeability, chemical resistance, and often abrasion resistance of the polymeric film. Combinations of polymer and paper provide outstanding performance in counter tops, table tops, and cabinets; and in electrical laminates. Combinations of polymer and wood produce plywood.

Sandwich Structures

Sandwich construction combines a high-modulus solid skin with a low-density highly-structured interior, such as a foam or honeycomb, to produce a maximum ratio of rigidity to weight. This is particularly useful in aircraft, and is finding growing application in building (ref 123).

In these and many other ways, the macrostructure of polymers may be combined with other structures to produce properties and performance for specific applications which could not have been obtained in any other way. The development of such complex multicomponent total macrostructures may offer the greatest promise for continued technical development and commercial growth, in the application of polymers in industrial and consumer markets.

REFERENCES

1. Billmeyer, F. W., Jr., *Textbook of Polymer Science,* Interscience, 1964, p 148.
2. Miller, M. L., *The Structure of Polymers,* Reinhold, 1966, p 512.
3. Alfrey, Turner, and Gurnee, E. F., *Organic Polymers,* Prentice-Hall, 1967, p 11.
4. *Ibid.,* p 9.
5. Miller, *op. cit.,* pp 512–15.
6. Billmeyer, *op. cit.,* p 150.
7. Baer, Eric, *Engineering Design for Plastics,* Reinhold, 1964, pp 774–75.
8. *Modern Plastics, 45* (1A), 19 (September, 1967).
9. Miller, *op. cit.,* pp 516–17.
10. *Ibid.,* p 515.
11. Nielsen, L. E., *Mechanical Properties of Polymers,* Reinhold, 1962, p 117.
12. *Ibid.,* p 187.
13. Billmeyer, *op. cit.,* p 166.
14. Morton, W. E., & Hearle, J. W. S., *Physical Properties of Textile Fibers,* Butterworths, 1962, pp 44–45.
15. Modern Plastics, *op. cit.,* 20.
16. Nielsen, *op. cit.,* p 125.
17. Billmeyer, *op. cit.,* pp 222–24.
18. Brydson, J. A., *Plastics Materials,* Iliffe, 1966, p 59.

19. Tobolsky, A. V., *Properties and Structure of Polymers,* Wiley, 1960, p 199.
20. Billmeyer, *op. cit.,* p 232.
21. Brydson, *p. cit.,* p 96.
22. Erath, E. H., & Spurr, R. A., *J. Polymer Sci., 35,* 391 (1959).
23. Erath, E. H., & Robinson, Morton, American Chemical Society Division of Organic Coatings & Plastics, *Preprints 23* (1), 395 (1963).
24. Alfrey, *op. cit.,* p 115.
25. Solomon, D. H., *The Chemistry of Organic Film Formers,* Wiley, 1967, pp 282–85.
26. Society of Plastics Industry, *Plastics Engineering Handbook,* 3rd Edition, Reinhold, 1960, Chapter 16.
27. Modern Plastics, *op. cit.,* 825–26.
28. *Ibid.,* 836–37.
29. Hercules, Inc., Technical Bulletins on Hi-Fax 1900.
30. Billmeyer, *op. cit.,* pp 422–23.
31. Nielsen, *op. cit.,* pp 255–56.
32. *Ibid.,* p 256.
33. Alfrey, *op. cit.,* Chapter 7.
34. *Ibid.,* p 99.
35. Nielsen, *op. cit.,* p 118.
36. Tobolsky, *op. cit.,* pp 81–82.
37. Miller, *op. cit.,* p 308.
38. *Ibid.,* p 483.
39. *Modern Plastics, op. cit.,* pp 20–21.
40. Nielsen, *op. cit.,* pp 131–34.
41. *Ibid.,* pp 172–73.
42. Tobolsky, *op. cit.,* p 82.
43. Nielsen, *op. cit.,* p 175.
44. *Ibid.,* p 174.
45. Allied Chemical Corporation, Technical Bulletins on "Source."
46. *Modern Plastics, op. cit.,* p 370.
47. Baer, *op. cit.,* p 995.
48. *Ibid.,* pp 1050–51.
49. *Ibid.,* p 997.
50. *Ibid.,* p 1012.
51. *Ibid.,* p 1010.
52. *Ibid.,* p 1011.
53. Alfrey, *op. cit.,* p 103.
54. Baer, *op. cit.,* p 1008.
55. *Ibid.,* p 1021.
56. *Ibid.,* p. 1013.
57. *Ibid.,* p 1023.
58. *Ibid.,* pp 1057–59.

59. Alfrey, *op. cit.*, p 105.
60. Baer, *op. cit.*, p 1025.
61. *Ibid.*, p 1030.
62. *Ibid.*, p 1048.
63. *Ibid.*, p 1014.
64. *Ibid.*, p 1034.
65. *Ibid.*, p 1038.
66. *Ibid.*, p 1037.
67. Brydson, *op. cit.*, p 96.
68. Baer, *op, cit.*, p 1043.
69. *Ibid.*, p 1044.
70. *Ibid.*, pp 1041–42.
71. *Ibid.*, pp 1039–41.
72. *Ibid.*, p 996.
73. Alfrey, *op. cit.*, pp 98–99.
74. Nielsen, *op. cit.*, p 190.
75. *Ibid.*, pp 189–91.
76. *Ibid.*, p 191.
77. *Ibid.*, p 189.
78. Volk, M. C., Lefforge, J. W., & Stetson, Russell, *Electrical Encapsulation,* Reinhold, 1962, p 23.
79. Alfrey, *op. cit.*, p 119.
80. Miller, *op. cit.*, p 671.
81. Schmidt, A. X., & Marlies, C. A., *Principles of High Polymer Theory & Practice,* McGraw-Hill, 1948, p 618.
82. Baer, *op. cit.*, pp 962–66.
83. Billmeyer, *op. cit.*, p 225.
84. Morton, Maurice, *Introduction to Rubber Technology,* Reinhold, 1959, pp 37–38.
85. Schmidt, *op. cit.*, pp 568–71.
86. Meals, R. N., & Lewis, F. M., *Silicones,* Reinhold, 1959, pp 36–37.
87. Morton, Maurice, *op. cit.*, pp 234–43.
88. Schmidt, *op. cit.*, pp 581–82.
89. Billmeyer, *op. cit.*, p 545.
90. Meares, Patrick, *Polymers: Structure and Bulk Properties,* Van Nostrand, 1965, pp 232–33.
91. Alfrey, *op. cit.*, p 88.
92. *Ibid.*, pp 100–102.
93. Baer, *op. cit.*, p 871.
94. Barton, R. S., *SPE Journal, 24* (5), 31 (May, 1968).
95. Baer, *op. cit.*, p 962.
96. *Ibid.*, pp 966–69.
97. Fiberfil Company, Technical Bulletins.
98. Liquid Nitrogen Processing Company, Technical Bulletins.

99. *Modern Plastics, op. cit.,* 599.
100. Schmidt, *op. cit.,* pp 366–68.
101. Dietz, A. G. H., *International Science & Technology,* August, 1964, p 62.
102. Morton, & Hearle, *op. cit.,* p 43.
103. *Ibid.,* pp 120–21.
104. Billmeyer, *op. cit.,* p 514.
105. McFarlane, S. B., *Technology of Synthetic Fibers,* Fairchild Publications, 1953, p 279.
106. Morton, & Hearle, *op. cit.,* pp 76–77.
107. *Ibid.,* pp 401–02.
108. McFarlane, *op. cit.,* p 278.
109. Billmeyer, *op. cit.,* p 512.
110. McFarlane, *op. cit.,* p 75.
111. *Ibid.,* pp 256–60.
112. Miller, *op. cit.,* p 305.
113. Baer, *op. cit.,* p 459.
114. *Ibid.,* p 452.
115. Miller, *op. cit.,* pp 305–08.
116. Nielsen, *op. cit.,* pp 126–30.
117. *Ibid.,* pp 124–25.
118. Merten, Ulrich, *Desalination by Reverse Osmosis,* MIT Press, 1966.
119. Baer, *op. cit.,* p 623.
120. Rogers Corporation, Technical Bulletins on Poron.
121. National Cash Register Company, Technical Bulletins and Patents on Microencapsulation.
122. Golding, Brage, *Polymers and Resins,* Van Nostrand, 1959, pp 696–98.
123. Baer, *op. cit.,* p 1052.
124. *Modern Plastics, op. cit.,* 540.
125. U.S. Department of Interior, Office of Saline Water, *Saline Water Conversion Report for 1964,* p 32.

Part Six

CONCLUSION

Chapter Eight

Commercial Polymers: Their Properties and Applications

Up to this point, the relationship between polymer structure and properties has been analyzed most conveniently by studying the effect of each individual structural feature upon each individual property. In this analysis it has become obvious that often a number of structural features interact to determine a single property. It is equally obvious that the choice of a particular polymer for a particular application depends upon more than a single critical property, but rather upon a balance of important practical properties. Each of these subjects deserves a separate volume for itself. The present study may very well be summarized, however, by considering briefly how the structures of the major commercial polymers interact to determine their properties, and how their balance of important properties determines their practical applications.

For this review, the commercial polymers are conveniently divided into a series of fundamental structures, beginning with polyethylene, proceeding to the side-chain substituents of the vinyl polymers, and then to the main-chain substitution of oxygen, sulfur, nitrogen, and silicon. In each group it is best to study carefully the structure and properties of the most general and important member of the group, and then consider more briefly the effects of structural variations upon those properties. For the purposes of the present study, the commercial polymers have been grouped as follows:

1. Polyolefins
2. Diene elastomers
3. Styrene polymers
4. Vinyl halide polymers
5. Polyvinyl alcohol and its derivatives
6. Acrylic polymers
7. Polyvinyl pyrrolidone
8. Aliphatic polyethers
9. Aromatic polyethers
10. Polysulfide elastomers
11. Acetals
12. Polyesters
13. Polyurethanes and ureas
14. Polyamides
15. High-temperature heterocycles
16. Amino resins
17. Phenolic resins
18. Silicones

8-1 POLYOLEFINS

High-Density Polyethylene

The study begins most conveniently with linear high-density polyethylene, the most general and fundamental model for organic polymer structures. The simple chemical composition, derived directly from very low-cost petrochemical sources, provides a polymer whose low price may be its most important property in the tremendous markets where other properties are relatively unimportant (Figure 8-1) (ref 1). The saturated hydrocarbon composition, once it has been purified to remove catalyst impurities, is so nonpolar that it offers excellent electrical resistance and very low dielectric constant and loss; this finds major application in electrical wire, cable, and conduit (ref 2).

Next to its simple hydrocarbon composition, the most important structural feature of high-density polyethylene is the balance between (*a*) its high crystallinity and (*b*) its lack of polarity or any other type of intermolecular bonding. Thus the polymer melts sharply at a moderate temperature, providing easy processing by many techniques. Crystallinity gives the solid polymer reasonably good rigidity and strength, while weakness of any other intermolecular bonds still makes it somewhat softer and weaker than more polar or sterically hindered structures. The heat deflection temperature is reasonable but not high. Crystalline growth into spherulites can produce translucency and opacity, unless they are kept small by quenching or stretching during processing. Crystallinity produces high solvent resistance; but rigidity accentuates the danger of stress-cracking, unless a little comonomer is used to soften and flexibilize the structure somewhat. This balance of properties finds major application in bottles, closures, piping, and toys.

The flexibility of the polyethylene molecule is potentially an important feature of polymer structure, but its effect on solid properties is masked into insignificance by packing it into the highly-ordered crystalline lattice. In the molten state during processing, on the other hand, molecular flexibility plays an important role in providing the easy processing which, along with low price, makes polyethylene the optimum

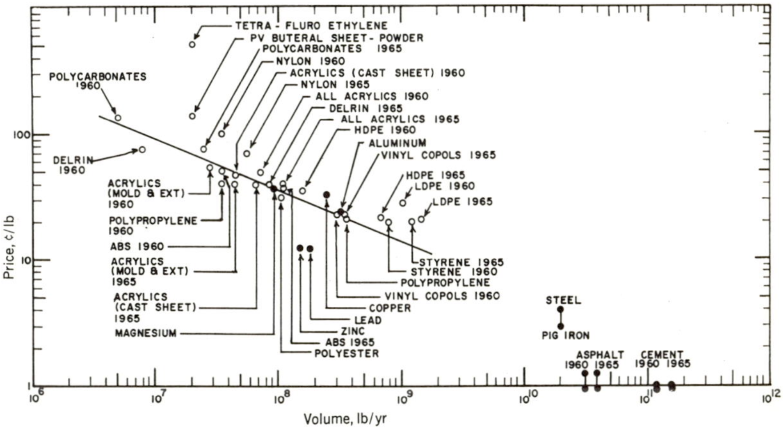

Figure 8-1. Price vs production of thermoplastic molding pellets (ref 1)

material for many major applications. From the theoretical point of view, the high molecular flexibility becomes apparent when the crystallinity of the solid state is reduced or eliminated by substitution, as discussed in the paragraphs which follow.

Low-Density Polyethylene

High-pressure free-radical polymerization of ethylene produces rather frequent short-chain branching, about 15 to 20 ethyl (and a few butyl) side-chains per 1000 carbon atoms in the main-chain (ref 3). These structural irregularities reduce crystallinity to intermediate values, permitting the main-chain to show more of its inherent flexibility, and thus making the polymer stiffly flexible in a leathery sort of way, greatly improving its impact strength, and retaining these properties down to the very low glass transition temperature of the large amorphous phase. Lower crystallinity, however, also lowers the melting and heat deflection temperatures to a level that limits usefulness in hot applications. Other than the effect of branching to reduce crystallinity, the structure and properties of low-density polyethylene generally resemble those of high-density polyethylene.

The other major difference is the lower price of the low-density polymer, primarily a result of its longer commercial history and larger production volume. Wherever price is a more important property than rigidity and hot strength, the low-density polymer is chosen first.

Thus it finds major applications in film form for packaging foods (produce, baked goods, meat, poultry, frozen foods, candy) and other products (clothing, crumb rubber), in diaper liners, as a vapor barrier in construction, and as liners for irrigation ditches and ponds in agriculture. Extruded coatings on paper and paperboard find major use in milk cartons and other packaging applications; and extruded coatings on other films provide heat-sealability and barrier qualities for many types of packaging. Extruded wire and cable insulation, pipe, and conduit are other major applications. Leading molded applications include bottles, closures, and toys.

Ionomers

Copolymerization of ethylene with moderate amounts of ionic comonomer, such as sodium methacrylate, introduces two major new structural features: (*a*) random comonomer units and (*b*) ionic intermolecular bonding. The random comonomer units reduce crystallinity sufficiently to produce transparency. Low crystallinity also reduces rigid packing of polymer molecules; but high intermolecular bonding takes its place to continue to restrict molecular mobility, thus retaining some of the rigidity and all of the strength of the highly crystalline homopolymer. The ionic cross-links have surprising thermal lability, permitting thermoplastic processing; and they offer high adhesion, particularly to polar substrates such as metals. Introduction of more expensive comonomer and post-treatment naturally raise the price, restricting such materials to applications where their improved properties have practical significance. Initial applications have developed in skin packaging and blister packaging, extrusion-coated foil packaging, adhesives for laminating metals, and blow-molded bottles (ref 4).

Ethylene Copolymers

Copolymerization of ethylene with moderate to considerable amounts of ester comonomers, particularly vinyl acetate and ethyl acrylate, reduces crystallinity markedly and increases polarity moderately. The low crystallinity makes the copolymers more transparent, soft, and resilient than polyethylene, without the undesirable fugitive plasticizers of vinyls, and thus finds growing application in food, beverage, and medical tubing, wire and cable insulation, syringes, squeeze bulbs and droppers, and squeeze toys. The ester groups introduce polarity, which produces sufficient dielectric loss for high-frequency sealing, and sufficient compatibility and adhesion for blending with waxes for coating paper for packaging. While price is naturally higher than for polyethylene due to more expensive comonomers, economics are sometimes competitive with plasticized vinyls (ref 5).

Chlorosulfonated Polyethylene

Treatment of polyethylene with chlorine and sulfur dioxide introduces occasional chlorine and sulfonyl chloride groups on the polymer chain. Random substitution destroys the crystallinity, permitting the polymer chains to exhibit maximum mobility and rubberiness. Highly reactive sulfonyl chloride groups permit vulcanization without carbon black, producing a wider color range than most elastomers, and without the main-chain unsaturation which causes poor aging resistance in most elastomers. Chlorine groups also contribute flame and oil resistance. But the chlorosulfonation reaction adds considerably to the price, limiting it to specialty applications. The resulting elastomer is used in some wire and cable insulations; liquid roof coatings; hose, gaskets, seals and diaphragms for handling oxidizing chemicals; and consumer items such as white sidewall tires, floor tile, coated fabrics, and shoe soles and heels (ref 6).

Ethylene/Propylene Rubber

Random copolymerization of ethylene and propylene, plus random atactic configuration of the propylene units in the polymer chain, produces a completely amorphous polymer with high molecular flexibility, resulting in typical rubbery properties of low modulus and high extensibility. The saturated aliphatic hydrocarbon structure of this rubber provides good electrical properties and high age resistance and color stability; and the extremely low cost of the monomers offers low ultimate price in large-scale production. The hydrocarbon polymer is attacked by petroleum hydrocarbons such as fuels and oils. The saturated hydrocarbon copolymer is so unreactive that it can be vulcanized only by difficult peroxide systems; but incorporation of a small amount of diene termonomer permits conventional sulfur vulcanization systems. Initial applications are mainly in special age- and heat-resistant products such as auto window and windshield channels and seals, radiator hose, pedal pads, and floor mats; washer-dryer hose, seals, and gaskets; high-voltage accessories for electrical cable and equipment; high-pressure steam hose; and hot materials-handling belts (ref 6); but ultimately low price may be expected to stimulate much greater technical and market development.

Polypropylene

Going from polyethylene to polypropylene, the regular substitution of a methyl group on every second carbon atom produces considerable restriction of rotation, producing a much less flexible molecule, with much less rubberiness even in the amorphous state; and much higher glass transition temperature, causing serious low-temperature brittleness in the solid polymer. Without stereospecific control of the polymerization reaction, the random atactic configuration of successive methyl branches produces an amorphous rubbery gum which has found some applications in compounding heat-seal adhesives for packaging (ref 7,8).

Most polypropylene, however, is produced under strict stereospecific control to give a highly stereoregular isotactic configuration, in which the steric hindrance of successive methyl groups produces a very regular coiled helical conformation of the polymer molecule, which crystallizes rapidly and thoroughly from the melt during processing. This high crystallinity, along with the steric hindrance of the methyl groups, reduces molecular mobility sharply, producing greater rigidity and equal or greater strength than linear polyethylene; as well as lower impact strength, and higher melting and heat deflection temperatures. As in the polyethylenes, crystallinity also produces translucency or opacity, unless crystal and spherulite growth are restrained by nucleation, quenching, or stretching during processing; and crystallinity also produces high solvent resistance. One unfortunate effect of methyl group substitution is the creation of active α-hydrogen atoms on every second carbon along the main-chain, making the polymer sensitive to oxidative degradation during aging, and requiring development of sophisticated stabilizer systems to retard such degradation, during both thermal processing and long-term use.

Finally, straightforward synthesis from low-cost petrochemical raw materials makes polypropylene inherently almost as low-price as polyethylene, stimulating rapid market growth wherever price is one of the critical properties; and where rigidity, strength, and heat deflection temperature requirements are somewhat more stringent than polyethylene can satisfy. Current major applications include fiber for carpeting, cordage, and sandbags; packaging film; and molded closures and toys.

Polypropylene Blends and Copolymers

The greatest weakness of polypropylene is its brittleness, particularly at low temperatures, producing early failure in many applications. This was originally improved somewhat by blending with softer polymers such as atactic polypropylene and especially polyisobutylene and butyl rubber. More recently impact strength has been improved still further by copolymerization with small amounts of ethylene, especially block copolymerization, to insert occasional flexible links into the stiff polypropylene chain without excessive loss of its regularity and crystallinity. Both blending and copolymerization incur some sacrifice of rigidity, strength, and heat deflection temperature; but they do improve impact strength sufficiently to make the polymer more satisfactory in many of its applications, particularly in thick sheet and molded form.

Poly-1-Butene

The ethyl side-chain on every second carbon atom in the main-chain of poly-1-butene creates as much restriction of rotation as the methyl side-chain in polypropylene; but being linear and flexible, its effects are more complex. The isotactic polymer coils into a regular helix, and crystallizes on cooling from the melt; but the longer flexible side-chain causes looser packing, first into a meta-stable crystalline form and then changing gradually into a stable crystalline form. Even the final stable form has somewhat looser packing than polypropylene, resulting in lower density, melting point, and especially modulus, and producing a stiffly flexible material of high impact strength and excellent stress-crack resistance. At the same time, the high crystallinity produces good strength, and especially high resistance to long-term cold flow, and to impact and tearing. This combination of resistance of cold flow, stress-cracking, impact, and tearing has led to initial commercial applications in heavy-duty water pipe and in lay-flat collapsible irrigation tubing;

but wider use in larger volume will have to go hand-in-hand with a marked decrease from development to fully commercial price.

Polyisobutylene and Butyl Rubber

Substitution of two methyl groups on every second carbon atom in the main-chain of polyisobutylene creates steric hindrance just as the methyl group in polypropylene; but whereas asymmetric substitution in isotactic polypropylene creates a preferred helical coil structure which is regular and crystalline, symmetrical substitution in polyisobutylene creates three equally probable conformations at every second atom in the main-chain, producing a randomly-coiled amorphous molecule of high flexibility (ref 9). With neither crystallinity nor polarity to create intermolecular attraction, polyisobutylene molecules have high mobility. Low-molecular-weight polymers are viscous liquids of broad compatibility, which find use in compounding adhesives, caulks, and sealants (ref 10,11). High-molecular-weight polymers are age-resistant elastomers, but require copolymerization with small amounts of isoprene to introduce the residual unsaturation required for conventional vulcanization. Their unusually low permeability to air made them the optimum material for auto tire inner tubes; but since the advent of tubeless tires, they have found more specialized applications such as high-voltage power cable, conveyor belts for hot materials, dairy hose, auto components, and some general-purpose use in auto tires (ref 5,12).

Poly(4-Methylpentene-1)

The isobutyl side-chain on every second carbon atom in the main chain of poly(4-methylpentene-1) produces sufficient steric hindrance to give the stereoregular isotactic polymer a uniform helical coil conformation which crystallizes readily; and the bulk of the side-chain requires a high melting tempreature to produce complete molten mobility. On the other hand, the bulk of the side-chain causes looser packing, producing the lowest-density polymer known, and thus permitting high permeability to gases. The balance of high crystallinity vs loose packing produces a fairly rigid polymer, but suffers from low impact strength on the one hand and low heat deflection temperature on the other. Loose packing of the crystalline phase makes its density as low as the amorphous phase, thus giving equal refractive indexes and clarity despite the crystallinity. The two tertiary hydrogen atoms in every monomer unit produce serious problems of oxidation resistance, requiring careful stabilization. And problems in monomer synthesis plus small-scale specialty production result in high initial development price for the polymer, which is not conducive to an energetic search for large-volume applications. Initial specialty applications include hospital and laboratory ware in the United States; plus lighting fixtures, milking machine and liquor dispenser sight glasses, sink traps, slot closures for electric motors, and packages for reheatable foods in Great Britain.

8-2 DIENE ELASTOMERS (Ref 5,6,13)

Polybutadiene

Cis-polybutadiene is one of the newer and faster-growing of the commercial diene elastomers, and its structure represents the best starting point for a study of the relationships between their structures, properties, and applications. The —CH_2—CH_2— groups have very little steric hindrance and no polar repulsion to restrict rotation; and the C═C bond, while itself planar and rigid, increases the freedom of rotation on the adjacent C—C bonds even further. Thus the polymer molecule is very flexible, and can respond to either external or internal stresses by large and fast changes in chain conformation; external stresses produce high elongation and low-temperature flexibility, while internal stresses due to entropy produce rapid elastic retraction and high resilience.

The unsubstituted hydrocarbon chains of polybutadiene have no polarity to produce any significant intermolecular attraction, thus permitting each chain to exercise its flexibility freely. The C═C electrons in adjacent chains probably interact weakly by resonance or London dispersion forces, producing some slight intermolecular attractions and contributing somewhat to mechanical properties. The high-molecular-weight molecules are highly entangled, and these entanglements prevent liquid flow under stress, and help produce elastic retraction when external stress is removed; they do, however, disentangle gradually under sustained external stress, producing weak mechanical properties and permitting high creep. Vulcanization produces a few widely-spaced permanent covalent cross-links, which provide much more dependable strength, creep resistance, recovery, and hot strength. The polymer chains have a very regular configuration, and when they are highly stretched they become nearly parallel in conformation as well, crystallizing temporarily and increasing strength still further; yet when the high strain is released, the probability (entropy) of returning to a random coiled structure is so much greater than the weak forces holding the cis chains together in the nonpolar crystal, that the crystals "melt" instantly and permit rapid elastic retraction and rebound. Addition of a reinforcing filler like carbon black, which forms fairly strong intermolecular bonds between polymer chains, greatly enhances properties still further.

Molecular flexibility and intermolecular attraction together produce intermediate strength properties but high resistance to abrasive wear.

The unsaturated C═C bond, and the adjacent allylic α-hydrogen which it in turn activates, produce a fairly high reactivity, which is useful in vulcanization. On the other hand, they also cause sensitivity to oxygen and ozone (especially when stretched) and to oxidizing chemicals; and this sensitization is increased by heat and light, producing rather poor aging properties. Addition of stabilizers is thus very important, and still not entirely satisfactory, especially under conditions of heat, light, and oxidizing agents.

Table 8-1. Rubber Usage: Nontire Products (ref 14)

Belts and belting	Other shoe products
Hose	Cements
Packing and gaskets	Proofing and coating fabrics
Aircraft equipment	Drug sundries
Auto equipment	Bullet-sealing fuel cells
Household products	Athletic goods
Mats and matting (exclud. auto.)	Toys and balloons
Hard rubber	Sponge rubber products
Other misc. mechanical goods	Pressure-sensitive tape
Wire and cable	Thread and related products
Rubber footwear	Rubber flooring
Heels and soles	Miscellaneous products
Inner shoe cushions	Latex foam
Cements for shoes	All other rubber products

The use of carbon black for high strength of course limits color possibilities severely. Even with light-colored, less effective reinforcing agents, the poor stability of the C═C bond during aging produces poor color stability.

Finally, the aliphatic hydrocarbon structure, low cross-linking, and fugitive crystallinity all combine to leave the polymer with poor resistance to fuels and oils, and thus restrict it from applications requiring exposure to such solvent materials.

Overall, diene elastomers in general find greatest use in pneumatic tires, plus a tremendous variety of applications in other soft, flexible, and rubbery products (Table 8-1). While some of these applications may prefer one diene elastomer rather than another, the great majority of them may be performed by any of the general-purpose diene elastomers.

Polyisoprene

Natural rubber and synthetic polyisoprene are essentially similar in structure, although they differ in small percentages of other configurations and impurities. Introduction of the methyl side group on every fourth main-chain carbon atom produces some restriction of rotation and reduction of molecular flexibility, which results in somewhat less resilience and more limited low-temperature flexibility. Otherwise technical properties are similar to cis-polybutadiene, and any superiority for natural rubber probably results from longer and more thorough experience with it in industry. Economically, the price and availability of natural rubber fluctuate erratically due to political and military conditions, whereas the price and availability of synthetic elastomers improve steadily with commercial growth; controlled production also produces more economical processing. Ultimately the synthetic elastomer which achieves lowest price will probably capture the vast majority of general-purpose applications, leaving the smaller-volume higher-priced materials to meet the needs of more demanding specialized applications.

Butadiene/Styrene Rubber

Random copolymerization produces an irregular structure which cannot crystallize on stretching, and thus gives very poor unreinforced "gum" stocks; here carbon black reinforcement is absolutely essential to good mechanical properties. Both the phenyl side groups of styrene, and the frequent branching due to chain-transfer during free radical polymerization, produce considerable restriction of rotation and increase in molecular stiffness, resulting in less resilience, more compression set, and more limited low-temperature flexibility; and especially higher hysteresis conversion of mechanical into heat energy during repeated flexing, resulting most notably in serious overheating of large bus and truck tires.

Historically the copolymer of butadiene/styrene 72/28 was commercialized long before the new stereoregular diene homopolymers, and already controls many markets and applications which may eventually have to be redivided on a more rational and permanent basis.

Butadiene/Styrene Block Copolymer

Sandwich molecules, containing long polybutadiene centers between shorter polystyrene ends, combine the elastomeric extensibility, resilience, and low-temperature flexibility of polybutadiene with the thermoplastic fusibility and strength of polystyrene (ref 15). Thus the polystyrene ends take the place of both vulcanization and reinforcement in conventional diene elastomers, providing "fusible cross-linking." This structure combines thermoplastic processability and wide color range with most of the properties of conventional diene elastomers. The major weakness is hot creep, where the polystyrene blocks no longer provide sufficient intermolecular attraction to prevent disentangling and flow. Some initial applications have appeared in rubber bands, toys, sporting goods, and footwear.

Nitrile Rubber

Copolymers of butadiene with 18%–50% of acrylonitrile contain increasing degrees of polarity and hydrogen-bonding between adjacent polymer chains. These produce good resistance to aliphatic and aromatic fuels, oils, and solvents, as well as improved abrasion and heat resistance. On the other hand, they also reduce molecular flexibility, giving lower resilience, higher hysteresis, and poorer low-temperature flexibility. High polarity also produces poor electrical properties. And acrylonitrile comonomer adds to the final price, limiting these copolymers to solvent-resistant specialty applications such as oil-well parts, fuel-cell liners, fuel hose, aircraft hose and gaskets, and carburetor and fuel-pump diaphragms.

Neoprene

The chlorine atom on every fourth main-chain carbon atom in neoprene makes it one of the most flame-resistant elastomers, and also increases the resistance of the unsaturated aliphatic hydrocarbon structure toward attack by oxygen, ozone, heat, light, and weather. In addition the polarity produced by the chlorine atom contributes to high tensile strength and elongation, as well as fair fuel and oil resistance. The regular structure and polarity permit temporary crystallization when the polymer is stretched, thus contributing further increase in strength, even without the use of reinforcing agents. Absence of carbon black reinforcement, and stabilization of the unsaturated structure, provide broad color range and good color stability. On the negative side, polarity produces poor low-temperature flexibility, fair electrical resistance, and higher price. As a result, neoprene is used mainly in specialty applications such as wire and cable jackets, industrial hose and belting, automotive seals, boots and diaphragms, golf ball covers, and curtain-wall gasketing.

8-3 STYRENE POLYMERS

Polystyrene

The phenyl group on every second carbon atom in the main-chain of polystyrene has a flat, rigid, bulky configuration, producing sufficient steric hindrance to make the polymer molecule very inflexible at room temperature. Resonance of electrons in the benzene ring also probably contributes some polarity and London dispersion forces, to create some intermolecular attraction between adjacent polymer molecules. This high inflexibility and mild intermolecular attraction combine to make polystyrene rigid and strong, and give it moderately high glass transition and heat deflection temperatures. On the other hand, molecular inflexibility leaves the pure homopolymer unable to withstand high rates of mechanical loading, making it brittle under impact; or to withstand surface attack by surfactants and solvents, causing stress-cracking.

At higher temperatures, weak intermolecular attractions become insignificant, and atomic vibrations increase sufficiently to make the molecule quite flexible. This produces easy processing by molding and other conventional thermal processing techniques. It also produces rather limited heat deflection temperatures, causing serious distortion at the boiling point of water.

Commercial free-radical polymerization produces a random atactic sequence of successive asymmetric carbon atoms, making the polymer an amorphous glass. This produces transparent solid products, but again limits heat deflection temperature to the glass transition temperature of the amorphous phase. Research has indicated that stereospecific polymerization to crystalline polymer can raise heat deflection temperature, but excessive brittleness prevents commercial use.

As a pure hydrocarbon polymer of low molecular mobility, polystyrene has excellent electrical insulating properties. As an amorphous hydrocarbon, it has poor resistance to organic solvents.

Two-step synthesis of styrene monomer from benzene and ethylene makes it a reasonably low-cost raw material, and public support of initial development during World War II has made polystyrene a large-volume low-price plastic in the United States. Lack of such historical factors in Europe and Asia has left polystyrene a much less important commercial material there than it is here.

Easy processability, rigidity, clarity, and low price combine to support major applications in molded and thermoformed packaging containers and lids, housewares, and toys; as well as radio, television, and other appliances, lighting and signs, closures, combs and brushes, and interior wall tiles. Unfortunately, many early applications over-enthusiastically stressed low price at the expense of poor impact strength and limited heat deflection temperature, and contributed heavily to the popular opinion of plastics as cheap substitutes during the 1940s and 1950s. More recently the development of improved copolymers, as discussed below, has overcome these limitations and helped establish styrene polymer properties and applications on a sounder commercial basis.

Polystyrene Foam

The thermoplasticity and solvent sensitivity of polystyrene combine to support the easy molding and extrusion of rigid, low-cost foams. The low-density closed-cell structure, containing primarily air, provides very good thermal insulation, and is widely used in construction for this reason, as well as in insulated drinking cups for both over-the-counter and vending-machine service. Ability to absorb energy gradually when crushed by impact provides excellent protection for delicate mechanisms and food stuffs, and is heavily used in packaging; in addition, nontoxicity and water resistance support major use in packaging of many foods, particularly in meat-trays. Light weight and rigidty account for large usage in housewares and toys; while low density, closed-cell structure, and water resistance are responsible for large usage in flotation applications. In all of these, of course, easy molding and extrusion, and low cost remain major determining factors.

Styrene/Acrylonitrile Copolymer

Copolymerization of styrene with a moderate amount of acrylonitrile provides polarity and hydrogen-bonding, increasing intermolecular attraction sufficiently to increase heat deflection temperature up to the boiling point of water, and also to increase chemical resistance somewhat. On the other hand, reactivity of the nitrile side-groups causes thermal discoloration which is masked only with difficulty; and addition of higher-cost comonomer increases price of the copolymer into the range of smaller-volume specialty resins. Leading applications are typically in heat-resistant products, particularly housewares such as dishwasher-proof drinking glasses, and appliances such as radio and television parts.

Impact Styrene

Copolymerization of styrene with flexibilizing comonomer such as butadiene can reduce brittleness, but only at the expense of rigidity and heat deflection temperature, as noted below. On the other hand, dispersion of a semicompatible rubbery phase within the rigid polystyrene matrix can provide discrete flexible areas which can absorb and disperse impact energy harmlessly, while the continuous polystyrene phase still retains most of its rigidity and heat deflection temperature relatively unchanged. Early commercial products, based upon simple dispersion of rubber in polystyrene, were too incompatible, and failed to transmit impact energy smoothly from the polystyrene matrix, across the jagged microscopic interface, into the dispersed rubber particles; so impact resistance was only slightly improved. Later development of graft copolymers produced compositions containing mainly rubber in the finely-dispersed particles, mainly polystyrene in the continuous rigid phase, and an intermediate graft copolymer to provide a gradual transition from one to the other at the interface. In such impact styrene compositions, the improved balance of rigidity and heat deflection temperature vs impact resistance permitted these improved materials to perform more satisfactorily than polystyrene homopolymer in many of its applications, and to open major new markets such as thermoformed refrigerator and door liners. Impact styrene technology still suffers several major limitations, primarily lower rigidity and heat deflection temperature than polystyrene, loss of transparency due to the different refractive index of the incompatible dispersed phase, poor weathering due to the unsaturated elastomer component, and gradually increasing cost with increased rubber content and impact resistance.

ABS

Some of the limitations of impact styrene are overcome, while others are accentuated, by the incorporation of acrylonitrile into the graft copolymer. When the continuous matrix phase is primarily a styrene/acrylonitrile copolymer, it retains more of polystyrene's rigidity and heat deflection temperature, despite the presence of a larger concentration of dispersed rubber particles, and also improves chemical resistance.

Meanwhile the higher concentration of rubber particles produces even higher impact strength. Opacity remains a limitation, foaming becomes more difficult, and weathering and cost become even worse. The desirable properties, however, are sufficient to produce major growth in a number of markets, particularly automotive parts; major appliances such as refrigerator door liners and inner liners; small appliances, business machines, and telephones; luggage; shoe heels; and pipe, fittings, and conduit, particularly drain-waste-and-vent pipe in building.

Styrene/Butadiene Copolymers

Copolymerization of styrene with moderate amounts ($\leq 40\%$) of butadiene increases molecular flexibility and lowers glass transition temperature, providing a polymer which is widely used for latex coatings on paper, latex paints on masonry, and solution-applied corrosion-resistant maintenance coatings on steel.

Poly-*p*-Xylylene

It is interesting to note that, when the benzene ring is in the main-chain instead of the side-chain, the higher symmetry produces a structure which is crystalline, high-melting, and solvent-resistant, and retains the excellent electrical and water-resistant properties of polystyrene, but is very sensitive to hot atmospheric oxidation. Present difficult polymerization and application technology limits it to very high-cost thin-film and coating applications in select specialties such as electrical equipment and instrument construction.

8-4 VINYL HALIDE POLYMERS

Polyvinyl and Vinylidene Chlorides

The chlorine atom on every second carbon in the main-chain of polyvinyl chloride produces polarity, which reinforces a low degree of crystallinity in the otherwise rather random, weakly syndiotactic structure. The large negative chlorine atoms also cause some steric hindrance and electrostatic repulsion, which reduce the flexibility of the polymer chain. All together, these structural features cause molecular immobility, resulting in difficult processing by either melt or solution techniques; high rigidity and strength coupled with brittleness; fair heat deflection temperature; good electrical resistance; and high solvent resistance.

The chlorine atoms tend to split out by dehydrohalogenation under the influence of heat during processing, and heat and light during use of the finished products, producing undesirable discoloration and embrittlement; this requires addition of special stabilizer systems to retard degradation, and many of them in turn introduce problems of opacity and toxicity. On the other hand, the high chlorine content of the polymer produces definite advantages in flame resistance and chemical resistance.

The rather random asymmetry of the polymer produced by conventional free radical polymerization limits the crystallinity to a low degree, permitting good transparency in the polymer itself; use of additives of course may cause opacity, particularly in rigid and filled compositions.

Synthesis from ethylene and chlorine, and easy polymerization reactions, produce a low-cost polymer whose relatively long commercial history has also contributed to very large volume production and use in a wide variety of applications.

Rigid PVC

Polyvinyl chloride alone is a fairly good rigid plastic, and suffers mainly from difficult processability and low impact strength. Both of these properties are greatly improved by addition of semicompatible elastomers or graft copolymers, particularly ABS and impact acrylic materials. These improve melt flow during processing; and they remain dispersed as small particles of fairly soft rubbery material in the continuous rigid PVC matrix, serving to absorb and disperse impact energy harmlessly without fracture, without seriously lowering the rigidity or heat deflection temperature of the PVC matrix. With this improved balance of properties, rigid vinyls are finding large and increasing application in building panels, siding, windows, and rainwater gutters and downspouts; pipe, fittings, and conduit, particularly for water and for chemical processing; blow-molded bottles; thermoformed sheet for packaging; and in Europe for magnetic tape.

Plasticized Polyvinyl Chloride

In most of its applications, polyvinyl chloride is softened by the addition of major proportions ($\geq$20 PHR) of compatible, nonvolatile, liquid plasticizers. These are primarily high-molecular-weight monomeric esters, selected for similar polarity, hydrogen-bonding, and solubility parameter to give complete miscibility and compatibility with the polymer. They function by intercepting and weakening the polarity and crystallinity which attracted adjacent chains to each other and held them immobile; and also by providing small free molecules as internal lubricants which can flow between adjacent polymer chains. Careful choice of compatible plasticizers provides both homogeneity and clarity, and reduces escaping-tendency to improve permanence in the composition. Small flexible plasticizer molecules, with moderate hydrogen-bonding by ester groups, provide efficient plasticization, which produces easier processing by both melt and paste technologies, and produces softness, flexibility, and even extensibility and gradual elasticity, which can be varied over a wide range by simple adjustment of polymer/plasticizer ratio; low-temperature flexibility depends mainly on the flexibility of the plasticizer molecule. Plasticizer permanence depends upon maximum compatibility, and also upon low volatility, solvent resistance, and minimum migration into adjacent materials; for maximum permanence, medium-molecular-weight aliphatic polyesters must be used, at some sacrifice in plasticizing efficiency. With increased plasticizer content and molecular mobility, ionic conduction becomes feasible and electrical resistance decreases considerably.

Plasticized polyvinyl chloride outperforms rubber in processability, wide range of mechanical properties, flame resistance, transparency and color possibilities, resistance to chemicals and aging, and cost. It outperforms textiles and paper in processability, wear resistance, flame resistance, impermeability, resistance to soiling, and often in cost. And it outperforms leather in easy processing and quality control, wear resistance, color, moisture resistance, and cost. For these reasons it has often replaced these conventional materials and developed very large applications in furniture and auto upholstery and interior wall coverings; in outdoor clothing, rainwear, shoes, and baby pants; in tablecloths, shower curtains, and closet accessories; in electrical wire and cable insulation; in packaging film, especially for meat, poultry, fresh fruit, and vegetables; in dolls and inflatable toys; in sporting goods such as balls and swimming pool liners; in garden hose and medical tubing, in weatherstripping and waterstops, in auto mats, and in stationery supplies.

Polyvinyl Chloride Foam

Plasticized polyvinyl chloride is readily foamed to fairly low density, most often by incorporating a thermally-decomposable azo compound in a plastisol formulation. The resulting foams add greatly to the softness and energy absorption already inherent in the plasticized polymer, giving richness, warmth, and elegance to upholstery, clothing, boots, handbags, luggage, dolls, and auto door panels; energy absorption for quiet and comfort in flooring, shoe soles and heels; energy absorption and crash-padding in auto headliners and sunvisors, and in packaging auto bumpers and bezels for shipments; and easy compressibility and recovery in caulking tapes, sealants, gaskets, spline, welting, stripping, bumper stock, and bottle and jar caps. In all of these applications, foam is replacing solid plasticized polyvinyl chloride and providing higher quality performance.

Vinyl Chloride/Acetate Copolymers

Copolymerization of vinyl chloride with 10%–15% of vinyl acetate (1) provides chain-transfer action which lowers molecular weight, (2) introduces flexible points into the rather stiff chain, (3) lowers regularity and crystallinity still further, and (4) lowers polarity somewhat. All of these modifications in structure make the copolymers easier to process in molten and solution form, without detracting seriously from the rigidity and heat deflection temperature. These copolymers find primary application in phonograph records, in plasticized and highly-filled flooring, and in solution-coating the interior of cans and other metal containers to prevent corrosion and protect their contents.

Chlorinated Polyvinyl Chloride

Partial chlorination to vinylidene chloride structures was commercialized in Germany 30 years ago; the random copolymer structure was less crystalline and more soluble in common solvents like acetone, permitting easy spinning of fibers. More recently, controlled chlorination of the methylene groups in polyvinyl chloride has produced a more regular structure, with more steric hindrance and electrostatic repulsion contributing to increase heat deflection temperature, making the new material serviceable beyond the boiling point of water. This polymer is finding application in pipe, fittings, and valves for handling hot water and corrosive chemicals; exhaust systems and scrubbers for handling hot acid fumes in steel-pickling lines; and chemical etching machines, tanks, liners, and hoods.

Polyvinylidene Chloride

Symmetrical substitution of two chlorine atoms on every second carbon in the main-chain provides a regular zigzag chain structure (ref

16) which orients and crystallizes readily (ref 17,18), producing high strength, high melting point, and especially outstanding impermeability to oil, grease, water vapor, oxygen, and carbon dioxide (ref 19,20). When the molten polymer is extruded in film form, quenched, and then oriented (ref 18), the crystallinity is fine enough to produce high clarity and flexibility. High polarity also produces high surface tension and friction (ref 21), making the film cling firmly when wrapped around contents or even on itself (ref 18). These properties contribute to widespread use in packaging film, particularly for food products which require the impermeable barrier protection (ref 20,22). Since the cost is rather high, vinylidene chloride is often first copolymerized with moderate amounts of vinyl chloride, alkyl acrylate, or acrylonitrile, to make it more processable; and then solution- or emulsion-coated onto paper, paperboard, or other films to provide the impermeability as economically as possible (ref 20,23–26). A smaller amount of the polymer is also extruded into monofilament and tape for outdoor furniture upholstery (ref 17,20), but is encountering increasing competition from newer polymers in such applications.

Fluoropolymers (ref 27–30)

Polytetrafluoroethylene

The fluorine atoms in polytetrafluoroethylene account for most of its unusual properties. As halogens, they produce flame resistance. They form very strong covalent bonds with the carbon atoms in the mainchain, producing high thermal and chemical stability. Their high electronegativity, small diameter, and high density produce a polymer of high density and low surface tension, low adhesion, and low coefficient of friction. And the difficulty and relative novelty of fluorochemical synthesis make the price of the polymer so high that it is largely restricted to highly critical specialty applications.

Complete substitution by fluorine atoms also makes the polymer molecule a stiff rod-like helical coil, producing very high melt viscosity and difficult processability. These rods crystallize with difficulty, leaving a partially amorphous structure which produces a low glass transition temperature and good low-temperature flexibility, but also permits considerable cold flow under mechanical stress. The immobility of the rods in the crystalline portion, on the other hand, produces a high crystalline melting point and good retention of form at high temperatures.

Finally, while the fluorine atoms are highly electronegative, complete substitution balances them and produces a nonpolar molecule with high electrical resistance and low dielectric constant. In addition, the immobility of the stiff rod-like molecules prevents any tendency toward motion under the influence of an electrical field, producing low and constant dielectric loss.

This combination of properties leads to a number of valuable specialty applications. Thermal stability and electrical properties combine to favor use for insulating electrical wiring in high-temperature environments; motors, locomotives, aircraft engines, missiles, and spacecraft; lighting fixtures, stoves, and ovens; switches, controls, and computers; and heating cable for pipe tracing in chemical plants and refineries. Low adhesion and low coefficient of friction, high heat stability and chemical resistance are used in gaskets, seals, packings, bearings, and cooking utensils. Low constant dielectric constant and loss are used in electrical insulators for radar and television. And chemical resistance is used in lining pipe, fittings, valves, and pumps in industrial plants, and hydraulic and fuel hose in aircraft, trucks, buses, and trains.

Fluorinated Ethylene/Propylene Copolymer

Copolymerization of tetrafluoroethylene with some hexafluoropropylene produces irregularity in the polymer molecule, making it less rod-like and less crystalline, lowering melting point and melt viscosity, and thus improving processability. Whereas the homopolymer must be sintered at high temperature and pressure to reduce voids and weak points to a minimum, the copolymer can actually be molded and extruded by conventional techniques. This change lowers thermal stability, but improves impact strength and clarity. Thus the copolymer can replace the homopolymer in applications requiring easier and more conventional processing. Specific applications include capacitors, printed circuits, multiconductor cable, mold liners, flexible belting for food and chemical plants, and shrink tubing on rolls for conveyors, textile finishing, offset printing, paper-making, and wax handling.

Polychlorotrifluoroethylene

Substitution of a chlorine atom in place of every fourth fluorine atom also produces irregularity in the polymer molecule, making it less rod-like and less crystalline, and thus moldable to products of higher rigidity and strength. The combination of high density and low porosity produces an unusual resistance to transmission of water vapor. On the other hand, lower regularity produces lower resistance to heat and solvents. Applications include electrical insulation for hook-up wire, cable jacketing, electrical and electronic coil forms, tube sockets, terminal insulators, and printed circuit boards; valves, fittings, pipe seals, O-rings, and coatings for corrosive process industries; caps for containers holding corrosive liquids; electrolytic diaphragms, and microporous chemical filters.

Polyvinylidene Fluoride

With alternating hydrogen and fluorine atoms, polyvinylidene fluoride has better processability but less thermal and chemical resistance than the previous fluoropolymers. Strength is higher, melting point lower, dielectric constant and loss higher, and resistance to solvents much less. Applications include chemical piping, valves, pumps, spargers, dutchmen, and large ducts; electrical wire insulation in aircraft, computers, oil wells, and heat-shrinkable insulating tubing for electronics resistors and diodes, soldered joints, and strain relief in high-density connectors; and in pigmented finishes for exterior metal siding such as aluminum and galvanized panels for stores and industries.

Polyvinyl Fluoride

Films of this polymer have demonstrated excellent outdoor durability, leading to applications in greenhouse and poultry glazing and on polystyrene lighting panels, and particularly in presurfaced exterior building panels of aluminum, plywood, reinforced plastic, ABS, asbestos cement, hardboard, doors and windows.

Vinylidene Fluoride/Hexafluoropropylene Elastomer

Copolymerization of these two fluoromonomers produces a random, amorphous elastomer which is most notable for its high-temperature resistance, as well as its resistance to chemicals, oxygen, sunlight, weather, and burning. Weaknesses are strength properties and low-temperature embrittlement. Primary applications are in gaskets, O-rings, seals, diaphragms, equipment linings, coated fabrics, hose, tubing, and carburetor needle valves, for appliances, automobiles, aircraft, and industrial equipment (ref 6).

8-5 POLYVINYL ALCOHOL AND ITS DERIVATIVES

Polyvinyl Alcohol

The hydroxyl groups on every second carbon atom in the main-chain of polyvinyl alcohol provide high polarity and hydrogen-bonding, which make it infusible but soluble in water, and resistant to oils, fats, and greases; and also provide strong adhesion, particularly to cellulosic substrates such as paper and textiles. Although the tertiary asymmetric carbon atoms are random and atactic, the hydroxyl groups are not much larger than simple hydrogen atoms, and do not seriously impair the symmetry of the molecule, permitting it to crystallize readily, producing high strength, solvent resistance, and impermeability to gases. Leaving a moderate proportion of acetate groups in the polymer molecule is sufficient to decrease regularity and packing, and increase flexibility and solubility. Plasticization with compatible liquids like glycerol also reduces bonding between adjacent polymer molecules, increases flexibility, and even produces sufficient thermoplasticity for molding and extrusion. Production from polyvinyl acetate by hydrolysis unfortunately adds considerably to its cost. Major applications are adhesives for paper and paperboard in packaging, bookbinding, and envelopes; textile sizing for strength and abrasion resistance during processing; aqueous thickeners and dispersing agents in emulsion and suspension polymerization; and film for vacuum-bag molding of reinforced plastics, polarizing instruments, and packaging (ref 31). In Japan it is also spun into fibers, then stabilized by cross-linking with formaldehyde to improve its water resistance.

Polyvinyl Acetate

The acetate groups on alternate carbon atoms in the main-chain of polyvinyl acetate are sufficient to push the polymer chains apart and make them readily fusible and soluble. Random orientation of the tertiary asymmetric carbon atoms makes the polymer atactic and amorphous, also favoring low softening temperature and high solubility. Considerable branching during polymerization also contributes to this amorphous structure. The size and shape of the acetate side-chains provide sufficient steric hindrance to make the polymer stiff at room temperature, but permit it to be softened readily by warming or plasticization. The ester groups and active hydrogen atoms provide moderate polarity and hydrogen-bonding, permitting easy fusion, solubility, plasticization, emulsification, and adhesion to a wide range of substrates. And easy low-cost application of emulsions to porous substrates provides attractive economics for adhesives and coatings. Molecular mobility and polarity lead to major use in adhesives for bookbinding, paper bags, milk cartons, drinking straws, envelopes, gummed tapes, convolute tubes, folding boxes, multiwall shipping bags, labels, and many other materials and products. Addition of monomeric ester plasticizers or flexibilizing comonomers increases molecular mobility at room temperature sufficiently to provide good wetting and film formation, and polarity and hydrogen-bonding provide good wetting and adhesion to pigments and substrates, leading to wide use in latex paints. Polarity and hydrogen-bonding for adhesion, easy fusibility for processing, and moderate stiffness at room temperature all contribute to considerable use in textile finishing (ref 32).

Polyvinyl Acetals

The acetal and residual hydroxyl groups in polyvinyl formal and butyral provide considerable polarity and hydrogen-bonding which produce solubility in polar solvents, compatibility with polar plasticizers, and high adhesion to a wide variety of substrates. The acetal and hydroxyl groups also provide reactivity for post-cure and for further reaction with many thermosetting resins to form copolymers. And the acetal rings on the main-chain restrict rotation sufficiently to provide considerable strength. Unfortunately the two-step synthesis from polyvinyl acetate adds considerably to their cost and restricts their use somewhat. Major use of polyvinyl butyral, resulting from its plasticizability, adhesion, and strength, is as the interlayer in laminated safety glass. Major use of polyvinyl formal, resulting from its solubility, reactivity, adhesion, and strength, is in heat-resistant nonconductive electrical wire enamels. Important coating applications, due to flexibility, reactivity, and adhesion, include wash primers for conditioning metal surfaces to prevent corrosion; and plasticizing components for epoxy and phenolic can linings (ref 33).

Polyvinyl Ethyl Ether

The short flexible side-chain is attached through a highly flexible oxygen link at every second carbon atom in the main-chain, pushing the polymer molecules apart without significantly restricting rotation, and thus produces high molecular flexibility. The random atactic configuration of successive asymmetric carbon atoms prevents crystallization and produces a random amorphous coil. Molecular flexibility and amorphous structure together produce high mobility of the polymer molecules.

Polarity of the oxygen atoms, and hydrogen-bonding of the oxygen and α-hydrogen atoms, provides good adhesion to a variety of surfaces. The combination of molecular mobility and high adhesion makes the polymer primarily useful in the formulation of pressure-sensitive adhesives (ref 34,35).

8-6 ACRYLIC POLYMERS

Ethyl Acrylate Polymers

Polyethyl acrylate is one of the important commercial acrylics, and provides a convenient starting point for the consideration of their structure, properties, and applications. The flexible carboethoxy side-chains push the molecules apart without significantly restricting rotation, and thus produce high molecular flexibility. The random atactic configuration of successive asymmetric carbon atoms prevents crystallization and produces a random amorphous coil. Molecular flexibility and amorphous coiled structure together produce a soft extensible material. At the same time, the carboxyl groups and α-hydrogen atoms provide considerable hydrogen-bonding, both between molecules to produce strength and elasticity, and to substrates to produce good adhesion.

The polarity and hydrogen-bonding are sufficient to provide easy solution in coating solvents or preparation in stable emulsion, both of which provide easy application of coatings and adhesives. Since the homopolymer itself is a rather soft, inert rubber, ethyl acrylate is generally copolymerized with a variety of other modifying acrylic comonomers to provide specific properties for specific applications. Most commonly the comonomer is methyl methacrylate, which hardens and strengthens the structure. Smaller amounts of hydroxyl, carboxyl, amine, and amide comonomers are often used to provide adhesion and thermosetting cross-linking. The major application is in high-quality latex paints for wood, wallboard, and masonry in homes. A variety of smaller and more specialized applications have developed in automotive and appliance coatings, finishes and adhesives for textiles and leather, floor polishes, and other specialty coatings and adhesives applications.

Polymethyl Methacrylate

Quaternary substitution on every second carbon atom in the main-chain provides considerable steric hindrance, making the polymer molecule quite rigid at ordinary temperatures, and providing good rigidity and strength in the solid plastic. Random configuration of the asymmetric carbon atoms produces a completely amorphous structure which provides high transparency to visible light. Absence of an α-hydrogen atom adjacent to the carbonyl group prevents oxidative free-radical reactions which would form conjugated double bonds and consequent differential light absorption and discoloration; therefore the polymer retains its white color readily throughout processing and aging, even when exposed to exterior ultraviolet light for long periods of time. The major degradative mechanism is depolymerization, causing gas bubbles during processing and high flammability during use.

Major applications are cast sheet which can be thermoformed into glazing for aircraft and institutional use, skylights, and outdoor signs, as well as decorative architectural panels. Other major markets for solid moldings include auto lenses and lighting fixtures for both home and industrial use. Solution-cast coatings find major use in automotive lacquers. And smaller amounts are used in specialty applications such as dentures, embedments, and fiber optics. In all of these, rigidity, clarity, and color stability are of prime importance.

Polyacrylonitrile

The highly electronegative nitrile groups on every second carbon atom in the main-chain, even though they occur in random atactic configuration on the asymmetric carbon atoms, exert so much mutual electronic repulsion that they force the molecule into an extended, stiff, rod-like conformation. The regularity of these rod-like structures permits sufficient orientation and crystallization to produce strong fibers, and the high hydrogen-bonding between adjacent chains produces high strength, infusibility, and resistance to moisture and solvents. To improve spinnability and dyeability of fibers, a moderate amount of a comonomer is generally added, such as vinyl acetate or methyl acrylate, to reduce regularity and crystallinity somewhat, and thus increase solubility and dye-affinity. Major use is in wool-like applications such as sweaters, blankets, and carpeting, where its strength and chemical stability are far superior.

8-7 POLYVINYL PYRROLIDONE

High polarity, hydrogen-bonding, and protein-like peptide groups produce water-solubility, film-forming ability, and biological compatibility. Despite rather high price, these properties have led to specialty applications such as blood plasma substitute, hair-sprays, pharmaceutical tablet binders, beverage clarification, and dye-receptors in synthetic fibers (ref 36).

8-8 ALIPHATIC POLYETHERS

Polyethylene Oxide

The oxygen atoms which alternate regularly with ethylene units in polyethylene oxide provide such polarity and hydrogen-bonding that the polymers are completely soluble in water as well as a range of organic solvents. The regular structure crystallizes readily, but the highly flexible ether links provide so much molecular mobility that the polymers remain flexible and low-melting. While the low-molecular-weight polymers are used as fluids and thickeners, the solid high-molecular-weight polymers find use in textile warp sizes, pharmaceutical binders, and water-soluble films (ref 37).

Polypropylene Oxide

The additional methyl group on every monomer unit makes these polymers less hydrophilic, and creates asymmetric carbon atoms whose configuration can lead to either amorphous or crystalline materials. Primary uses are for low-molecular-weight polymers as surfactants and as prepolymer blocks for polyurethanes (ref 38,39).

Polyepichlorohydrin

Alternating ethylene units and ether links in the main-chain provide molecular flexibility, and substitution of one chloromethyl group on each monomer unit separates the polymer molecules from each other and thus increases their mobility without adding any serious steric hindrance. Random atactic configuration of the asymmetric carbon atoms prevents crystallization and keeps the molecule in the form of a random amorphous coil. All these structural features make the polymer soft, flexible, and rubbery.

The chlorine atoms provide reactive points for vulcanization to give strength and elasticity; and they also produce flame resistance, a rather uncommon quality in elastomers. The polarity of both oxygen and chlorine atoms produces good resistance to fuels and oils. The absence of residual diene unsaturation permits good resistance to heat aging. And the chlorine provides high density which contributes to impermeability to air and other gases. Potential applications include fuel-pump diaphragms, pipe gaskets, fuel and oil hose, oil well equipment, printing press blankets, and food jar gaskets (ref 40).

Poly[3,3-Bis(Chloromethyl)Oxetane]

The chlorine atoms provide chemical and flame resistance, and the absence of hydrogen atoms on adjacent carbons prevents dehydrochlorination, providing thermal stability. The symmetry permits easy crystallization, giving good hot strength. Applications are primarily in coating steel or replacing stainless steel for hot corrosive conditions, particularly in valves, pumps, meters, and pipe fittings, and as linings for steel pipe and tanks subjected to such corrosive conditions (ref 41,42).

8-9 AROMATIC POLYETHERS

Phenoxy Resins

Phenoxy resins are linear and therefore fusible and soluble for processing. The aromatic rings provide restricted rotation and electronic attraction, and the hydroxyl groups provide hydrogen-bonding, all of which contribute rigidity, creep resistance, and fair hot strength. The aliphatic ether portions provide molecular flexibility which contributes processability, ductility, and impact resistance. These qualities have proven useful in molded and blown bottles for food and drug packaging.

The combination of aliphatic, aromatic, ether, and hydroxyl groups provides adhesion to a wide variety of surfaces, and the hydroxyl groups provide thermosetting reactivity, all of which are useful in coating primers and adhesives. These can be applied both from melt and solution (ref 43).

Epoxy Resins

The low molecular weight of epoxy resins, generally monomeric to oligomeric in size, gives them exceptional molecular mobility during processing: all of them are either liquid at room temperature, or on warming, and all are readily dissolved to form fluid solutions. This high mobility during processing permits fast thorough wetting of surfaces, which is particularly important to their use in coatings, reinforced structures, and adhesives; and also easy pouring into final form, which is particularly important in potting, encapsulation, flooring, and decorative laminates.

The high reactivity of the epoxide groups with amines, anhydrides, and other curing agents, provides fast easy conversion of the liquid monomers into high-molecular-weight highly-cross-linked materials. Since these curing reactions generally occur with little or no change in the number of bonds, and are generally addition reactions with little or no evolution of volatile condensation by-products, they cause little shrinkage or evolution of gases, and permit easy conversion of liquid raw materials directly into finished solid end products. This is particularly useful in reinforced structures, potting, encapsulation, flooring, and decorative laminates.

The combination of aliphatic, aromatic, ether, and hydroxyl groups in the original epoxy resins, plus the amine, ester, and other groups formed by reaction with the curing agent, all provide a range of structures, polarities, and hydrogen-bonding which is compatible with and attracted to a wide variety of surfaces, producing good wetting and penetration which is particularly important in coatings, reinforced structures, and adhesives. In fact, the curing reactions may even form ionic or covalent bonds directly with these surfaces.

The high degree of cross-linking in the completely cured structure produces hardness, strength, heat resistance, electrical resistance, and broad chemical resistance. These properties are important in coatings, reinforced structures, adhesives, and electrical equipment.

The major use of epoxies is in protective coatings for plant maintenance, auto primers, can and drum linings, pipes, appliances, floor toppings, and bridges. Second in importance are reinforced plastics, particularly filament-wound pipe, tanks, tank wagons, pressure vessels, hydrospace vehicles, rocket cases, and other simple, hollow-form, high strength to weight ratio constructions. Third in importance are bonding and adhesive applications: (1) *metal to metal* adhesives such as aluminum honeycomb for aircraft, metal laminating solders to replace riveting, and auto body repairs; (2) *metal to plastic* adhesives for bonding polyester film to aluminum foil in heat-sealable impermeable packaging, copper to phenolic boards in printed circuits, and polytetrafluoroethylene to steel or stainless steel in nonstick cook-ware; (3) *plastic to plastic* adhesives such as reinforced polyester constructions and boat repairing; and (4) bonding *concrete* in repair of highways and buildings. Other important applications include potting and encapsulation of electrical and electronic components such as coils, resistors, capacitors, semiconductors, miniature circuits, transformers, switchgear, and motor windings; printed circuit boards and wire enamels; terrazzo flooring; and cast aggregate decorative laminates for wall panels.

Poly(2,6-Dimethylphenylene Oxide)

The repeating phenylene rings create both steric hindrance to rotation within the molecule and electronic attraction of resonating electron systems in adjacent molecules; and both contribute to a high degree of molecular immobility, producing high rigidity, strength, creep resistance, dimensional stability, and heat deflection temperature. The symmetrical structure has no unbalanced polarity, and the high molecular rigidity further contributes, to produce low and constant dielectric constant and dissipation factor. The low polarity also produces very low water absorption.

Molecular rigidity has produced problems in melt processing, which have been solved by blending with impact styrene. While the phenyl ether structure is very stable, the methyl groups appear sensitive to thermal oxidation at high use temperatures (ref 49). The aromatic and ether groups in the linear polymer permit easy solubility in a range of common solvents.

Initial applications are largely in electrical equipment, business machines, appliances, and water distribution equipment.

Polysulfones

Repeating phenylene rings again create both steric hindrance to rotation within the molecule, and electronic attraction of resonating electron systems between adjacent molecules; and both contribute to a high degree of molecular immobility, producing high rigidity, strength, creep resistance, dimensional stability, and heat deflection temperature. Phenyl ether and phenyl sulfone groups have high thermal and oxidative stability, producing long-term high-temperature stability during use. Difficult processing, resonance color, sensitivity to solvents, and high development prices are limiting properties. Initial applications are developing in electronic connectors and integrated circuits; electrical circuit breakers, meter and timer housings; automotive under-the-hood parts; aircraft cabin interior parts; kitchen range hardware; alkaline battery cases; and film for transparencies.

8-10 POLYSULFIDE ELASTOMERS

The aliphatic ether sequences provide rubbery flexibility, while the polysulfide links produce high resistance to fuel, oil, and solvents, as well as impermeability to gases. The polysulfide links also are chemically labile, which is useful in processing but produces low strength and high creep during aging. They also cause undesirable odor in finished products.

Solid polymers are used mainly as noncuring putty sealers in aircraft cabins and gasoline-handling systems, as hose liners in paint-spraying equipment, in printing rolls, and in coated paper and felt gaskets for automotive use. Low-molecular-weight liquid polymers, with reactive mercaptan end groups, are copolymerized with liquid epoxy resins to produce heat-and-impact resistant plastic tooling and electrical potting compounds; laminates, and adhesives; and chemically-resistant coatings (ref 44).

8-11 ACETALS

Polyoxymethylene

The regularity, symmetry, and flexibility of the polymer molecule permit rapid high crystallization which produces high rigidity, strength, and heat deflection temperature. At the same time, the flexibility of the alternating methylene and especially ether links produces good resilience, even in the rigid total structure. While these properties approach those of nonferrous metals such as copper, brass, and zinc, other properties are actually superior. Thus thermoplasticity provides easier and more economical processing. Methylene groups provide lower polarity and higher lubricity at sliding surfaces. And organic composition provides better resistance to most corrosive environments, while crystalline structure provides very high resistance to organic solvents. Major weakness is the sensitivity of the aliphatic ether link toward heat oxidation and ionic hydrolysis, all of which initiate depolymerization during thermal processing or during extended use in corrosive conditions.

Major applications are in business machines as wheels, bearings, cams, and gears, for computers and typewriters, and as pushbuttons for telephones; in automotive gears, bearings, turn signal switches, carburetor components, fender extensions, and trim clips; in plumbing fittings such as tub assemblies, shower heads, lawn sprinklers, valves, and pump housings; in hardware such as furniture casters and handles; in industrial parts such as gears, cams, bushings, clips, lugs, truck trailer connectors, and table-top conveyors; in appliance housings, electric clock parts, shaver cartridges, lawn mower wheels, and gears for toys; in electric devices in construction; and in aerosol containers, pens and pencils, and bra buckles.

Cellulose and Its Derivatives

Cellulose

The bulky ring structure of the cellulose molecule is inherently quite stiff and inflexible. It is also highly regular, and when oriented—by natural growth or by man's technology—it becomes partly crystalline. It is also highly polar and hydrogen-bonded. Crystallinity, polarity, and hydrogen-bonding add to its molecular immobility, making it rigid, strong, infusible, and insoluble. On the other hand, its multiplicity of hydroxyl groups makes it absorb moisture readily in its amorphous phase and decompose by dehydration when heated; and its hemiacetal groups make it sensitive to hydrolysis and degradation.

The pure fibers from cotton have long been one of man's favorites for textiles, because the bulky rings, crystallinity, polarity, and hydrogen-bonding produce good mechanical properties; and because the hydroxyl groups in the amorphous phase have a reactivity which permits good dyeability, and water absorption and desorption which help man establish more constant and comfortable temperature and humidity conditions against the variations in his surroundings. The impure fibers from wood have long been the basis of most paper-making, because of the mechanical properties of the fibers; because their hydroxyl groups permit them to be first dispersed in water, then hydrogen-bonded to each other by removal of the water; and because the resulting paper has sufficient polarity and reactivity to accept inks readily.

When cellulose is dissolved by the xanthate process, extruded and recoagulated to produce rayon fiber, it is stretched in the coagulated wet state before drying in order to produce maximum orientation and crystallization to generate maximum mechanical properties for use in tire cord. When the cellulose xanthate is instead extruded and recoagulated to produce cellophane film, a plasticizing polyol such as glycerol is added to improve flexibility and clarity, producing man's first major transparent packaging film; but the hydroxyl groups cause high water-vapor permeability and poor heat-sealability, both of which must be remedied by coatings such as vinylidene chloride copolymers and nitrocellulose.

Alternatively, sheets of cellulose paper can be swollen by aqueous zinc chloride and leached to recoagulate the cellulose in sheet form. The resulting "vulcanized fiber" is rigid and strong due to the bulky rings, crystallinity, polarity, hydrogen-bonding, and partly fibrous structure; impact resistant because of its ether links, amorphous portions, absorbed moisture, and partly fibrous structure; flame resistant because of density and absorbed moisture; and solvent resistant because of crystallinity, polarity, and hydrogen-bonding. Being infusible and insoluble, it can only be processed by machining operations; because of its hydroxyl groups which are highly polar and hydrogen-bonded, it tends to absorb considerable water. Applications are largely in electrical insulation such as switches, track insulation, motors, and terminal blocks; and in materials-handling equipment such as shuttles, bobbins, and tote boxes.

Cellulose Esters

Replacement of the hydroxyl groups by aliphatic ester groups reduces crystallinity, polarity, and hydrogen-bonding, making the polymers more compatible with plasticizers, fusible, soluble, stiffly flexible, transparent, and glossy. In cellulose triacetate, regularity is still too high to permit thermoplastic processing, but solution-cast films are strong, fold resistant, dimensionally stable, and solvent and water resistant; they are used largely in sound motion-picture film and magnetic recording tape.

In the other esters, the mixed structure of ester and residual hydroxyl groups is less regular, more compatible with plasticizers, and more easily fusible and melt-processable. Major applications in general are stiffly-flexible transparent film and sheet for packaging, especially blister packaging; personal items such as brushes, pens, and pencils;

industrial sheeting for signs and machine guards; optical goods; and lighting fixtures, tool handles, toys, and tubing. The acetate is hardest in properties and lowest in cost, and is used most widely for blister and skin packaging, windows of folding and set-up boxes, and packaging of fresh fruit and produce. The propionate is easier-processing, more impact resistant, more weather resistant, and compatible with less volatile plasticizers which give it greater dimensional stability; it is used widely in thermoformed blister packaging. The butyrate is still eaiser-processing, more impact resistant, and more weather resistant, and likewise compatible with less volatile plasticizers which give it greater dimensional stability; it finds considerable use in building panels and siding, vacuum-metallized foil for woven fabrics, and coatings on metals.

Cellulose nitrate contains a mixture of highly-polar residual hydroxyl groups and less-polar nitrate ester groups, making it amorphous, compatible with plasticizers, tough, flexible, dimensionally stable, and water resistant. Its thermal instability and high flammability prevent thermoplastic processing, and limit it to solution-cast lacquers and cements, and personal accessories fabricated from sheet.

Cellulose Ethers

Replacement of most of the hydroxyl groups in cellulose by ethoxyl groups reduces regularity, crystallinity, polarity, and hydrogen-bonding, making the polymer soluble, fusible, and compatible with a wide range of plasticizers, and thus gives it superior low-temperature impact strength. Applications are in football helmets, bowling pin bases, auto armrests, flashlight cases, table edging, fire extinguisher components, tool handles, electrical appliance parts, printing inks, lacquers, and hot-melt coatings.

Partial replacement of hydroxyl by methoxyl groups likewise reduces regularity and crystallinity, but produces a much more polar polymer which is soluble in water, and is used mainly as an aqueous thickening and suspending agent.

Partial replacement of hydroxyl groups by sodium carboxymethyl groups decreases regularity and crystallinity, increases polarity by actually introducing ionic groups, and produces complete solubility in water. Applications are primarily in thickening and stabilizing aqueous systems, such as ice cream, gelatin, dietetic foods, pharmaceutical suspensions and tablets, shampoo, toothpaste, laundry finishes, textile printing pastes, paper and paperboard manufacture, latex paints, and adhesives.

8-12 POLYESTERS

Polycarbonate

The phenyl and methyl groups at the quaternary carbon atoms produce considerable steric hindrance, making a very stiff molecular structure. The regularity of the polymer molecule makes it crystallizable. And the overlap of resonant phenyl groups from adjacent chains produces some degree of intermolecular attraction. These factors together produce high melting point and high melt viscosity which cause difficult melt processing, high rigidity and strength, high heat deflection temperature, and good electrical properties.

The flexibility of the carbon-oxygen single bonds permits some molecular flexbility, which produces high impact strength; but the molecular mobility is not sufficient to insure this unambiguously, and measured impact strengths are very sensitive to thickness and other specific design considerations.

Restricted molecular motion impedes the ability of the molecules to move into the crystalline lattice, and restricts crystallinity to rather low degree and small size. The largely amorphous material is therefore transparent to visible light.

Intermediate polarity and lack of major intermolecular attraction permit the polymer to dissolve in a range of common solvents. The organic carbonate group, though generally considered rather unstable, is stabilized by aromatic resonance, and is hydrolyzed only by alkaline conditions. At higher temperatures encountered during some processing, and when exposed to ultraviolet light, the carbonate groups may undergo color-forming side reactions.

The balance of rigidity, impact strength, heat deflection temperature, electrical properties, and clarity lead to electrical and electronic applications such as electric knife handles and other appliances, electric devices in construction, lighting fixtures and glazing and skylights, and plumbing and bathroom fixtures.

Poly(Ethylene Terephthalate)

The benzene rings are bulky and rigid, and resonate with the carbonyl groups to produce large flat rigid bulky units in the polymer chain. Their resonance creates some overlap and attraction between adjacent chains. The regular structure crystallizes quite readily despite its bulkiness. All these factors comprise molecular immobility, producing high modulus, strength, melting point, electrical quality, and moisture and solvent resistance.

The ethylene glycol units are very flexible, and provide sufficient molecular mobility for melt processing, stretching and orientation above the glass transition temperature, and flexibility in fibers and films down to very low temperatures.

Moderate polarity of aromatic esters provides good resistance to moisture and staining.

Primary application is in fiber blends with cotton or wool for wash-and-wear and wrinkle-resistant fabrics. The other important form is in thin films, which are of major importance in food packaging, particularly boil-in-bag, heat-and-serve, and vacuum and inert gas packaging of meats and cheeses; as electrical insulation in capacitors, cable-wrapping, flexible printed circuits, slot liners, motors, and transformers; magnetic recording tape; office supplies such as typewriter ribbon and graphic arts; and weather balloons and satellites.

Unsaturated Polyesters

Low-molecular-weight polyesters are viscous liquids or soluble brittle solids, which can be dissolved in liquid monomers like styrene to produce any viscosity desired for impregnation and lamination of glass fabrics. The fumarate/maleate ester units in the polyester provide easy reactivity with styrene monomer, permitting easy cure to glass-fiber-reinforced plastics. Propylene glycol can be replaced by butylene or diethylene glycols for greater flexibility, or by hydrogenated or ethoxylated bisphenols for greater heat and chemical resistance. Phthalic anhydride is used to control unsaturated reactivity, to increase hardness, and to lower cost; isophthalic acid is sometimes used for improved properties, adipic acid for flexibility, or chlorinated or brominated acids for flame resistance. Styrene may be replaced by methyl methacrylate for improved color and weather stability, or by triallyl cyanurate for improved heat stability. The glass fibers may be filament-wound for simple hollow products of maximum radial strength; woven fabric for simple shapes of maximum isotropic strength; or chopped mat for more complex shapes and fast versatile moldability. The final structure is highly cross-linked and reinforced, providing high rigidity, strength, and impact resistance, hot strength, and chemical resistance. While the raw materials are low to moderate in price, the production techniques are more complex than other plastics processing, increasing the cost of the finished products, and thus satisfying primarily markets which require maximum rigidity, strength, and impact resistance, along with light color and chemical resistance.

The largest single market is in small to medium size boats and associated marine equipment such as decks, superstructures, and smaller components. Transportation markets in general are very large, and include bodies and body components for autos, trucks, trailers, railway cars, farm machinery, buses, subway cars, aircraft, and motorcycles, as well as many interior components. Another very large market is building and construction, including building panels, siding, glazing, and skylights. Other large markets include bathtubs and shower stalls; pipes, ducts, and tanks for chemical processing, mining, petroleum, and water distribution; appliance housings and components; industrial housings, containers, and trays; structural electrical insulation; and in lighting fixtures and decorative cast flooring.

Diallyl Phthalate

The moderate reactivity of allyl esters permits partial prepolymerization to easily-molded low-molecular-weight prepolymers, which are reinforced with short glass fibers and then compression-molded into highly-cross-linked completely-cured products. The benzene rings, high degree of cross-linking, and glass-fiber reinforcement provide high rigidity and strength. The hydrocarbon chains, low polarity, and high cross-linking provide good thermal, electrical, and chemical resistance. And the allyl prepolymer provides final cure during molding with no byproducts and little shrinkage, thus permitting precision molding.

Primary applications are in molded structural electrical insulation such as housings, circuit boards, junction boxes, radomes, potentiometers, encapsulating shells, and multipronged connectors. The resin itself is also used for surfacing decorative wood laminates in furniture and panelling, and in vacuum impregnation of porous nonferrous metal castings.

Drying Oils

Conventional oil paints are primarily dispersions of colored pigments in polyunsaturated glyceryl esters. The liquid esters have low viscosity and moderate polarity, which favor wetting and dispersion of pigments, wetting and penetration of wood and other substrates, and coherent film formation. Their reactivity permits gradual polymerization by atmospheric oxygen, converting them into cross-linked structures of some flexibility and good adhesion and water resistance. Thus they have been the primary basis of most paints throughout history until quite recently.

Alkyd Coatings

Branched prepolymers of glycerol and phthalic anhydride, copolymerized with glyceryl esters of fatty acids, combine the high cross-linking and strength of the glyceryl phthalate with the flexibility of the fatty acids, permitting control of curing conditions and cured properties over a wide range of compositions. For air-drying coatings, high concentrations of drying-oil acids provide the reactivity of conventional drying oils with atmospheric oxygen; for oven-baked coatings, lower concentrations of less unsaturated acids, and residual alcohol and acid groups in the glyceryl phthalate prepolymer, provide high-temperature reactivity. These have found increasing use in place of simple drying oils, both in air-dried coatings on wood and large structures, and in oven-baked coatings on automobiles and appliances.

8-13 POLYURETHANES

Polyurethanes form very rapidly when their low-viscosity liquid reactants are mixed at room temperature, and thus are easily poured or sprayed from low-cost equipment and converted directly into finished form. Use of a variety of polyether and polyester diols and higher-functionality polyols, diisocyanates and polyisocyanates, permits easy control of structure and properties over a wide range. Reaction between excess isocyanate and water, or evaporation of low-boiling fluorocarbons, produce convenient foaming of the liquid reactants, and the immediate cure reaction which follows captures this foam, solidifies, and cross-links it permanently.

In general, the aliphatic polyol, and especially polyether, portions of the finished polymer contribute flexibility and elasticity; the urethane groups contribute polarity and hydrogen-bonding to prevent creep and improve strength and elasticity; the controlled cross-links provide increased strength and elasticity, and at high concentration can actually produce rigidity; and the foam structure provides cushioning, insulating, and low-density properties in the finished products.

Flexible Foams

Largest use of polyurethanes is in flexible foams, where long aliphatic polyether blocks provide high flexibility; urethane group polarity and hydrogen-bonding, and controlled mild cross-linking, provide strength and resilience; and the low density and high air content of the open-celled foam provides softness and resilience. Major application is in furniture cushioning. Second largest is in auto seat cushioning and safety padding. Third is in bedding, both mattresses and pillows. Other large applications are textile lamination for body warmth, and carpet underlay for softness and resilience; while smaller amounts are used for cushioning in packing delicate products, and in other specialities.

Rigid Foams

Second largest use of polyurethanes is in rigid foams. Here high-functionality polyethers, and often also high-functionality polyisocyanates as well, produce sufficient cross-linking to immobilize the molecules and make the structure entirely rigid at room and even elevated temperatures. When foamed with low-boiling fluorocarbon gases, contained in a closed-cell foam structure, thermal conductivity is exceptionally low, permitting use of thinner layers to produce equal insulating capacity, and thus producing superior design possibilities; fortunately the polarity and cross-linking of the polymer produce sufficient chemical resistance to contain the fluorocarbon gases, making the use of these foams possible. In addition, the closed cell structure, easily poured into any type of cavity, has great buoyancy and considerable rigidity/weight ratio, making it useful in flotation and structural applications as well. Largest applications are refrigerated railroad cars and trucks; household and commercial coolers, refrigerators, and freezers; foam sandwich panels which combine thermal insulation and light-weight rigidity, in building roofs, ceilings, walls, and siding panels and in aircraft construction; and marine flotation applications including boats, pontoons, floats, buoys, docks, and raising sunken ships. Smaller applications include insulation of industrial tanks and piping in chemical, gas, and petroleum plants; fire seals in mining; and smaller specialities.

Elastomers

Long flexible aliphatic diols, with moderate trifunctionality for cross-linking, can be mixed with diisocyanate, in the absence of foaming agents, and cast into elastomeric products using simple equipment and techniques. The long aliphatic polyethers and polyesters provide softness and flexibility; the urethane groups provide polarity and hydrogen-bonding, and the mild cross-links provide stronger intermolecular bonding, together providing high strength, elasticity, and especially abrasion resistance. And the urethane polarity and hydrogen-bonding plus the mild cross-linking provide chemical resistance. Some formulations further use diamines to produce polyurea groups, which provide even higher polarity and hydrogen-bonding, as well as faster cure. Major applications are in cast flooring; solid tires and wheels for industrial trucks, aircraft, and materials-handling equipment; shoe heels and soles; friction drives and power transmission belts; oil-resistant gaskets; shock and vibration dampers; wire and cable jacketing; and metal-forming pads and punches.

Fibers

Fibers prepared from block copolymers of aliphatic polyurethanes and aromatic polyureas combine the softness and flexibility of the aliphatic polyether or polyester polyurethanes with the polarity, hydrogen-bonding, and microcrystallization of the aromatic polyureas, and sometimes add covalent cross-linking as well, which make them strong and elastic. As compared with rubber thread, these "spandex" fibers are superior in strength and in resistance to abrasion, heat, light, and ozone. Spandex thread has therefore rapidly replaced rubber thread in the stretch fabrics used in foundation garments, swimming suits, hosiery, surgical applications, outerwear, dresswear, and upholstery.

Coatings

Coatings made of polyurethane have all the same advantages of flexibility, toughness, abrasion resistance, and chemical resistance. Their main limitations are processing, light-stability, and cost. All-urethane coatings can be deposited by reaction directly on the substrate, but the reaction is sensitive to moisture and other accidental ingredients, and the cost is high; while copolymers with drying oils can be applied like drying-oil coatings, but sacrifice much of the advantages of the polyurethanes. Light-stability is achieved either by pigmentation or by restricting the coatings to indoor use. Typical uses are in wire enamels, industrial and marine maintenance paints to combat corrosion, clear gloss finishes on institutional wood floors, and clear finishes on furniture and concrete floors.

Adhesives

Adhesive applications derive from the low viscosity and high polarity of the starting materials, which provide good wetting and penetration of the substrate; easy room-temperature polymerization and cure; reactivity in situ which may involve reactive sites on the substrate as well; varied structure and polarity of the finished polymer, providing many sources of bonding to the substrate; and high cohesive strength of the final polyurethane layer itself. Typical uses are in shoemaking and in bonding rubber to fabrics and to metals.

8-14 POLYAMIDES

The regular symmetrical structure and molecular flexibility of nylons produce high crystallinity, and the polarity and hydrogen-bonding of the regularly repeating amide groups greatly enhance the crystalline intermolecular forces, producing good rigidity, and high strength, heat deflection temperature, and chemical resistance. The flexibility of the aliphatic chains produces low melt viscosity and easy processability, and also permits good orientation by stretching of fibers and films, and further enhances strength in this way. The flexibility of the aliphatic chains also contributes impact resistance, and their low polarity undoubtedly contributes to high lubricity, low surface friction, and good abrasion resistance. Polarity and hydrogen-bonding of amide groups favors dyeability and creates water sensitivity, which softens nylons 6 and 66 and causes poor dimensional stability, but becomes less noticeable in the higher nylons 610, 11, and 12, with their greater proportion of aliphatic hydrocarbon chains.

Largest application is in high-strength tire-cord. Second largest application is in fine strong fibers for sheer textiles used in stockings, undergarments, and dresses, and in blends with wool, rayon, and cotton. Large amounts of fiber are also used in top-quality long-wearing carpeting.

Molded plastic applications replace metals in mechanical applications by virtue of superior lubricity, and abrasion, electrical, and chemical resistance. Leading markets are automotive components such as timing sprockets, speedometer take-off gears, door latches, steering linkage bushings and column brackets, hand-brake brackets, carburetor linkage parts, fuel and oil tubing, fuel line filters, gas tank floats, dome light covers, and fender extensions. Related applications in machinery in general include gears, bearings, bushings, cams, pulleys, rollers, fasteners, gaskets, valve seats, printing roll covers, boat propellors, coatings on metal gas meter shafts, fertilizer distributor heads, alkaline battery cases, valves, conveyor rollers, hand-wheels, tanks. Second largest class of plastic applications is in electrical insulation such as abrasion-resistant wire-and-cable-jacketing, auto wire connectors, and alternator coil forms. Considerable amounts are also used in molded appliance parts, plumbing and bath fixtures; in extruded brush bristles and fishing-line filaments; and extruded film for food-packaging such as gas-flush packaging of cheese, vacuum-packaging of bacon and frankfurters and luncheon meats, boil-in-bag frozen foods, bake-in-bag bread products, and roast-in-bag meat and poultry. In addition nylons 11 and 12 in Europe have been finding film markets in sausage casing, packaging of live lobsters, and other food-packaging markets; and powder coatings on metal handrails, hospital beds, bus-stop shelters, street signs, and institutional furniture.

8-15 HIGH-TEMPERATURE HETEROCYCLES (Ref 45,46)

The most general failure of most plastics and other polymeric materials is their softening and degradation at high temperatures. In instrument and aerospace applications, where such high-temperature rigidity and stability are worth the cost of research and development, these properties are gradually approaching the 500°C level, mainly through the development of heterocyclic polymers and ladder structures. The strength and stability of the carbon-nitrogen, carbon-oxygen, and carbon-sulfur bonds in these polymers are greater than simple carbon-carbon and carbon-hydrogen bonds, and their many and extended possibilities for equivalent resonance structures adds still further to their thermal stability. As they approach completely ladder-like structures, the breakage of any single bond by thermal and/or oxidative attack is less and less likely to break the entire molecule. Thermal stabilities at 500°C and higher have been observed for many of the newer structures of this type.

The multiple ring structures built into the main-chain of the polymer molecule make it quite inflexible, and resonance between series of adjacent rings produces long flat planar structures. The rings attached in the side-chains of many of them add further steric hindrance to rotation. The regular structures may also be crystalline, and their resonance, polarity, and hydrogen-bonding then add to the strength of the crystalline intermolecular forces, making them very high-melting. Some cross-linking also undoubtedly occurs during the formation of the predominantly ring structures, causing true thermoset qualities. All in all, these polymers are generally very rigid and creep resistant, and almost infusible and insoluble. This produces excellent final properties, but creates great problems in fabrication and even in the original polymerization reaction.

Polyimides, the first to develop commercial importance, are finding application in magnet-wire enamels, impregnated glass cloth and wrapping film for high-temperature motors and capacitors, and molded or machined high-temperature bearings, gaskets, diamond grinding wheels, and structural applications. These are all used primarily in aerospace development and special instrumentation.

8-16 AMINO RESINS

Urea-Formaldehyde

The high reactivity of urea, formaldehyde, and their low prepolymers, with each other and probably with the cellulosic hydroxyls of wood as well, produce rapid cross-linking and cure, even at room temperature, to highly cross-linked thermoset products. The high polarity and hydrogen-bonding of the intermediates also permits their simple and low-cost application from aqueous solutions or dispersions. The cured products combine the high cross-linking of the urea-formaldehyde and the fibrous and cellular structure of the wood, to produce low-cost products of good rigidity, strength, and impact resistance. Their major drawback is sensitivity to moisture.

Largest application is in bonding wood chips and fibers into particleboard for furniture and for flooring underlay. Other major applications include lamination of indoor plywood, treatment of paper for wet-strength, treatment of textiles to reduce shrinkage and improve wash-and-wear and permanent-press qualities, and copolymerization in alkyd baking enamels to improve color and chemical resistance of appliance and kitchen cabinet finishes. Molding applications include electrical wiring devices in construction, bottle and jar closures, and smaller amounts in lighting fixtures, housings for mixers and electric shavers, and hardware such as knobs and handles for stoves and refrigerators (ref 47,48).

Melamine-Formaldehyde

Melamine is superior to urea in resins because of its higher cross-linking, and because the heterocyclic rings and methylene bridges are stabler toward heat, light, moisture, and chemicals. The highly cross-linked structure gives high hardness, rigidity, strength, heat, abrasion electrical, and chemical resistance. The fibrous cellulose reinforcement provides still greater rigidity and strength, and especially impact strength.

Primary application is in molded dinnerware. The other major molding application is in buttons. Hardness, heat stability, and good color are important in decorative surfaces on plastic and wood laminates used for counter-tops, tabletops, furniture surfacing, and wall panelling. Copolymerization with alkyd coatings improves their retention of color, gloss, and resistance to moisture, chemicals, and checking, particularly in automotive enamels and coil coatings. Glass-fabric laminates rely on rigidity, strength, impact, heat, flame, arc, and alkali resistance in switchboard panels, circuit breaker parts, arc barriers, and armature and slot wedges. Melamine-resin treatment of wool controls shrinkage, while treatment of cotton improves strength and abrasion resistance, and treatment of paper improves strength and water resistance.

8-17 PHENOLIC RESINS

Phenolic resins combine the high reactivity of phenol, formaldehyde, their prepolymers, and probably fibrous cellulosic fillers as well, with each other, in aqueous solution, organic solution, or molten form, providing rapid complete cross-linking to highly-cured structures. The high cross-linking and aromatic structure produce high hardness, rigidity, strength, heat resistance, electrical insulation, and chemical resistance. The fibrous reinforcement produces still higher rigidity and strength and especially impact strength. The low-cost raw materials and good reactivity produce low-cost polymers and finished products.

On the other hand, the rapid thermosetting reaction is only gradually being adapted to continuous processes like injection-molding and extrusion. Long-fiber and fabric reinforcement for maximum mechanical properties involve more difficult and expensive processing. And aromatic quinoid structures generally limit color to opaque dark browns and blacks.

Plastic moldings find major applications in many fields: appliance and automotive parts; business machines; bottle and jar closures; electrical devices, wiring devices, controls, and switchgear; abrasives, friction materials, and brake linings; machine parts; telephones and intercoms; utensils and their handles; and washing-machine agitators. Laminating adhesive applications provide maximum moisture resistance at lowest cost wherever color is not important: exterior plywood, resin-bonded fibrous and granulated wood, and all but the decorative surface of most paper and wood decorative laminates in building and furniture. Paper and fabric laminates find considerable use in gears, cams, bearings, switchgear, and electrical assemblies. Tremendous amounts of phenolic resin adhesive are used in bonding fiberglass mats into thermal insulation for building. Large amounts are also used as binder for sand in foundry and shell molding, and for weather-resistant varnishes as protective coatings on wood and metals.

8-18 SILICONES

The high energy and strength of the silicon-oxygen bond provide high heat and chemical stability; while the free rotation about this bond provides soft, flexible quality down to very low temperature, and—with a little cross-linking to prevent creep—also provides good rubbery elasticity as well. The methyl groups present a surface of low polarity, which provides low surface tension, high lubricity, and water repellency. The low polarity and balanced structure also provide good electrical resistance. And the ionic lability of the silicone bonds provides room-temperature conversion of liquid prepolymers into solid rubbers for easy processing under limited conditions and with little equipment.

These properties lead to a great variety of specialty applications: Low-molecular-weight oils provide their low surface tension and low polarity for antifoams, mold releases, and waterproofing agents. Coatings on paper are nonblocking and release readily from contents packed in them. Elastomers are cast for potting and encapsulating electrical and electronic components, and to apply sealants, roof coatings, concrete-patio toppings, and curtain-wall sealants; and they are molded into a wide variety of elastomeric specialties for use at very high and very low temperatures, such as spark plug connectors, oven door seals, oxygen hoses, space suits, medical equipment and body implants. Trifunctional resins are used to laminate glass fabrics for engine heat barriers, motor slot liners, class *H* transformers, radio and radar insulators.

This survey of commercial polymer properties and applications has illustrated briefly how the structural features of each polymer interact to produce a different characteristic balance of properties, and how these different combinations of properties make each polymer particularly useful in certain specific applications. The subject is obviously too complex to be treated thoroughly in such a brief survey, organized simply from the point of view of structure, and really deserves separate treatment in separate volumes organized from the point of view of properties and applications respectively.

REFERENCES

1. Fulmer, G. E., *SPE Journal, 24* (2), 22 (February, 1968).
2. Most of the applications in this chapter were taken from: (*a*) *Modern Plastics, 45* (5), (January, 1968), and (*b*) *Modern Plastics Encyclopedia 1968*, McGraw-Hill, 1967. These references apply to the entire chapter. When other sources are used for individual polymers, they will be listed individually.
3. *Modern Plastics Encyclopedia op. cit.*, p 208.
4. DuPont, Technical Bulletins on Surlyn Ionomers.
5. Larsen, P. J., *Product Engineering*, January 8, 1962, p 54.
6. *Materials in Design Engineering*, August, 1965, p 119.
7. Avisun, Technical Bulletins on Oletac Amorphous Polypropylene.
8. Eastman, Technical Bulletins on Epolene Amorphous Polypropylene.
9. Baer, Eric, *Engineering Design for Plastics*, Reinhold, 1964, p 65.
10. Amoco, Technical Bulletins on Indopol Polybutenes.
11. Chevron, Technical Bulletins on Oronite Polybutenes.
12. Morton, Maurice, *Introduction to Rubber Technology*, Reinhold, 1959, Chapter 12.
13. *Ibid.*, Chapters 1, 3, 10, 11, & 13.
14. *Ibid.*, p 12.
15. Shell, Technical Bulletins on Kraton.
16. Baer, *op. cit.*, pp 119, 122.
17. Schildknecht, C. E., *Vinyl and Related Polymers*, Wiley, 1952, Chapter 8.
18. Winding, C. C., & Hiatt, G. D., *Polymeric Materials*, McGraw-Hill, 1961, pp 312–15.
19. Baer, *op. cit.*, pp 660, 682, 684.
20. Smith, W. M., *Manufacture of Plastics*, Reinhold, 1964, pp 337–40.
21. Baer, *op. cit.*, pp 701, 702, 736.
22. *Modern Plastics, 45* (5), 93–94 (1968).
23. Dow, Technical Bulletins on Saran vinylidene chloride copolymers.
24. Grace, Technical Bulletins on Daran vinylidene chloride copolymers.
25. National Starch & Chemical, Technical Bulletins on Resyn vinyldene chloride copolymers.
26. Rohm & Haas, Technical Bulletins on Rhoplex vinylidene chloride copolymers.
27. DuPont, Technical Bulletins on Teflon, Viton, and Tedlar.

28. 3M, Technical Bulletins on Kel-F.
29. Allied Chemical, Technical Bulletins on Halon.
30. Pennsalt, Technical Bulletins on Kynar.
31. Skeist, Irving, *Handbook of Adhesives,* Reinhold, 1962, pp 366–73.
32. *Ibid.,* pp 349–50.
33. *Ibid.,* pp 395–99.
34. *Ibid.,* pp 406, 586.
35. Union Carbide, Technical Bulletins on Bakelite vinyl ethyl ether resins.
36. GAF, Technical Bulletins on PVP Polyvinyl Pyrrolidone.
37. Gaylord, N. G., *Polyethers,* Part I, Interscience, 1963, Chapters 3 & 4.
38. Wyandotte, Technical Bulletins on Pluracols and Pluronics.
39. Saunders, J. H., & Frisch, K. C., *Polyurethanes,* Interscience, 1962.
40. Goodrich, Technical Bulletins on Hydrin Elastomers.
41. Hercules, Technical Bulletins on Penton Chlorinated Polyether.
42. Smith, W. M., *op. cit.,* Chapter 15.
43. Union Carbide, Technical Bulletins on Phenoxy Resins.
44. Morton, *op. cit.,* Chapter 15.
45. Frazer, A. H., *High Temperature Resistant Polymers,* Interscience, 1968.
46. Lee, Henry; Stoffey, Donald; & Neville, Kris, *New Linear Polymers,* McGraw-Hill, 1967, Chapters 7–10.
47. Skeist, *op. cit.,* Chapter 24.
48. Parker, D. H., *Principles of Surface Coating Technology,* Interscience, 1965, Chapter 17.
49. Lee, *op. cit.,* pp 72–75.

Author Index

Subject Index